Wilczyński

Rheology in Polymer Processing

Krzysztof Wilczyński

Rheology in Polymer Processing

Modeling and Simulation

HANSER

Hanser Publishers, Munich Hanser Publications, Cincinnati

The Author:

Prof. Krzysztof Wilczyński,
Warsaw University of Technology, Warsaw, Poland

MIX
Papier aus verantwortungsvollen Quellen
FSC® C083411
FSC
www.fsc.org

Distributed in the Americas by:
Hanser Publications
414 Walnut Street, Cincinnati, OH 45202 USA
Phone: (800) 950-8977
www.hanserpublications.com

Distributed in all other countries by:
Carl Hanser Verlag
Postfach 86 04 20, 81631 Munich, Germany
Fax: +49 (89) 98 48 09
www.hanser-fachbuch.de

Library of Congress Control Number: 2020946060

Editor: Dr. Mark Smith
Production Management: Jörg Strohbach
Coverconcept: Marc Müller-Bremer, www.rebranding.de, Munich
Coverdesign: Max Kostopoulos
Coverphoto: © shutterstock.com/Mr.1
Typesetting: Kösel Media, Krugzell, Germany
Printed and bound by CPI books GmbH, Leck
Printed in Germany

ISBN: 978-1-56990-660-6
E-Book ISBN: 978-1-56990-661-3

Dedicated to my parents and my wife Gertruda

Preface

This book is designed to provide a polymer rheology background for polymer processing to both engineering students and practicing engineers. It is written on an intermediate, sometimes advanced level, required to enable the reader to understand the complex rheological behavior of polymers and its far reaching consequences in polymer processing.

There are many excellent books on polymer rheology as well as polymer processing and process modeling. However, this book is comprehensive with content that allows the reader to understand how the engineering CAD/CAE software for process modeling works, and also how this is developed, which makes its use more conscious and effective. This book is written to bridge the gap between a science and an engineering practice. It begins with fundamentals of rheology and ends on using and applications of CAD/CAE software.

The first chapter covers the basics of rheology, i.e. the basics of continuum mechanics, including the concepts of stress, rate of strain, constitutive equations, as well as the concepts of viscosity and viscoelasticity. Tensor notation is used here for a better understanding of these concepts. This notation allows us to generalize simple one-dimensional problems to three-dimensional problems, as well as to consider simple problems as special cases of complex problems.

The second chapter covers the basics of rheometry, which is a science concerning the measurement of rheological properties of polymers, and thus on the measurements of stress and strain. These are made using rheometers. Good rheological material data are the basis for the correct modeling of polymer processing. Understanding the issues of measuring the rheological properties of polymers, including their limitations, is important for the informed use of CAD/CAE systems.

The third chapter describes the basic processing techniques, and three steps of processing are distinguished: pre-processing or compounding in which chemically pure polymers are prepared for processing, such as mixing and pelletizing, polymer processing in which products are formed, e.g. extrusion, injection molding, thermoforming, and post-processing or complementary processing, i.e. finishing treatment of manufactured products.

The fourth chapter covers the modeling of polymer processing, which is applied to better understand the manufacturing processes. Analytical solutions for some simple flow problems are presented, which can be used to solve a lot of engineering problems. However, most of the polymer processing models are represented by sets of non-linear partial differential equations and/or integral equations, which do not have analytical solutions. These problems can be solved using methods of numerical analysis.

The fifth chapter covers the modeling of polymer extrusion, which is the most important technology in the polymer processing industry since more than half of all polymers are extruded. Moreover, extrusion is fundamental for compounding, i. e. mixing, pelletizing, filling, and reinforcing. The first and fundamentals models of polymer processing were developed for single screw extrusion, and these early concepts are still highly important in modeling of polymer processing.

The sixth chapter describes the software for computer modeling of polymer processing. The selected flow problems are solved using ANSYS Polyflow for CFD modeling, Autodesk Moldflow for injection molding, and MULTI-SCREW for extrusion. These programs enable the modeling of most problems in polymer processing.

In this book, the references are limited to books, with the exception of the sixth chapter, which contains also the author's papers. The historically first and fundamental books on rheology and fluid mechanics are referenced, i. e. those by Reiner (1949), Eirich et al. (1956), and Bird et al. (1960), then chronologically, e. g. by Tanner (1985), Bird et al. (1987), White (1990), Macosko (1994), Carreau et al. (1997), until the current books, e. g. by Vlachopoulos (2016). The historically first and fundamental books on polymer processing and process modeling are also referenced, i. e. by Bernhardt et al. (1959), Schenkel (1959), and McKelvey (1962), then chronologically, e. g. by Tucker (1987), Hensen et al. (1989), White and Potente (2002), Tadmor and Gogos (2006), Rauwendaal (2013), until the current books e. g. by Agassant et al. (2017).

The author would like to acknowldge the invaluable help of his colleagues at the Polymer Processing Department at the Warsaw University of Technology during the preparation of this manuscript. In particular I would like to thank Dr. Andrzej Nastaj for co-authorship of the chapter on modeling of extrusion and for development of figures throughout the book, Dr. Adrian Lewandowski for co-authorship of the chapter on CFD modeling, Dr. Krzysztof J. Wilczyński for co-authorship of the chapter on modeling of injection molding, and my PhD student MSc Kamila Buziak for preparing the manuscript. I thank Dr. Mark Smith of Hanser Publishers for his co-operation during the production of this book.

I would like to dedicate this book to the memory of Professor James Lindsay White of the University of Akron, who passed away a little over 10 years ago (1938–2009), and who creatively inspired me to research work during my Fulbright scholarship at The University of Akron, and later. Professor John Vlachopoulos of McMaster University and Professor Avraam Isayev of The University of Akron are cordially thanked for their support throughout this project.

Summer 2020 *Krzysztof Wilczyński*

Warsaw, Poland

Contents

Preface .... **VII**

1 Rheology **1**

1.1 Fundamentals of Continuum Mechanics 1
1.1.1 Stress 1
1.1.2 Rate of Strain 5
1.1.3 The Conservation Laws 8
1.1.3.1 Conservation of Mass 8
1.1.3.2 Conservation of Momentum 11
1.1.3.3 Conservation of Energy 16
1.1.4 Constitutive Equations 21
1.1.5 The Basic Problem of Fluid Mechanics 22

1.2 Viscosity 23
1.2.1 The Concept of Viscosity 23
1.2.2 Newtonian and Non-Newtonian Fluids 25
1.2.3 Viscosity of Polymers 28
1.2.3.1 Effect of Temperature on Viscosity 35
1.2.3.2 Effect of Pressure on Viscosity 37
1.2.3.3 Effect of Molecular Weight on Viscosity 38
1.2.3.4 Effect of Time of Deformation on Viscosity 39
1.2.3.5 Effect of Crosslinking on Viscosity 40
1.2.4 Rheological Models of Viscous Fluids 40
1.2.4.1 Newtonian Fluid 40
1.2.4.2 Generalized Newtonian Fluids 47
1.2.5 Multiphase Systems of Polymers 53

1.3 Viscoelasticity 55
1.3.1 The Concept of Viscoelasticity 55
1.3.2 Characteristic Phenomena of Viscoelasticity 57
1.3.3 Linear Viscoelasticity 59
1.3.3.1 The Concept of Linear Viscoelasticity 59

1.3.3.2 Mechanical Rheological Models 61
1.3.3.3 Time Effects of Viscoelasticity 66
1.3.3.4 The General Differential Model of Linear Viscoelasticity 75
1.3.4 Nonlinear Viscoelasticity 76
1.3.4.1 The Concept of Nonlinear Viscoelasticity 76
1.3.4.2 Normal Stress Differences 77
1.3.4.3 Normal Stress Effects 79
1.3.5 Rheological Models of Viscoelastic Liquids 83
1.3.5.1 The Rivlin-Ericksen Model of Second Order .. 84
1.3.5.2 The Criminale-Ericksen-Filbey Model 86
1.3.5.3 The Maxwell Convective Model 86
1.3.5.4 The White-Metzner Model 86

2 Rheometry 89

2.1 The Concept of Rheometry 89
2.2 Classification of Rheometric Methods 90
2.3 Single-Point Methods 95
2.3.1 Melt Flow Index 95
2.3.2 The Single-Point Method of Determination of Viscosity Curve 98
2.4 Capillary Rheometers 99
2.4.1 Principle of Operation 99
2.4.2 Theoretical Basics 99
2.4.3 Errors of Capillary Rheometry 106
2.4.4 Determination of Viscosity 112
2.4.5 Determination of Extensional Viscosity 115
2.4.6 Determination of Normal Stresses 116
2.5 Cone-Plate Rheometers 117
2.5.1 Principle of Operation 117
2.5.2 Theoretical Basics 118
2.6 Extensional Rheometers 123
2.6.1 Principle of Operation 123
2.6.2 Theoretical Basics 124

3 Polymer Processing 129

3.1 Extrusion 129
3.1.1 Introduction 129
3.1.2 Single Screw Extrusion 133
3.1.3 Twin Screw Extrusion 138

3.1.3.1 Co-Rotating Twin Screw Extrusion 139
3.1.3.2 Counter-Rotating Twin Screw Extrusion 140
3.1.4 Extrusion Dies 142

3.2 Injection Molding 149
3.2.1 Introduction 149
3.2.2 Injection Molding Process 150
3.2.3 Injection Molds 153
3.2.4 Special Injection Molding Processes 156

3.3 Blow Molding 157
3.3.1 Introduction 157
3.3.2 Film Blowing 158
3.3.3 Extrusion Blow Molding 159
3.3.4 Injection Blow Molding 160

3.4 Thermoforming 161
3.4.1 Introduction 161
3.4.2 Negative Thermoforming 162
3.4.3 Positive Thermoforming 163

3.5 Calendering 164

3.6 Compression Molding 166

4 Process Modeling 171

4.1 Introduction 171

4.2 Simple Flows 172
4.2.1 Pressure Flows 172
4.2.1.1 Flow between Parallel Plates 173
4.2.1.2 Flow through a Circular Tube 184
4.2.1.3 Flow through a Tapered Channel 192
4.2.1.4 Flow through a Cone 193
4.2.1.5 Flow through an Annulus 194
4.2.1.6 Disk Flow 199
4.2.1.7 Hele-Shaw Flow 204
4.2.2 Drag Flows 207
4.2.2.1 Isothermal Flow between Parallel Plates 208
4.2.2.2 Non-Isothermal Flow between Parallel Plates . 210
4.2.3 Pressure-Drag Flow between Parallel Plates 212

4.3 Numerical Methods 214
4.3.1 Introduction 214
4.3.2 Finite Difference Method 215
4.3.2.1 Basic Formulations 215
4.3.2.2 Example of Computations 217

5 Modeling of Extrusion 223

5.1 Introduction 223

5.2 Single Screw Extrusion 224
5.2.1 Physical Model of Extrusion 224
5.2.2 Basic Assumptions of Extrusion Theory 225
5.2.3 Solid Conveying 228
5.2.3.1 Solid Conveying Mechanism 228
5.2.3.2 Flow Rate 230
5.2.3.3 Pressure Development 231
5.2.4 Plasticating 234
5.2.4.1 Pre-Melting 234
5.2.4.2 Melting 236
5.2.5 Melt Conveying in Conventional Screws 245
5.2.5.1 Classification of Models 245
5.2.5.2 Newtonian Model 246
5.2.5.3 Non-Newtonian Model 255
5.2.6 Melt Conveying in Non-Conventional Screws 257
5.2.6.1 Dispersive Mixing Elements 257
5.2.6.2 Distributive Mixing Elements 260
5.2.7 Characteristics of Extruder Operation 262

5.3 Extrusion Dies 265
5.3.1 Classification of Dies 265
5.3.2 Methodology of Modeling 266
5.3.2.1 General Assumptions 266
5.3.2.2 Newtonian Model 268
5.3.2.3 Non-Newtonian Model 270
5.3.2.4 The Concept of Representative Viscosity 271
5.3.3 Circular Dies 273
5.3.4 Slit Dies 273
5.3.5 Annular Dies 278
5.3.5.1 Center Fed Mandrel Dies 279
5.3.5.2 Side Fed Mandrel Dies 279
5.3.6 Profile Dies 286
5.3.6.1 Computation by Cross-Section Division 286
5.3.6.2 Computation by Shape Correction 289

5.4 Global Modeling 292

6 Computer Modeling for Polymer Processing 295

6.1 Overview of Computer Modeling Software 295

6.2 CFD Modeling 297
Co-authored by Adrian Lewandowski
6.2.1 ANSYS Polyflow – Program Overview 297
6.2.2 Modeling Procedure 298
6.2.3 Examples of Modeling 300
6.2.3.1 Pressure Flow 300
6.2.3.2 Extrudate Swell 317
6.2.3.3 Extrudate Swell Inverse Problem 321
6.2.3.4 Single Screw Extrusion 325
6.2.3.5 Co-Rotating Twin Screw Extrusion 332
6.2.3.6 Counter-Rotating Twin Screw Extrusion 339

6.3 Injection Molding 348
Co-authored by Krzysztof J. Wilczyński
6.3.1 Autodesk Moldflow – Program Overview 348
6.3.2 Modeling Procedure 349
6.3.3 Examples of Modeling 350
6.3.3.1 Gate Location 350
6.3.3.2 Filling Analysis 351
6.3.3.3 Flow Balancing 352
6.3.3.4 Melt Flipper 354

6.4 Extrusion 359
Co-authored by Andrzej Nastaj
6.4.1 MULTI-SCREW System – Program Overview 359
6.4.2 Modeling Procedure 361
6.4.3 Examples of Modeling 361
6.4.3.1 Single Screw Extrusion 361
6.4.3.2 Twin Screw Extrusion 368

Index 373

1 Rheology

1.1 Fundamentals of Continuum Mechanics

1.1.1 Stress

If external forces act on a material medium (body), they induce forces in that medium that affect each cross-section of the body. The surface density of these forces is called the **stress.**

The forces acting on a material medium can be characterized as body forces or surface forces. **Body forces** act on every mass element of the medium and are proportional to its total mass. They do not affect the geometrical form of the body. Examples of body forces are gravity forces, inertia forces, or electromagnetic forces. These forces affect any body; however, a physical contact with the surroundings in which these forces occur is not necessary. On the other hand, in the case of **surface forces,** contact with the surroundings is necessary. These forces do not depend on the mass of the medium, but on the size of the surface they operate on. They affect the geometrical form of the body, causing it to deform. Examples of surface forces are pressure forces or viscosity forces.

At the state of equilibrium, the mass and surface forces balance each other. But in motion, the algebraic sum of these forces is proportional to the product of mass and acceleration of the body, according to the second principle of Newton's dynamics.

In order to define the concept of stress, we make an analysis of forces acting on the body, as shown in Figure 1.1. In this body, a surface element ΔS is selected, on which an elementary force $\Delta \underset{\sim}{\boldsymbol{F}}$ acts at a point P. The position of an element ΔS is defined by the vector $\underset{\sim}{\boldsymbol{r}}$, and its direction by the normal unit vector $\underset{\sim}{\boldsymbol{n}}$. The unit force acting on this elementary surface is defined as the stress vector $\underset{\sim}{\boldsymbol{f}}$.

$$\underset{\sim}{\boldsymbol{f}} = \lim_{\Delta S \to 0} \left(\frac{\Delta \underset{\sim}{\boldsymbol{F}}}{\Delta S} \right) \tag{1.1}$$

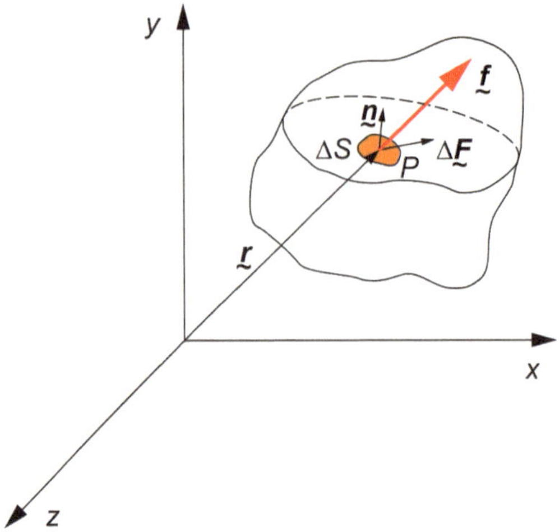

Figure 1.1
The stress vector $\underset{\sim}{f}$

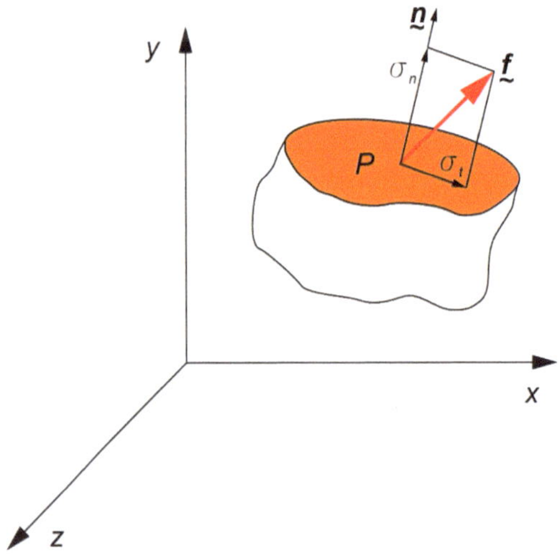

Figure 1.2
Components of the stress vector $\underset{\sim}{f}$:
σ_n - normal stress, σ_t - shear stress

The component of the stress vector $\underset{\sim}{f}$ normal to the surface dS is defined as the **normal stress** σ_n, while the component parallel to that surface is the **tangential (shear) stress** σ_t (Figure 1.2).

The stress vector $\underset{\sim}{f}$ cannot characterize the state of stress at a given point in the body because it depends on the orientation of the surface element dS, i.e., on the vector $\underset{\sim}{n}$. The state of stress at a point P is characterized by the relation of vectors $\underset{\sim}{f}$ and $\underset{\sim}{n}$, that is, by the **stress tensor** $\underset{\sim}{\sigma}(\underset{\sim}{r})$. It is defined by the product

$$\underset{\sim}{f} = \underset{\sim}{n}\underset{\sim}{\sigma} \tag{1.2}$$

The state of stress at a point P in a given coordinate system is uniquely determined by three stress vectors. They correspond to three base vectors of this system which represent three planes perpendicular to each other, and intersecting at a point P. Thus, in the Cartesian coordinate system x, y, z, defined by the base vectors $\underset{\sim}{i}_x$, $\underset{\sim}{i}_y$, $\underset{\sim}{i}_z$, the stress vectors are

$$\underset{\sim}{f}\left(\underset{\sim}{i}_x\right)=\sigma_{xx}\underset{\sim}{i}_x+\sigma_{xy}\underset{\sim}{i}_y+\sigma_{xz}\underset{\sim}{i}_z \tag{1.3}$$

$$\underset{\sim}{f}\left(\underset{\sim}{i}_y\right)=\sigma_{yx}\underset{\sim}{i}_x+\sigma_{yy}\underset{\sim}{i}_y+\sigma_{yz}\underset{\sim}{i}_z \tag{1.4}$$

$$\underset{\sim}{f}\left(\underset{\sim}{i}_z\right)=\sigma_{zx}\underset{\sim}{i}_x+\sigma_{zy}\underset{\sim}{i}_y+\sigma_{zz}\underset{\sim}{i}_z \tag{1.5}$$

The stress vectors $\underset{\sim}{f}\left(\underset{\sim}{i}_x\right)$, $\underset{\sim}{f}\left(\underset{\sim}{i}_y\right)$, $\underset{\sim}{f}\left(\underset{\sim}{i}_z\right)$, respectively, act on the surfaces perpendicular to the axis of the coordinate system x, y, z (Figure 1.3).

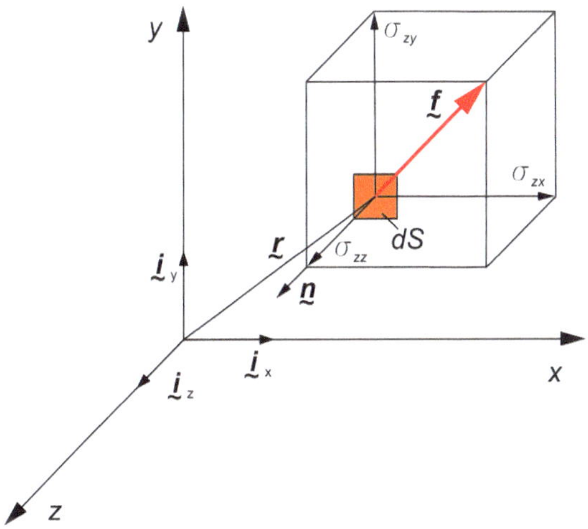

Figure 1.3
Components of the stress vector $\underset{\sim}{f}$ in the Cartesian coordinate system

As a result, the state of stress is determined by the components σ_{ij} of the corresponding stress vectors $\underset{\sim}{f}\left(\underset{\sim}{i}_x\right)$, $\underset{\sim}{f}\left(\underset{\sim}{i}_y\right)$, $\underset{\sim}{f}\left(\underset{\sim}{i}_z\right)$ and can be represented in the form of matrix:

$$\underset{\sim}{\sigma}\equiv\sigma_{ij}=\begin{vmatrix}\sigma_{xx} & \sigma_{xy} & \sigma_{xz}\\ \sigma_{yx} & \sigma_{yy} & \sigma_{yz}\\ \sigma_{zx} & \sigma_{zy} & \sigma_{zz}\end{vmatrix} \tag{1.6}$$

The first subscript i refers to the coordinate which is normal to the plane on which the stress is applied. The second subscript j defines the direction of the stress (the corresponding component of the stress vector). Thus, σ_{xy} is a shear stress acting on a plane normal to the x-axis and in the y-direction, whereas σ_{xx} is a normal stress acting on a plane normal to the x-axis and in the x-direction (Figure 1.4). It is assumed that normal stresses are positive if tensile and negative if compressive.

Based on the balance of forces and momentum (mass and surface) acting on the selected element of the body, it can be shown that for isotropic bodies the stress is a symmetric tensor, and therefore $\sigma_{ij} = \sigma_{ji}$, which means $\sigma_{xy} = \sigma_{yx}$, $\sigma_{xz} = \sigma_{zx}$, and $\sigma_{yz} = \sigma_{zy}$. Thus, the state of stress is uniquely determined by six components of the stress tensor.

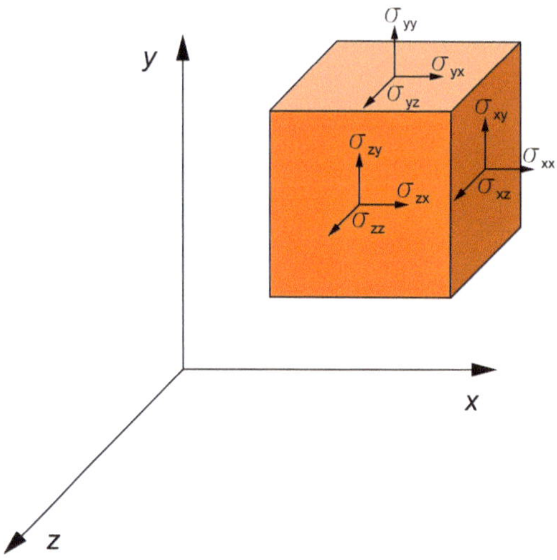

Figure 1.4
Stresses exerted on the planes of an elemental cube

The stress tensor can be divided into isotropic (spherical) and deviatoric parts:

$$\underset{\sim}{\sigma} = -p\underset{\sim}{1} + \underset{\sim}{\tau} \tag{1.7}$$

where p is the hydrostatic pressure; $\underset{\sim}{1}$ is the tensor unity; $\underset{\sim}{\tau}$ is the extra-stress tensor (deviatoric stress tensor).

The **hydrostatic pressure** p is defined as the mean of normal components of the stress tensor:

$$p = -\frac{\sigma_{xx} + \sigma_{yy} + \sigma_{zz}}{3} \tag{1.8}$$

The **isotropic part** of the stress tensor ($-p\underset{\sim}{1}$) is interpreted as the part that causes the change in the volume of the body (volume deformation). On the other hand, the **deviatoric part** ($\underset{\sim}{\tau}$) is the part that causes the change in the shape of the body (deformation) at constant volume.

The stress tensor components, when decomposed into isotropic and deviatoric parts, assume in a Cartesian system the following form:

$$\sigma_{ij} = -p\begin{vmatrix} 1 & 0 & 0 \\ 0 & 1 & 0 \\ 0 & 0 & 1 \end{vmatrix} + \begin{vmatrix} \tau_{xx} & \tau_{xy} & \tau_{xz} \\ \tau_{yx} & \tau_{yy} & \tau_{yz} \\ \tau_{zx} & \tau_{zy} & \tau_{zz} \end{vmatrix} \tag{1.9}$$

and, in the cylindrical coordinate system r, φ, z:

$$\sigma_{ij} = -p\begin{vmatrix} 1 & 0 & 0 \\ 0 & 1 & 0 \\ 0 & 0 & 1 \end{vmatrix} + \begin{vmatrix} \tau_{rr} & \tau_{r\varphi} & \tau_{rz} \\ \tau_{\varphi r} & \tau_{\varphi\varphi} & \tau_{\varphi z} \\ \tau_{zr} & \tau_{z\varphi} & \tau_{zz} \end{vmatrix} \tag{1.10}$$

Tangential components τ_{ij} of the deviatoric stress tensor are equal to the tangential components σ_{ij} of the total stress tensor, i.e. $\tau_{ij} = \sigma_{ij}$, when $i \neq j$. However, the normal components of these tensors are not equal, because $\sigma_{ii} = -p + \tau_{ii}$. Therefore, in order to extract the pure deformation in the normal directions, the concept of difference of normal stresses is introduced:

$$N_1 = \sigma_{11} - \sigma_{22} = \tau_{11} - \tau_{22} \tag{1.11}$$

$$N_2 = \sigma_{22} - \sigma_{33} = \tau_{22} - \tau_{33} \tag{1.12}$$

where N_1 is the first normal stress difference; N_2 is the second normal stress difference; 1, 2, 3 are the axes of the coordinate system, e.g., Cartesian system x, y, z or cylindrical system r, φ, z.

1.1.2 Rate of Strain

Flow of the liquid is generally a continuous movement of the material, which is accompanied by large displacements at relatively short time intervals. In this case, **Euler's motion analysis** (description of motion at fixed point of space) is used. According to this, the temporary state of liquid motion is described by its velocity field, which in the generalization can be presented as a function of position and time

$$\underset{\sim}{v} = \underset{\sim}{v}\left(\underset{\sim}{x}, t\right) \tag{1.13}$$

where $\underset{\sim}{v}$ is the vector of velocity; $\underset{\sim}{x}$ is the vector of position; t is the time.

In the discussed flow, two points P and Q are separated by an infinitesimal distance $\mathrm{d}\underset{\sim}{x}$. The velocity at the point P is $\underset{\sim}{v}$, and at the point Q it is $\underset{\sim}{v} + \mathrm{d}\underset{\sim}{v}$ (Figure 1.5).

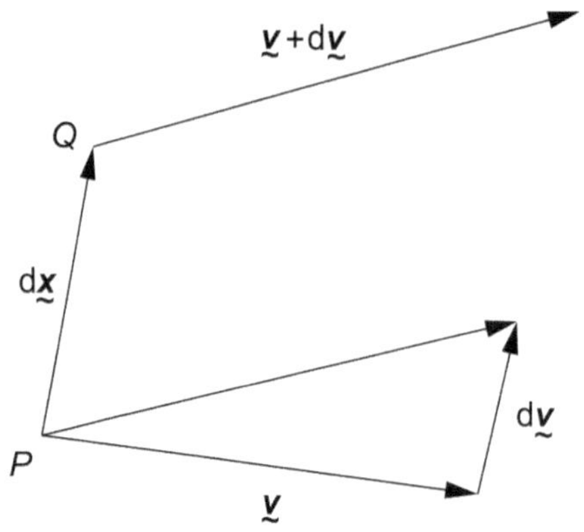

Figure 1.5
The rate of strain tensor

What is interesting here is not the absolute velocity of points P and Q but rather their difference $\mathrm{d}\underset{\sim}{v}$ (relative velocity). It can be specified at any point of the flow based on the gradient of the velocity field as

$$\mathrm{d}\underset{\sim}{v} = \underset{\sim}{L}\mathrm{d}\underset{\sim}{x} \tag{1.14}$$

where $\underset{\sim}{L}$ is the velocity gradient tensor equal to

$$\underset{\sim}{L} = \operatorname{grad} \underset{\sim}{v} = \frac{\partial \underset{\sim}{v}}{\partial \underset{\sim}{x}} \tag{1.15}$$

The velocity gradient $\underset{\sim}{L}$ is a second order tensor that assigns the local vector of the velocity change $\mathrm{d}\underset{\sim}{v}$ to the local vector of displacement $\mathrm{d}\underset{\sim}{x}$. In the index notation, in the Cartesian system, it has the following form:

$$\underset{\sim}{L} \equiv L_{ij} = \frac{\partial v_i}{\partial x_j} = \begin{vmatrix} \frac{\partial v_x}{\partial x} & \frac{\partial v_x}{\partial y} & \frac{\partial v_x}{\partial z} \\ \frac{\partial v_y}{\partial x} & \frac{\partial v_y}{\partial y} & \frac{\partial v_y}{\partial z} \\ \frac{\partial v_z}{\partial x} & \frac{\partial v_z}{\partial y} & \frac{\partial v_z}{\partial z} \end{vmatrix} \tag{1.16}$$

where the first subscript i is the direction of velocity, and the second subscript j is the gradient direction. Thus, e.g. $L_{xy} = \frac{\partial v_x}{\partial y}$ is the gradient of the velocity component v_x in the direction y.

The **velocity gradient tensor** $\underset{\sim}{L}$ can be decomposed into symmetric and antisymmetric parts as follows:

$$\frac{\partial v_i}{\partial x_j} = \frac{1}{2}\left(\frac{\partial v_i}{\partial x_j} + \frac{\partial v_j}{\partial x_i}\right) + \frac{1}{2}\left(\frac{\partial v_i}{\partial x_j} - \frac{\partial v_j}{\partial x_i}\right) \tag{1.17}$$

The symmetric part of the velocity gradient tensor is called the **rate of deformation (rate of strain) tensor $\underset{\sim}{\boldsymbol{D}}$**, and its components are defined as

$$D_{ij} = \frac{1}{2}\left(\frac{\partial v_i}{\partial x_j} + \frac{\partial v_j}{\partial x_i}\right) \tag{1.18}$$

whereas the antisymmetric part is called the **rate of rotation tensor $\underset{\sim}{\boldsymbol{W}}$**, and its components are defined as

$$W_{ij} = \frac{1}{2}\left(\frac{\partial v_i}{\partial x_j} - \frac{\partial v_j}{\partial x_i}\right) \tag{1.19}$$

The rate of deformation tensor $\underset{\sim}{\boldsymbol{D}}$ describes the deformation of the body whereas the rate of rotation tensor $\underset{\sim}{\boldsymbol{W}}$ describes the rigid rotation of the body (there is not any deformation of the body). In the Cartesian system, the rate of deformation tensor and the rate of rotation tensor are defined by the following matrices:

$$D_{ij} = \begin{bmatrix} \dfrac{\partial \nu_x}{\partial x} & \dfrac{1}{2}\left(\dfrac{\partial \nu_x}{\partial y} + \dfrac{\partial \nu_y}{\partial x}\right) & \dfrac{1}{2}\left(\dfrac{\partial \nu_x}{\partial z} + \dfrac{\partial \nu_z}{\partial x}\right) \\ \dfrac{1}{2}\left(\dfrac{\partial \nu_x}{\partial y} + \dfrac{\partial \nu_y}{\partial x}\right) & \dfrac{\partial \nu_y}{\partial y} & \dfrac{1}{2}\left(\dfrac{\partial \nu_y}{\partial z} + \dfrac{\partial \nu_z}{\partial y}\right) \\ \dfrac{1}{2}\left(\dfrac{\partial \nu_x}{\partial z} + \dfrac{\partial \nu_z}{\partial x}\right) & \dfrac{1}{2}\left(\dfrac{\partial \nu_y}{\partial z} + \dfrac{\partial \nu_z}{\partial y}\right) & \dfrac{\partial \nu_z}{\partial z} \end{bmatrix} \tag{1.20}$$

$$W_{ij} = \begin{bmatrix} 0 & \dfrac{1}{2}\left(\dfrac{\partial \nu_x}{\partial y} - \dfrac{\partial \nu_y}{\partial x}\right) & \dfrac{1}{2}\left(\dfrac{\partial \nu_x}{\partial z} - \dfrac{\partial \nu_z}{\partial x}\right) \\ \dfrac{1}{2}\left(\dfrac{\partial \nu_y}{\partial x} - \dfrac{\partial \nu_x}{\partial y}\right) & 0 & \dfrac{1}{2}\left(\dfrac{\partial \nu_y}{\partial z} - \dfrac{\partial \nu_z}{\partial y}\right) \\ \dfrac{1}{2}\left(\dfrac{\partial \nu_z}{\partial x} - \dfrac{\partial \nu_x}{\partial z}\right) & \dfrac{1}{2}\left(\dfrac{\partial \nu_z}{\partial y} - \dfrac{\partial \nu_y}{\partial z}\right) & 0 \end{bmatrix} \tag{1.21}$$

The rate of deformation tensor is a symmetric tensor, and therefore $D_{ij} = D_{ji}$, which means $D_{xy} = D_{yx}$, $D_{xz} = D_{zx}$, and $D_{yz} = D_{zy}$. Thus, the state of deformation is determined by six components of the rate of deformation tensor.

Table 1.1 shows the components of the rate of deformation tensor in Cartesian and cylindrical systems.

Table 1.1 Components of the Rate of Deformation Tensor $\underset{\sim}{D}$

Cartesian system *xyz*		
$D_{xx} = \frac{\partial v_x}{\partial x}$	$D_{xy} = D_{yx}$	$D_{xz} = D_{zx}$
$D_{yx} = \frac{1}{2}\left(\frac{\partial v_x}{\partial y} + \frac{\partial v_y}{\partial x}\right)$	$D_{yy} = \frac{\partial v_y}{\partial y}$	$D_{yz} = D_{zy}$
$D_{zx} = \frac{1}{2}\left(\frac{\partial v_x}{\partial z} + \frac{\partial v_z}{\partial x}\right)$	$D_{zy} = \frac{1}{2}\left(\frac{\partial v_y}{\partial z} + \frac{\partial v_z}{\partial y}\right)$	$D_{zz} = \frac{\partial v_z}{\partial z}$
Cylindrical system $r\varphi z$		
$D_{rr} = \frac{\partial v_r}{\partial r}$	$D_{r\varphi} = D_{\varphi r}$	$D_{rz} = D_{zr}$
$D_{\varphi r} = \frac{1}{2}\left[r\frac{\partial}{\partial r}\left(\frac{v_\varphi}{r}\right) + \frac{1}{r}\frac{\partial v_r}{\partial \varphi}\right]$	$D_{\varphi\varphi} = \frac{1}{r}\frac{\partial v_\varphi}{\partial \varphi} + \frac{v_r}{r}$	$D_{\varphi z} = D_{z\varphi}$
$D_{zr} = \frac{1}{2}\left(\frac{\partial v_r}{\partial z} + \frac{\partial v_z}{\partial r}\right)$	$D_{z\varphi} = \frac{1}{2}\left(\frac{\partial v_\varphi}{\partial z} + \frac{1}{r}\frac{\partial v_z}{\partial \varphi}\right)$	$D_{zz} = \frac{\partial v_z}{\partial z}$

1.1.3 The Conservation Laws

Continuous media are subjected to certain laws of physics called the **conservation laws**. These are

- conservation of mass,
- conservation of momentum,
- conservation of energy.

The conservation laws determine relations between parameters describing the body which are related to position of the body, time, and temperature. They are valid for any continuum. However, there are also the **constitutive equations,** which distinguish one class of bodies from another.

1.1.3.1 Conservation of Mass

Classical continuum mechanics assumes that mass of the body does not change during its motion. In order to formulate the conservation law of mass, **Euler's motion analysis** is used [37]. In the continuum of density $\rho(x, y, z, t)$, we consider a control volume $\Delta x \Delta y \Delta z$ fixed in space in relation to the reference system x, y, z (Figure 1.6).

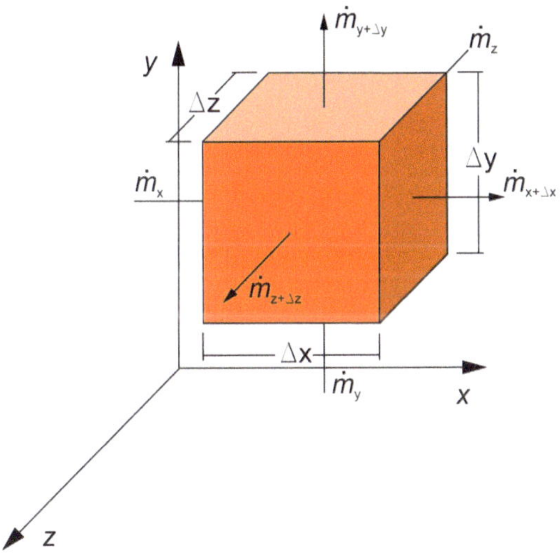

Figure 1.6
Control volume $\Delta x \Delta y \Delta z$;
the conservation law of mass

In this volume, the change of mass in time may result only from the difference of the rate of mass flow into the volume and the rate of mass flow out of the volume according to the equation

$$\frac{\partial m}{\partial t} = \dot{m}_1 - \dot{m}_2 \tag{1.22}$$

where m is the mass; t is the time; $\dot{m}_1$ is the rate of mass flow into the control volume (mass flow rate); $\dot{m}_2$ is the rate of mass flow out of the control volume.

The change of mass in time, in the control volume, is equal to

$$\frac{\partial m}{\partial t} = \frac{\partial \rho}{\partial t} \Delta x \Delta y \Delta z \tag{1.23}$$

The difference of the rate of mass flow into the control volume and the rate of mass out of the control volume results from the balance of the flow rates through the walls of surfaces $\Delta y \Delta z$, $\Delta x \Delta z$, $\Delta x \Delta y$, perpendicular to the respective axes of the coordinate system xyz, and expresses as

$$\dot{m}_1 - \dot{m}_2 = \left(\dot{m}_x - \dot{m}_{x+\Delta x}\right) + \left(\dot{m}_y - \dot{m}_{y+\Delta y}\right) + \left(\dot{m}_z - \dot{m}_{z+\Delta z}\right) \tag{1.24}$$

The rates of mass flow into the control volume are

$$\dot{m}_x = \rho v_x \Delta y \Delta z \tag{1.25}$$

$$\dot{m}_y = \rho v_y \Delta x \Delta z \tag{1.26}$$

$$\dot{m}_z = \rho v_z \Delta x \Delta y \tag{1.27}$$

and the rates of mass flow out of the volume can be expressed as

$$\dot{m}_{x+\Delta x} = \dot{m}_x + \frac{\partial}{\partial x}(\dot{m}_x)\Delta x \tag{1.28}$$

$$\dot{m}_{y+\Delta y} = \dot{m}_y + \frac{\partial}{\partial y}(\dot{m}_y)\Delta y \tag{1.29}$$

$$\dot{m}_{z+\Delta z} = \dot{m}_z + \frac{\partial}{\partial z}(\dot{m}_z)\Delta z \tag{1.30}$$

As a result, after substituting Eqs. (1.25), (1.26), (1.27) and (1.28), (1.29), (1.30) into Eq. (1.24), then Eqs. (1.23) and (1.24) into Eq. (1.22), and dividing this equation by $\Delta x\Delta y\Delta z$, we receive the equation of mass conservation in the differential form:

$$\frac{\partial\rho}{\partial t} + \frac{\partial}{\partial x}(\rho v_x) + \frac{\partial}{\partial y}(\rho v_y) + \frac{\partial}{\partial z}(\rho v_z) = 0 \tag{1.31}$$

which is often referred to as the **continuity equation.**

In the vector notation, it takes the form

$$\frac{\partial\rho}{\partial t} + \operatorname{div}(\rho \underset{\sim}{v}) = 0 \tag{1.32}$$

where $\underset{\sim}{v}$ is the velocity vector; $\operatorname{div}\rho(\underset{\sim}{v})$ is the divergence of vector $\rho\underset{\sim}{v}$.

While in the Cartesian index notation they can be represented as

$$\frac{\partial\rho}{\partial t} + \frac{\partial}{\partial x_i}(\rho v_i) = 0 \tag{1.33}$$

where the repeating index i, in accordance with Einstein's summation rule, means summing up with respect to this index.

Equation (1.31) can be transformed into the form

$$\frac{\partial\rho}{\partial t} + v_x\frac{\partial\rho}{\partial x} + v_y\frac{\partial\rho}{\partial y} + v_z\frac{\partial\rho}{\partial z} + \rho\left(\frac{\partial v_x}{\partial x} + \frac{\partial v_y}{\partial y} + \frac{\partial v_z}{\partial z}\right) = 0 \tag{1.34}$$

in which the material (substantial) derivative of density with respect to time can be extracted:

$$\frac{D\rho}{Dt} = \frac{\partial\rho}{\partial t} + v_x\frac{\partial\rho}{\partial x} + v_y\frac{\partial\rho}{\partial y} + v_z\frac{\partial\rho}{\partial z} \tag{1.35}$$

where $\frac{D\rho}{Dt}$ is the material time derivative of density.

Then, the continuity equation, in the vector notation, takes the form

$$\frac{D\rho}{Dt} + \rho\operatorname{div}\underset{\sim}{v} = 0 \tag{1.36}$$

For incompressible fluids ($\rho = \text{const}$), the continuity equation is reduced to

$$\operatorname{div} \underset{\sim}{v} = \frac{\partial v_x}{\partial x} + \frac{\partial v_y}{\partial y} + \frac{\partial v_z}{\partial z} = 0 \tag{1.37}$$

Table 1.2 shows the continuity equations in Cartesian and cylindrical systems.

Table 1.2 The Equations of Continuity in Different Coordinate Systems

Cartesian system *xyz*
$\frac{\partial \rho}{\partial t} + \frac{\partial}{\partial x}(\rho v_x) + \frac{\partial}{\partial y}(\rho v_y) + \frac{\partial}{\partial z}(\rho v_z) = 0$
Cylindrical system $r\varphi z$
$\frac{\partial \rho}{\partial t} + \frac{1}{r}\frac{\partial}{\partial r}(\rho r v_r) + \frac{1}{r}\frac{\partial}{\partial \varphi}(\rho v_\varphi) + \frac{\partial}{\partial z}(\rho v_z) = 0$

1.1.3.2 Conservation of Momentum

The conservation of momentum is based on the second law of Newton's dynamics. According to this law, the rate of change of the total momentum of a body (product of mass and velocity) at time t is equal to the resultant external forces acting at the same moment on that body.

In order to formulate the conservation law of momentum, **Euler's motion analysis** is used [37]. In the continuum of density $\rho(x, y, z, t)$, we consider a control volume $\Delta x \Delta y \Delta z$ fixed in space in relation to the reference system x, y, z (Figure 1.7).

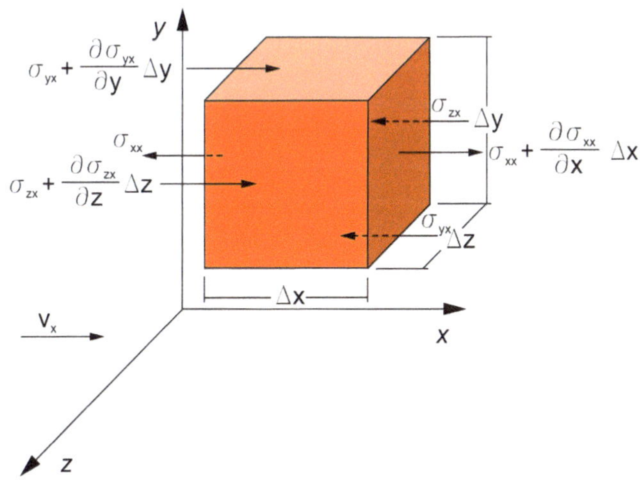

Figure 1.7 Control volume $\Delta x \Delta y \Delta z$; the conservation law of momentum

The rate of change of momentum in the control volume results from the difference in the rates of momentum into and out of the control volume, and from the influence of external forces, according to the equation

$$\frac{\partial}{\partial t}(m\underset{\sim}{v})=(\dot{m}_1\underset{\sim}{v}_1-\dot{m}_2\underset{\sim}{v}_2)+\Sigma\underset{\sim}{F} \tag{1.38}$$

where m is the mass; t is the time; $\underset{\sim}{v}$ is the velocity; $\dot{m}_1$ is the rate of mass flow into the control volume (mass flow rate); $\dot{m}_2$ is the rate of mass flow out of the control volume; $\underset{\sim}{v}_1$ is the velocity of mass flowing into the control volume; $\underset{\sim}{v}_2$ is the velocity of mass flowing out of the control volume, $\Sigma\underset{\sim}{F}$ is the sum of external forces (surface and mass) acting on the medium contained in the control volume.

Equation (1.38) is a vector equation and can be represented by three scalar equations, i. e. for a Cartesian system of coordinates we have

$$\frac{\partial}{\partial t}(mv_x)=\dot{m}_1(v_x)_1-\dot{m}_2(v_x)_2+\Sigma F_x \tag{1.39}$$

$$\frac{\partial}{\partial t}(mv_y)=\dot{m}_1(v_y)_1-\dot{m}_2(v_y)_2+\Sigma F_y \tag{1.40}$$

$$\frac{\partial}{\partial t}(mv_z)=\dot{m}_1(v_z)_1-\dot{m}_2(v_z)_2+\Sigma F_z \tag{1.41}$$

Let us consider the balance of momentum in the direction x. Then, the rate of momentum change in the control volume is

$$\frac{\partial}{\partial t}(mv_x)=\frac{\partial(\rho v_x)}{\partial t}\Delta x\Delta y\Delta z \tag{1.42}$$

The total rate of momentum in the direction x into the control volume $\dot{m}_1(v_x)_1$ is the sum of the momentum rates into that volume through planes perpendicular to the axes of the coordinate system xyz at point 0 (x, y, z), i. e.

$$\dot{m}_1(v_x)_1=(\dot{m}_x v_x)_x+(\dot{m}_y v_x)_y+(\dot{m}_z v_x)_z \tag{1.43}$$

On the other hand, the total rate of momentum in the direction x out of the control volume $\dot{m}_2(v_x)_2$ is the sum of the momentum rates out of that volume through planes perpendicular to the axes of the coordinate system xyz, respectively at points ($x+\Delta x$, y, z), (x, $y+\Delta y$, z), (x, y, $z+\Delta z$). Thus

$$\dot{m}_2(v_x)_2=(\dot{m}_x v_x)_{x+\Delta x}+(\dot{m}_y v_x)_{y+\Delta y}+(\dot{m}_z v_x)_{z+\Delta z} \tag{1.44}$$

In the case of flow through two planes perpendicular to the x-axis, the component of the rate momentum in the x-direction into the control volume through the plane perpendicular to that axis at the point x, y, z will be equal to

$$\left(\dot{m}_x v_x\right)_x = \rho v_x \Delta y \Delta z v_x \tag{1.45}$$

On the other hand, the component of the rate of momentum out of the control volume through the plane perpendicular to the x-axis at the point $x + \Delta x$, y, z can be represented by

$$\left(\dot{m}_x v_x\right)_{x+\Delta x} = \left(\dot{m}_x v_x\right) + \frac{\partial}{\partial x}\left(\dot{m}_x v_x\right)\Delta x \tag{1.46}$$

Thus, the difference in the rates of momentum, using Eqs. (1.45) and (1.46), will be equal to

$$\left(\dot{m}_x v_x\right)_x - \left(\dot{m}_x v_x\right)_{x+\Delta x} = -\frac{\partial}{\partial x}\left(\rho v_x v_x\right)\Delta x \Delta y \Delta z \tag{1.47}$$

And similarly, in the case of flow through the planes perpendicular to the axes y and z, we receive

$$\left(\dot{m}_y v_x\right)_y = \rho v_y \Delta x \Delta z v_x \tag{1.48}$$

$$\left(\dot{m}_z v_x\right)_z = \rho v_z \Delta x \Delta y v_x \tag{1.49}$$

and

$$\left(\dot{m}_y v_x\right)_y - \left(\dot{m}_y v_x\right)_{y+\Delta y} = -\frac{\partial}{\partial y}\left(\rho v_y v_x\right)\Delta x \Delta y \Delta z \tag{1.50}$$

$$\left(\dot{m}_z v_x\right)_z - \left(\dot{m}_z v_x\right)_{z+\Delta z} = -\frac{\partial}{\partial z}\left(\rho v_z v_x\right)\Delta x \Delta y \Delta z \tag{1.51}$$

As a result, with respect to the momentum balance in the x-direction, the difference in the rates of momentum into and out of the control volume is equal to

$$\dot{m}_1\left(v_x\right)_1 - \dot{m}_2\left(v_x\right)_2 = -\left[\frac{\partial}{\partial x}\left(\rho v_x v_x\right) + \frac{\partial}{\partial y}\left(\rho v_y v_x\right) + \frac{\partial}{\partial z}\left(\rho v_z v_x\right)\right]\Delta x \Delta y \Delta z \tag{1.52}$$

The surface forces (normal and tangential) and mass forces (gravity) act on the medium contained in the control volume.

The normal forces acting in the x-direction can be determined from Figure 1.7 as

$$F_{nx} = \left(\sigma_{xx} + \frac{\partial \sigma_{xx}}{\partial x}\Delta x\right)\Delta y \Delta z - \sigma_{xx}\Delta y \Delta z = \frac{\partial \sigma_{xx}}{\partial x}\Delta x \Delta y \Delta x \tag{1.53}$$

while the tangential (shear) forces acting in the same direction are

$$\begin{aligned} F_{tx} &= \left(\sigma_{yx} + \frac{\partial \sigma_{yx}}{\partial y}\Delta y\right)\Delta x \Delta z - \sigma_{yx}\Delta x \Delta z + \\ &+ \left(\sigma_{zx} + \frac{\partial \sigma_{zx}}{\partial z}\Delta z\right)\Delta x \Delta y - \sigma_{zx}\Delta x \Delta y = \frac{\partial \sigma_{yx}}{\partial y}\Delta x \Delta y \Delta z + \frac{\partial \sigma_{zx}}{\partial z}\Delta x \Delta y \Delta z \end{aligned} \tag{1.54}$$

The gravitational forces exerted on the fluid in the x-direction can be represented by

$$F_{gx} = \rho g_x \Delta x \Delta y \Delta z \tag{1.55}$$

where g_x is the component of the unit gravity force.

On the basis of Eqs. (1.53), (1.54), and (1.55), the sum of all external forces exerted on the fluid contained in the control volume in the direction of x-axis can be presented as

$$\Sigma F_x = \left(\frac{\partial \sigma_{xx}}{\partial x} + \frac{\partial \sigma_{yx}}{\partial y} + \frac{\partial \sigma_{zx}}{\partial z} + \rho g_x \right) \Delta x \Delta y \Delta z \tag{1.56}$$

As a result, after substituting Eqs. (1.42), (1.52), and (1.56) into Eq. (1.39) and dividing by $\Delta x \Delta y \Delta z$, we receive the equation of momentum in the x-direction as

$$\begin{gathered} \frac{\partial (\rho v_x)}{\partial t} = -\left[\frac{\partial}{\partial x}(\rho v_x v_x) + \frac{\partial}{\partial y}(\rho v_y v_x) + \frac{\partial}{\partial z}(\rho v_z v_x) \right] + \\ + \left(\frac{\partial \sigma_{xx}}{\partial x} + \frac{\partial \sigma_{yx}}{\partial y} + \frac{\partial \sigma_{zx}}{\partial z} \right) + \rho g_x \end{gathered} \tag{1.57}$$

and, after transformation

$$\begin{gathered} \rho \left(\frac{\partial v_x}{\partial t} + v_x \frac{\partial v_x}{\partial x} + v_y \frac{\partial v_x}{\partial y} + v_z \frac{\partial v_x}{\partial z} \right) + \\ + v_x \left[\frac{\partial \rho}{\partial t} + \frac{\partial}{\partial x}(\rho v_x) + \frac{\partial}{\partial y}(\rho v_y) + \frac{\partial}{\partial z}(\rho v_z) \right] = \\ = \left(\frac{\partial \sigma_{xx}}{\partial x} + \frac{\partial \sigma_{yx}}{\partial y} + \frac{\partial \sigma_{zx}}{\partial z} \right) + \rho g_x \end{gathered} \tag{1.58}$$

The term in square brackets in the second line of Eq. (1.58) represents the continuity equation, Eq. (1.31). Thus, the equation of momentum in the x-direction takes the form

$$\rho \left(\frac{\partial v_x}{\partial t} + v_x \frac{\partial v_x}{\partial x} + v_y \frac{\partial v_x}{\partial y} + v_z \frac{\partial v_x}{\partial z} \right) = \left(\frac{\partial \sigma_{xx}}{\partial x} + \frac{\partial \sigma_{yx}}{\partial y} + \frac{\partial \sigma_{zx}}{\partial z} \right) + \rho g_x \tag{1.59}$$

Similarly, by making the balance of momentum in the directions y and z, we have the following equations:

$$\rho \left(\frac{\partial v_y}{\partial t} + v_x \frac{\partial v_y}{\partial x} + v_y \frac{\partial v_y}{\partial y} + v_z \frac{\partial v_y}{\partial z} \right) = \left(\frac{\partial \sigma_{xy}}{\partial x} + \frac{\partial \sigma_{yy}}{\partial y} + \frac{\partial \sigma_{zy}}{\partial z} \right) + \rho g_y \tag{1.60}$$

$$\rho \left(\frac{\partial v_z}{\partial t} + v_x \frac{\partial v_z}{\partial x} + v_y \frac{\partial v_z}{\partial y} + v_z \frac{\partial v_z}{\partial z} \right) = \left(\frac{\partial \sigma_{xz}}{\partial x} + \frac{\partial \sigma_{yz}}{\partial y} + \frac{\partial \sigma_{zz}}{\partial z} \right) + \rho g_z \tag{1.61}$$

Then, after extracting the isotropic part of the stress tensor (hydrostatic pressure) on the basis of Eq. (1.9), the following equations of momentum can be formed:

$$\rho\left(\frac{\partial v_x}{\partial t}+v_x\frac{\partial v_x}{\partial x}+v_y\frac{\partial v_x}{\partial y}+v_z\frac{\partial v_x}{\partial z}\right)= \\ =-\frac{\partial p}{\partial x}+\frac{\partial \tau_{xx}}{\partial x}+\frac{\partial \tau_{yx}}{\partial y}+\frac{\partial \tau_{zx}}{\partial z}+\rho g_x \tag{1.62}$$

$$\rho\left(\frac{\partial v_y}{\partial t}+v_x\frac{\partial v_y}{\partial x}+v_y\frac{\partial v_y}{\partial y}+v_z\frac{\partial v_y}{\partial z}\right)= \\ =-\frac{\partial p}{\partial y}+\frac{\partial \tau_{xy}}{\partial x}+\frac{\partial \tau_{yy}}{\partial y}+\frac{\partial \tau_{zy}}{\partial z}+\rho g_y \tag{1.63}$$

$$\rho\left(\frac{\partial v_z}{\partial t}+v_x\frac{\partial v_z}{\partial x}+v_y\frac{\partial v_z}{\partial y}+v_z\frac{\partial v_z}{\partial z}\right)= \\ =-\frac{\partial p}{\partial z}+\frac{\partial \tau_{xz}}{\partial x}+\frac{\partial \tau_{yz}}{\partial y}+\frac{\partial \tau_{zz}}{\partial z}+\rho g_z \tag{1.64}$$

The left side of these equations corresponds to the material time derivative of the velocity vector. Therefore, the equation of momentum can be represented in the tensorial notation as

$$\rho\frac{D\underset{\sim}{\mathbf{v}}}{Dt}=-\text{grad }p+\text{div }\underset{\sim}{\boldsymbol{\tau}}+\rho\underset{\sim}{\mathbf{g}} \tag{1.65}$$

where $\frac{D\underset{\sim}{\mathbf{v}}}{Dt}$ is the material time derivative of velocity vector; grad p is the gradient of hydrostatic pressure (isotropic part of the stress tensor $\underset{\sim}{\boldsymbol{\sigma}}$); div $\underset{\sim}{\boldsymbol{\tau}}$ is the divergence of extra-stress tensor (deviatoric part of the stress tensor $\underset{\sim}{\boldsymbol{\sigma}}$).

In the index notation, in the Cartesian system, the equation of momentum can be represented as

$$\rho\left(\frac{\partial v_j}{\partial t}+v_i\frac{\partial v_j}{\partial x_i}\right)=-\frac{\partial p}{\partial x_j}+\frac{\partial \tau_{ij}}{\partial x_i}+\rho g_j \tag{1.66}$$

where the repeating indices i, according to Einstein's summation rule, mean summing up with respect to this index.

Table 1.3 shows the equations of momentum in Cartesian and cylindrical systems.

Table 1.3 The Equations of Momentum in Different Coordinate Systems

Cartesian system *xyz*
$\rho\left(\frac{\partial v_x}{\partial t}+v_x\frac{\partial v_x}{\partial x}+v_y\frac{\partial v_x}{\partial y}+v_z\frac{\partial v_x}{\partial z}\right)=-\frac{\partial p}{\partial x}+\frac{\partial \tau_{xx}}{\partial x}+\frac{\partial \tau_{yx}}{\partial y}+\frac{\partial \tau_{zx}}{\partial z}+\rho g_x$
$\rho\left(\frac{\partial v_y}{\partial t}+v_x\frac{\partial v_y}{\partial x}+v_y\frac{\partial v_y}{\partial y}+v_z\frac{\partial v_y}{\partial z}\right)=-\frac{\partial p}{\partial y}+\frac{\partial \tau_{xy}}{\partial x}+\frac{\partial \tau_{yy}}{\partial y}+\frac{\partial \tau_{zy}}{\partial z}+\rho g_y$
$\rho\left(\frac{\partial v_z}{\partial t}+v_x\frac{\partial v_z}{\partial x}+v_y\frac{\partial v_z}{\partial y}+v_z\frac{\partial v_z}{\partial z}\right)=-\frac{\partial p}{\partial z}+\frac{\partial \tau_{xz}}{\partial x}+\frac{\partial \tau_{yz}}{\partial y}+\frac{\partial \tau_{zz}}{\partial z}+\rho g_z$
Cylindrical system $r\varphi z$
$\rho\left(\frac{\partial v_r}{\partial t}+v_r\frac{\partial v_r}{\partial r}+\frac{v_\varphi}{r}\frac{\partial v_r}{\partial \varphi}-\frac{v_\varphi^2}{r}+v_z\frac{\partial v_r}{\partial z}\right)=-\frac{\partial p}{\partial r}+\frac{1}{r}\frac{\partial}{\partial r}(r\tau_{rr})+\frac{1}{r}\frac{\partial \tau_{r\varphi}}{\partial \varphi}-\frac{\tau_{\varphi\varphi}}{r}+\frac{\partial \tau_{rz}}{\partial z}+\rho g_r$
$\rho\left(\frac{\partial v_\varphi}{\partial t}+v_r\frac{\partial v_\varphi}{\partial r}+\frac{v_\varphi}{r}\frac{\partial v_\varphi}{\partial \varphi}-\frac{v_r v_\varphi}{r}+v_z\frac{\partial v_\varphi}{\partial z}\right)=-\frac{1}{r}\frac{\partial p}{\partial \varphi}+\frac{1}{r^2}\frac{\partial}{\partial r}\left(r^2\tau_{r\varphi}\right)+\frac{1}{r}\frac{\partial \tau_{\varphi\varphi}}{\partial \varphi}+\frac{\partial \tau_{\varphi z}}{\partial z}+\rho g_\varphi$
$\rho\left(\frac{\partial v_z}{\partial t}+v_r\frac{\partial v_z}{\partial r}+\frac{v_\varphi}{r}\frac{\partial v_z}{\partial \varphi}+v_z\frac{\partial v_z}{\partial z}\right)=-\frac{\partial p}{\partial z}+\frac{1}{r}\frac{\partial}{\partial r}(r\tau_{rz})+\frac{1}{r}\frac{\partial \tau_{\varphi z}}{\partial \varphi}+\frac{\partial \tau_{zz}}{\partial z}+\rho g_z$

1.1.3.3 Conservation of Energy

The conservation of energy is based on the first law of thermodynamics. According to this law, the rate of change of total energy of a closed system is equal to the difference of the heat transferred to the system and the work done by the system to the surroundings in the unit time. The total energy is a sum of mechanical energy and thermal energy. Usually, the thermal energy is important in polymer processing.

In order to formulate the conservation law of thermal energy [37], we consider in the continuum of density $\rho(x, y, z, t)$ a control volume $\Delta x\Delta y\Delta z$ fixed in space in relation to the reference system x, y, z (Figure 1.8).

For incompressible fluids (ρ = const) the rate of change of energy in the control volume results from the difference in the rates of energy into and out of the control volume, and from the rate of energy dissipation (heat generation), according to the equation

$$\frac{\partial E}{\partial t}=\left(\dot{Q}_1-\dot{Q}_2\right)+\dot{W} \tag{1.67}$$

where E is the energy; t is the time; $\dot{Q}_1$ is the rate of energy into the control volume; $\dot{Q}_2$ is the rate of energy out of the control volume; $\dot{W}$ is the rate of energy dissipation in the control volume.

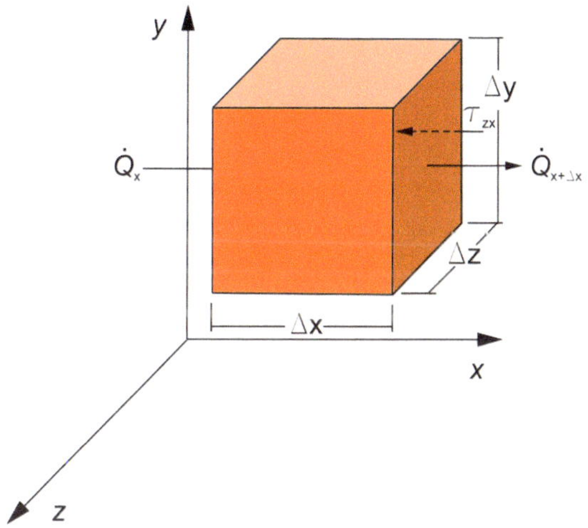

Figure 1.8
Control volume $\Delta x \Delta y \Delta z$;
the conservation law of energy

The rate of change of internal energy (energy accumulation) in the control volume is

$$\frac{\partial E}{\partial t} = \frac{\partial}{\partial t}(\rho e \Delta x \Delta y \Delta z) \tag{1.68}$$

where *e* is the unit internal energy of the fluid.

Since the density in the control volume is constant ($\rho = \text{const}$), Eq. (1.68) can be written as

$$\frac{\partial E}{\partial t} = \rho \frac{\partial e}{\partial t} \Delta x \Delta y \Delta z \tag{1.69}$$

The energy is transferred into the control volume or out of the control volume by heat conduction and convection. Thus

$$\dot{Q}_1 = \dot{Q}_1' + \dot{Q}_1'' \tag{1.70}$$

$$\dot{Q}_2 = \dot{Q}_2' + \dot{Q}_2'' \tag{1.71}$$

where $\dot{Q}_1'$, $\dot{Q}_2'$ are the rates of energy transferred by conduction; $\dot{Q}_1''$, $\dot{Q}_2''$ are the rates of energy transferred by convection.

Let us consider the energy transfer in the *x*-direction. Then, the rate of energy transferred to the control volume by conduction can be represented, using Fourier's law, as

$$\dot{Q}_x' = -k \frac{\partial T}{\partial x} \Delta y \Delta z \tag{1.72}$$

where *k* is the thermal conductivity; *T* is the temperature.

The rate of energy transferred to the volume by convection can be expressed as

$$\dot{Q}_x'' = \rho e v_x \Delta y \Delta z \tag{1.73}$$

The rates of energy transferred out of the control volume in the x-direction can be represented by

$$\dot{Q}_{x+\Delta x}' = \dot{Q}_x' + \frac{\partial}{\partial x}\left(\dot{Q}_x'\right)\Delta x \tag{1.74}$$

$$\dot{Q}_{x+\Delta x}'' = \dot{Q}_x'' + \frac{\partial}{\partial x}\left(\dot{Q}_x''\right)\Delta x \tag{1.75}$$

As a result, using Eqs. (1.72) to (1.75), we can write

$$\dot{Q}_x' - \dot{Q}_{x+\Delta x}' = k\frac{\partial^2 T}{\partial x^2}\Delta x \Delta y \Delta z \tag{1.76}$$

$$\dot{Q}_x'' - \dot{Q}_{x+\Delta x}'' = -\rho\frac{\partial}{\partial x}\left(e v_x\right)\Delta x \Delta y \Delta z \tag{1.77}$$

And similarly, with respect to the heat transfer in the direction of y- and z-axes, the following relationships can be formed:

$$\dot{Q}_y' - \dot{Q}_{y+\Delta y}' = k\frac{\partial^2 T}{\partial y^2}\Delta x \Delta y \Delta z \tag{1.78}$$

$$\dot{Q}_z' - \dot{Q}_{z+\Delta z}' = k\frac{\partial^2 T}{\partial z^2}\Delta x \Delta y \Delta z \tag{1.79}$$

$$\dot{Q}_y'' - \dot{Q}_{y+\Delta y}'' = -\rho\frac{\partial}{\partial y}\left(e v_y\right)\Delta x \Delta y \Delta z \tag{1.80}$$

$$\dot{Q}_z'' - \dot{Q}_{z+\Delta z}'' = -\rho\frac{\partial}{\partial z}\left(e v_z\right)\Delta x \Delta y \Delta z \tag{1.81}$$

As a result, the difference in the energy rates in Eq. (1.67) can be expressed by the formula

$$\dot{Q}_1 - \dot{Q}_2 = \\ = \left[k\left(\frac{\partial^2 T}{\partial x^2} + \frac{\partial^2 T}{\partial y^2} + \frac{\partial^2 T}{\partial z^2}\right) - \rho\left(\frac{\partial\left(e v_x\right)}{\partial x} + \frac{\partial\left(e v_y\right)}{\partial y} + \frac{\partial\left(e v_z\right)}{\partial z}\right)\right]\Delta x \Delta y \Delta z \tag{1.82}$$

After rearranging Eq. (1.82) and reducing the component equivalent to the continuity equation for incompressible fluids we receive

$$\dot{Q}_1 - \dot{Q}_2 = \\ = \left[k\left(\frac{\partial^2 T}{\partial x^2} + \frac{\partial^2 T}{\partial y^2} + \frac{\partial^2 T}{\partial z^2} \right) - \rho\left(v_x \frac{\partial e}{\partial x} + v_y \frac{\partial e}{\partial y} + v_z \frac{\partial e}{\partial z} \right) \right] \Delta x \Delta y \Delta z \tag{1.83}$$

The frictional energy production (energy dissipation) in the control volume is due to the stresses in a fluid. In general, the rate of energy (power) is the dot product of velocity and force. The rate of energy dissipation is therefore the corresponding product of stress, area on which the stress is exerted, and the velocity change in the direction of the stress.

Thus, e.g. the rate of energy produced by the stress component τ_{yx} of the tensor $\underset{\sim}{\boldsymbol{\tau}}$ can be expressed as

$$\dot{W}_{\tau_{yx}} = \tau_{yx} \Delta x \Delta z \Delta v_x \tag{1.84}$$

and, after rearranging

$$\dot{W}_{\tau_{yx}} = \tau_{yx} \frac{\partial v_x}{\partial y} \Delta x \Delta y \Delta z \tag{1.85}$$

The rate of energy produced by all nine components of the stress tensor is a sum of the rates produced by these components and is expressed as

$$\dot{W} = \left(\tau_{xx} \frac{\partial v_x}{\partial x} + \tau_{xy} \frac{\partial v_x}{\partial y} + \tau_{xz} \frac{\partial v_x}{\partial z} + \tau_{yx} \frac{\partial v_y}{\partial x} + \tau_{yy} \frac{\partial v_y}{\partial y} + \tau_{yz} \frac{\partial v_y}{\partial z} + \right. \\ \left. + \tau_{zx} \frac{\partial v_z}{\partial x} + \tau_{zy} \frac{\partial v_z}{\partial y} + \tau_{zz} \frac{\partial v_z}{\partial z} \right) \Delta x \Delta y \Delta z \tag{1.86}$$

Finally, after substituting Eqs. (1.69), (1.83), and (1.86) into Eq. (1.67), dividing by $\Delta x \Delta y \Delta z$, and taking into account the stress tensor symmetry, we receive the equation of energy in the following form:

$$\rho\left(\frac{\partial e}{\partial t} + v_x \frac{\partial e}{\partial x} + v_y \frac{\partial e}{\partial y} + v_z \frac{\partial e}{\partial z} \right) = \\ = k\left(\frac{\partial^2 T}{\partial x^2} + \frac{\partial^2 T}{\partial y^2} + \frac{\partial^2 T}{\partial z^2} \right) + \left(\tau_{xx} \frac{\partial v_x}{\partial x} + \tau_{yy} \frac{\partial v_y}{\partial y} + \tau_{zz} \frac{\partial v_z}{\partial z} \right) + \\ + \left[\tau_{xy} \left(\frac{\partial v_x}{\partial y} + \frac{\partial v_y}{\partial x} \right) + \tau_{xz} \left(\frac{\partial v_x}{\partial z} + \frac{\partial v_z}{\partial x} \right) + \tau_{yz} \left(\frac{\partial v_y}{\partial z} + \frac{\partial v_z}{\partial y} \right) \right] \tag{1.87}$$

The left side of Eq. (1.87) corresponds to the material time derivative of internal energy. Therefore, this equation can be represented in the tensorial notation as

$$\rho \frac{De}{Dt} = k \nabla^2 T + \underset{\sim}{\boldsymbol{\tau}} : \operatorname{grad} \underset{\sim}{\mathbf{v}} \tag{1.88}$$

where $\frac{De}{Dt}$ is the material time derivative of internal energy; $\underset{\sim}{\boldsymbol{\tau}}$ is the extra-stress tensor; grad $\underset{\sim}{\mathbf{v}}$ is the gradient of velocity vector.

Taking into account that for incompressible fluids the total internal energy can be expressed by

$$\frac{De}{Dt} = c_p \frac{DT}{Dt} \tag{1.89}$$

where c_p is the specific heat at constant pressure, the equation of energy can be represented in the tensorial notation as

$$\rho c_p \frac{DT}{Dt} = k\nabla^2 T + \underset{\sim}{\boldsymbol{\tau}} : \text{grad}\, \underset{\sim}{\mathbf{v}} \tag{1.90}$$

In the index notation, in the Cartesian system, the equation of energy can be represented in the form

$$\rho c_p \left(\frac{\partial T}{\partial t} + v_i \frac{\partial T}{\partial x_i} \right) = k \frac{\partial^2 T}{\partial x_i^2} + \tau_{ji} \frac{\partial v_j}{\partial x_i} \tag{1.91}$$

Table 1.4 shows the equations of energy in Cartesian and cylindrical systems.

Table 1.4 The Equations of Energy in Different Coordinate Systems

Cartesian system xyz
$\rho c_p \left(\frac{\partial T}{\partial t} + v_x \frac{\partial T}{\partial x} + v_y \frac{\partial T}{\partial y} + v_z \frac{\partial T}{\partial z} \right) =$ $= k \left(\frac{\partial^2 T}{\partial x^2} + \frac{\partial^2 T}{\partial y^2} + \frac{\partial^2 T}{\partial z^2} \right) + \left[\tau_{xx} \frac{\partial v_x}{\partial x} + \tau_{yy} \frac{\partial v_y}{\partial y} + \tau_{zz} \frac{\partial v_z}{\partial z} \right] +$ $+ \left[\tau_{xy} \left(\frac{\partial v_x}{\partial y} + \frac{\partial v_y}{\partial x} \right) + \tau_{xz} \left(\frac{\partial v_x}{\partial z} + \frac{\partial v_z}{\partial x} \right) + \tau_{yz} \left(\frac{\partial v_y}{\partial z} + \frac{\partial v_z}{\partial y} \right) \right]$
Cylindrical system $r\varphi z$
$\rho c_p \left(\frac{\partial T}{\partial t} + v_r \frac{\partial T}{\partial r} + \frac{v_\varphi}{r} \frac{\partial T}{\partial \varphi} + v_z \frac{\partial T}{\partial z} \right) = k \left[\frac{1}{r} \frac{\partial}{\partial r} \left(r \frac{\partial T}{\partial r} \right) + \frac{1}{r^2} \frac{\partial^2 T}{\partial \varphi^2} + \frac{\partial^2 T}{\partial z^2} \right] +$ $+ \left[\tau_{rr} \frac{\partial v_r}{\partial r} + \tau_{\varphi\varphi} \frac{1}{r} \left(\frac{\partial v_\varphi}{\partial \varphi} + v_r \right) + \tau_{zz} \frac{\partial v_z}{\partial z} \right] +$ $+ \tau_{r\varphi} \left[r \frac{\partial}{\partial r} \left(\frac{v_\varphi}{r} \right) + \frac{1}{r} \frac{\partial v_r}{\partial \varphi} \right] + \tau_{rz} \left(\frac{\partial v_z}{\partial r} + \frac{\partial v_r}{\partial z} \right) + \tau_{\varphi z} \left(\frac{1}{r} \frac{\partial v_z}{\partial \varphi} + \frac{\partial v_\varphi}{\partial z} \right)$

1.1.4 Constitutive Equations

In general, the presented conservation laws are valid (with reservations made at their derivation) for any continuum. They form the following set of equations:

- **equation of continuity** (1.36)

$$\frac{D\rho}{Dt}+\rho\,\mathrm{div}\,\underset{\sim}{\boldsymbol{v}}=0$$

- **equation of momentum** (1.65)

$$\rho\frac{D\underset{\sim}{\boldsymbol{v}}}{Dt}=-\mathrm{grad}\,p+\mathrm{div}\,\underset{\approx}{\boldsymbol{\tau}}+\rho\underset{\sim}{\boldsymbol{g}}$$

- **equation of energy** (1.90)

$$\rho c_p\frac{DT}{Dt}=k\nabla^2T+\underset{\approx}{\boldsymbol{\tau}}:\mathrm{grad}\,\underset{\sim}{\boldsymbol{v}}$$

These three equations are a set of five scalar equations, because the equation of momentum is a vector equation which is expressed by three scalar equations (Table 1.3). There are eleven unknowns, i.e. scalar functions of hydrostatic pressure p and temperature T, vector function of velocity $\underset{\sim}{\boldsymbol{v}}$, and function of symmetric stress tensor $\underset{\approx}{\boldsymbol{\tau}}$. This set of equations is inconsistent and requires six additional equations.

The missing equations are the constitutive equations of continuum. They describe any reaction of continuum to a mechanical load. These equations are relationships between deformations and stresses (or their rates) and temperature. Thus, they define a given continuum (material) and distinguish it from another continuum.

The density ρ might also be considered as an unknown. Then, one more equation is needed, namely the equation of state which determines the relationship between density and the state parameters, pressure p, and temperature T:

$$\rho=\rho(p,T) \tag{1.92}$$

Additional equations provided by the constitutive equation of continuum (and possibly by the state equation) complement the set of equations of conservation of continuity (Eq. (1.36)), momentum (Eq. (1.65)), and energy (Eq. (1.90)), and make this set consistent where the number of unknown functions equals the number of independent equations. The solution of this set of equations, with appropriate boundary and initial conditions, is the so-called basic problem of continuum mechanics (fluid dynamics).

1.1.5 The Basic Problem of Fluid Mechanics

The basic problem of fluid mechanics is to determine the motion of fluid in a given flow geometry. It consists in solving the set of differential equations of the conservation of continuity (Eq. (1.36)), momentum (Eq. (1.65)), and energy (Eq. (1.90)) with the corresponding constitutive equation of the material (and possibly the equation of state) and the corresponding boundary and initial conditions.

In order to solve the problem, one must define the appropriate field of physical quantities that describe the problem. Such a field is a space in which a specific value of a physical quantity, such as velocity, pressure, temperature, etc., is assigned to each point at any time.

If a physical quantity (flow parameter), e.g. velocity $\underset{\sim}{v}$, varies with the position only (e.g. in the Cartesian system) and does not vary with time t, the flow field is said to be steady *(the steady flow)*. Then

$$\frac{\partial \underset{\sim}{v}}{\partial t}=0 \text{ and } \underset{\sim}{v}=\underset{\sim}{v}(x, y, z) \tag{1.93}$$

On the other hand, if this parameter varies with time t, the flow field is said to be unsteady *(the unsteady flow)*, and then

$$\frac{\partial \underset{\sim}{v}}{\partial t}\neq 0 \text{ and } \underset{\sim}{v}=\underset{\sim}{v}(x, y, z, t) \tag{1.94}$$

Depending on the number of space coordinates defining the flow, one can distinguish the following:

- three-dimensional field (three-dimensional flow)

$$\underset{\sim}{v}=\underset{\sim}{v}(x, y, z) \tag{1.95}$$

- two-dimensional field (two-dimensional flow)

$$\underset{\sim}{v}=\underset{\sim}{v}(x, y) \tag{1.96}$$

- one-dimensional field (one-dimensional flow)

$$\underset{\sim}{v}=\underset{\sim}{v}(x) \tag{1.97}$$

Boundary and initial conditions are dependent on the problem to be solved. They often refer to the velocity and temperature field. Boundary conditions determine the flow parameters at selected points of the flow field, e.g. on the surface that confine the flow. On the other hand, initial conditions determine the flow parameters at a selected time, e.g. at the beginning of the flow.

Solving the basic problem of fluid mechanics using analytical methods is possible only in the case of relatively simple problems. Usually, it is necessary to use numerical methods, such as the finite difference method, finite element method, or boundary element method.

1.2 Viscosity

1.2.1 The Concept of Viscosity

Viscosity is a measure of the resistance of a material to flow. In order to define this concept, let us consider (in the Cartesian system) the fluid flow between two parallel plates of area A, one of which is fixed and the other is moved by the application of a force F with a velocity U (Figure 1.9). Assuming there is no slip on the plates, the velocity of the fluid adjacent to the fixed plate will be zero. However, the velocity of the fluid adjacent to the moving plate will be equal to the velocity U.

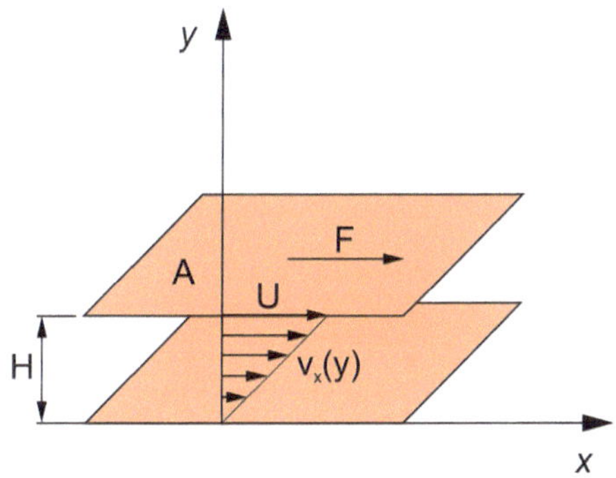

Figure 1.9
Simple shear flow

Experimental studies of such flows have shown that for many materials the velocity distribution in the gap between the plates is linear. It was also found that the force F is proportional to the velocity U and the area A of the plates, and inversely proportional to the distance H between the plates:

$$F \sim A\frac{U}{H} \tag{1.98}$$

The ratio of a force F to an area A expresses the shear stress $\tau = \tau_{yx}$, while the ratio of a velocity U to a distance H between the plates is equal to the gradient of velocity v_x in the direction y, i.e. $\frac{dv_x}{dy}$. So, we can write

$$\tau_{yx} \sim \frac{dv_x}{dy} \tag{1.99}$$

The proportionality (Eq. (1.99)) can be transformed into an equality by introducing the proportionality coefficient μ, called the **dynamic viscosity.** Then

$$\tau_{yx} = \mu \frac{dv_x}{dy} \tag{1.100}$$

Eq. (1.100) is referred to as **Newton's law of viscosity.** Newton first formulated the hypothesis of the proportionality of shear stress and velocity gradient in the fluid.

The velocity gradient is equivalent to the rate of deformation of a fluid caused by the shear stress $\tau = \tau_{yx}$. Hence, this gradient is often referred to as the **shear rate.** As shown in Figure 1.10, the ABCD fluid element, after the time dt, is deformed to $A'B'C'D'$. As a result, the triangle $AA'D$ shows that

$$d\gamma = \frac{AA'}{AD} = \frac{dx}{dy} \tag{1.101}$$

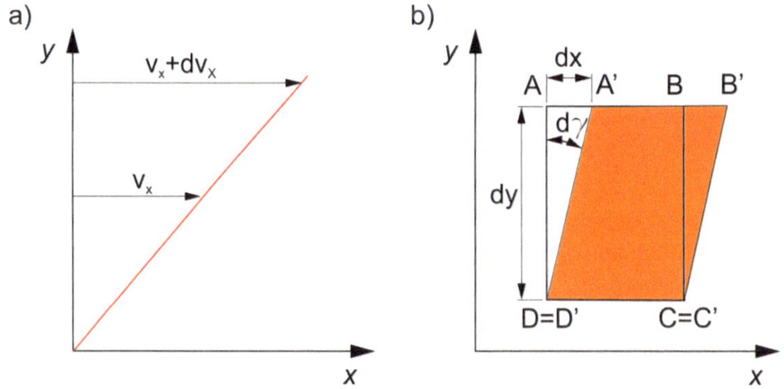

Figure 1.10 Deformation of a fluid element in a simple shear flow: a) velocity profile, b) deformation of a fluid element

Keeping in mind that dx = dv_xdt, we obtain an expression that defines the rate of deformation, i.e. the shear rate:

$$\frac{d\gamma}{dt} = \dot{\gamma} = \frac{dv_x}{dy} \tag{1.102}$$

The dimension of viscosity in SI units is Pa·s = (N·s)/m^2. Before the introduction of SI, the dimension of viscosity (in *c.g.s.* units) was poise (dyne·s/cm^2), and 1 poise = 10^{-1} Pa·s. It is interesting that the viscosity of water is 1 centipoise = 10^{-2} poise = 10^{-3} Pa·s. The dimension of shear rate is (s^{-1}).

The ratio of dynamic viscosity of the fluid to its density is referred to as the **kinematic viscosity:**

$$\nu = \frac{\mu}{\rho} \qquad (1.103)$$

where ν is the kinematic viscosity; μ is the dynamic viscosity; ρ is the density.

The dimension of kinematic viscosity in SI units is (m^2/s), while in *c. g. s.* units (cm^2/s).

Table 1.5 shows viscosity data for various substances.

Table 1.5 Viscosity of Various Substances

Substance	Dynamic Viscosity, Pa·s
Air (20 °C)	10^{-5}
Water (20 °C)	10^{-3}
Aluminum (700 °C)	$1.1 \cdot 10^{-3}$
Mercury (20 °C)	$1.5 \cdot 10^{-3}$
Zinc (700 °C)	$1.7 \cdot 10^{-3}$
Iron (1400 °C)	$6.0 \cdot 10^{-3}$
Blood (37 °C)	10^{-2}
Oil	10^{-2}–10^{0}
Honey	10^{1}
Molten polymers	10^{2}–10^{4}
Dough	10^{3}–10^{5}
Cheese	10^{7}–10^{8}
Asphalt	10^{8}
Molten glass (500 °C)	10^{12}
Glass	10^{40}

1.2.2 Newtonian and Non-Newtonian Fluids

Newton's law (Eq. (1.100)) expresses a linear relationship between shear stress and shear rate. It is applicable for many fluids, including air (and basically all gases), water, gasoline, molten metals, etc. In general, this relationship is valid for gases and liquids and solutions with low molecular mass. The graph of the function $\tau = f(\dot{\gamma})$ is called the **flow curve.** In the case of a Newtonian fluid, it is linear and passes through the origin of the coordinate system. The tangent of the slope angle α of this line relative to the abscissa equals the dynamic viscosity μ (Figure 1.11).

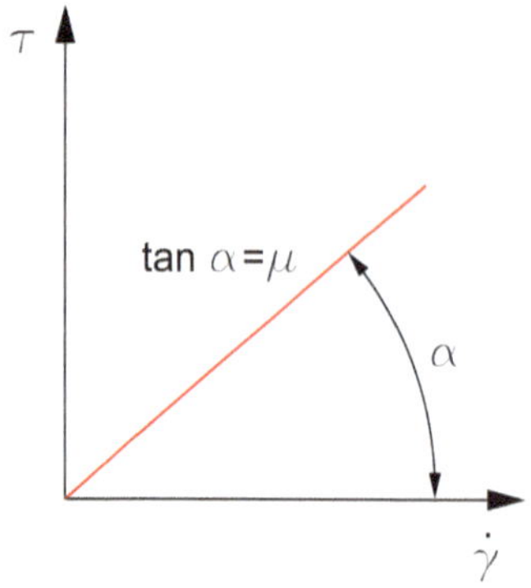

Figure 1.11
Flow curve of a Newtonian fluid

There are also many materials (fluids) that do not show such a simple relationship between shear stress and shear rate, e.g. molten polymers, paints and varnishes, building materials (clay, asphalt), foodstuffs (cheese, jam, chocolate), blood, etc. For fluids of this type, the flow curve is not a straight line or does not pass through the origin of the coordinate system (Figure 1.12). Fluids with such properties are called non-Newtonian fluids.

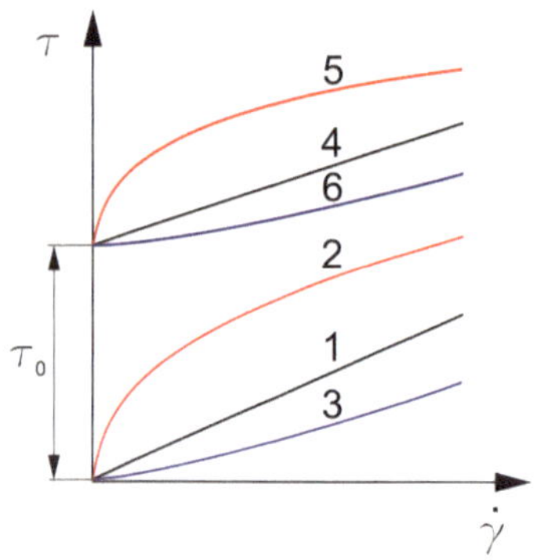

Figure 1.12
Flow curves of non-Newtonian fluids: 1, 2, 3 - fluids without yield stress, 1 - Newtonian fluid, 2 - shear thinning (pseudoplastic), 3 - shear thickening (dilatant); 4, 5, 6 - fluids with yield stress τ_0, 4 - Bingham fluid, 5 - shear thinning, 6 - shear thickening

The **non-Newtonian fluids** can be divided into two basic groups.

The first group includes fluids that have no yield stress (e.g. molten polymers) for which the flow curve passes through the origin of the coordinate system $\dot{\gamma}\,\tau$, but it is not a straight line, i.e. it does not meet Newton's law (Eq. (1.100)).

The second group includes fluids that have a yield stress, that is, the so-called viscoplastic fluids (e.g. clay, asphalt, margarine, mayonnaise, etc.) that flow only after exceeding a certain stress threshold τ_0, and at lower stress behave like elastic solids.

The fluids that belong to the first group can be divided into two types:

- **shear thinning fluids,** for which shear stress rises more slowly than linearly with increasing shear rate,
- **shear thickening fluids,** for which shear stress rises faster than linearly with increasing shear rate.

Shear thinning fluids (historically also called pseudoplastic) are the largest group of non-Newtonian fluids, while shear thickening fluids (also called dilatants) are relatively rare.

The dependence of shear stress on the shear rate for fluids that do not have a yield stress can be approximated by the **power law model** proposed by Ostwald and de Waele:

$$\tau = m\dot{\gamma}^n \tag{1.104}$$

where m, n are the equation parameters; m is the consistency coefficient, Pa·s^n; n is the power law exponent (dimensionless).

When $n < 1$, the law describes pseudoplastic fluids, while when $n > 1$ it describes dilatant fluids. For $n = 1$, Eq. (1.104) is transformed into a Newtonian equation for which $\mu = m$.

In the group of viscoplastic fluids that have a yield stress, it is possible to distinguish the fluids that have a linear relationship between shear stress and shear rate (these are called Bingham bodies) and those for which this relationship is nonlinear.

The Bingham flow curve is described by the following relationships:

$$\begin{aligned} &\tau = \tau_0 + \mu_0 \dot{\gamma} \ \text{ for } \tau \geq \tau_0 \\ &\dot{\gamma} = 0 \ \text{ for } \tau < \tau_0 \end{aligned} \tag{1.105}$$

where τ_0 is the yield stress; μ_0 is the proportionality factor, i. e. plastic viscosity.

The concept of viscosity defined by the Newton equation (1.100) as a constant value, determined by the slope angle of the Newton's flow curve in relation to the shear rate axis, is only valid for Newtonian fluids. For non-Newtonian fluids, the concept of apparent viscosity is introduced, which is defined by the ratio of shear stress to shear rate:

$$\eta = \frac{\tau}{\dot{\gamma}} \tag{1.106}$$

This ratio, for pseudoplastic and dilatant fluids, is a function of the shear rate, but for a Newtonian fluid it is constant and equal to the Newtonian viscosity, i. e. $\eta = \mu$ (Figure 1.13).

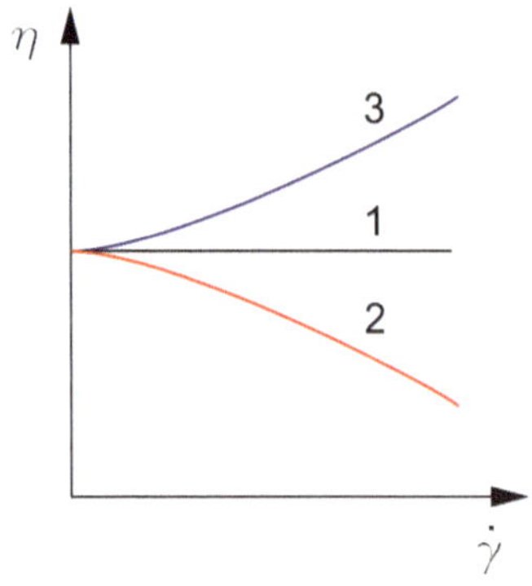

Figure 1.13
Apparent viscosity of fluids without yield stress:
1 - Newtonian fluid, 2 - shear thinning fluid,
3 - shear thickening fluid

1.2.3 Viscosity of Polymers

Polymers can be broadly divided according to their ability to crosslink. Thus, one can distinguish:

- thermoplastics that do not have the ability to crosslink,
- elastomers that form a weakly crosslinked structure,
- thermosets that form a strongly crosslinked structure in the solid state.

The basic group of polymers is thermoplastics, which in dependence on the temperature can reversibly change their state. When heated, they change from a solid to a liquid, and when cooled from a liquid to a solid. These transformations can take place many times, so that these polymers can be processed repeatedly.

Thermoplastics can have an amorphous structure (disordered) or partially crystalline (partially ordered), i.e. crystalline only in certain areas and in the remaining amorphous. The amorphous polymers harden at a temperature lower than the glass transition temperature T_g, whereas the partially crystalline polymers harden at a temperature lower than the crystallization (or melting) temperature T_m. However, the amorphous areas that occur in partially crystalline polymers harden only at a temperature lower than the glass transition temperature.

The glass transition temperature of amorphous polymers is higher than room temperature, while the glass transition temperature of semi-crystalline polymers is usually lower than 0 °C. It is generally assumed that the glass transition temperature of these materials is equal to 2/3 of the melting temperature (in Kelvin).

$$T_g = \frac{2}{3} T_m \tag{1.107}$$

where T_g is the glass transition temperature, T_m is the melting point.

Thermoplastics do not exhibit high-elastic deformation at room temperature.

Thermosets and **elastomers** have the ability to form a crosslinked structure, to a lesser extent (elastomers) or larger extent (thermosets, curable). The crosslinking process is irreversible, so that these materials cannot reversibly change their state depending on the temperature change. They cannot be processed multiple times.

Elastomers are characterized by the ability to undergo high-elastic deformations at room temperature. These are the materials whose glass transition temperature is lower than room temperature. Thermosets (cured materials) do not show high-elastic deformation at room temperature. The glass transition temperature of these materials is therefore higher than room temperature and sometimes close to the temperature of their thermal degradation. Elastomers and thermosets have an amorphous structure.

Polymers in a molten state are most often non-Newtonian pseudoplastic fluids, i. e. shear thinning fluids. The viscosity of polymers depends on the shear rate and decreases with increasing the shear rate. The viscosity of polymers is also dependent on the temperature and pressure. As the temperature increases the viscosity decreases, while as the pressure increases the viscosity also increases. For a given type of the material, the viscosity depends on the molecular weight (mass) and its distribution. It increases with an increase in molecular weight, but decreases with broadening of its distribution. The viscosity of some, relatively rare, polymeric materials depends on the time of deformation. These are the so-called **rheologically unstable fluids, thixotropic** or **antithixotropic.** The viscosity of crosslinking polymers depends on the degree of crosslinking. As it increases, the viscosity increases until the polymer reaches a solid state.

Polymers have a characteristic flow curve and a characteristic apparent viscosity function (Figure 1.14).

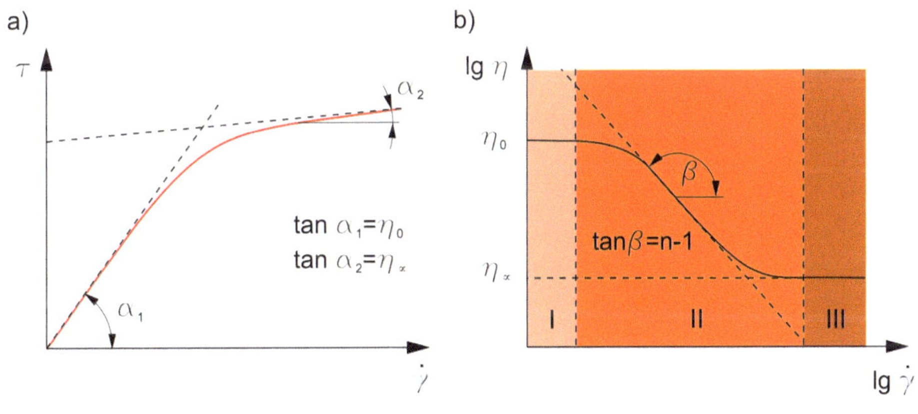

Figure 1.14 Flow characteristics of polymers: a) flow curve, b) viscosity curve; I - the first region of Newtonian flow, η_0 - the zero shear rate viscosity, II - the region of non-Newtonian flow, n - power law exponent, III - the second region of Newtonian flow, η_∞ - the infinite shear rate viscosity

In the range of low shear rate, there is a constant ratio of shear stress to shear rate, that is, the polymer behaves like a Newtonian fluid with a constant viscosity η_0 (it is equal to the tangent of the initial slope angle α_1 of the flow curve with abscissa). Thus, the viscosity η_0 is the limit viscosity of the polymer at a shear rate tending to zero. It is referred to as the **zero shear rate viscosity:**

$$\eta_0 = \eta(\dot{\gamma} \to 0) \tag{1.108}$$

Within the range of intermediate shear rates, the ratio of shear stress to shear rate is not constant and decreases with increasing shear rate, which characterizes the behavior of non-Newtonian pseudoplastic fluids. It is characteristic that this ratio in a double logarithmic system of the shear stress versus shear rate is usually approximately constant. So, in a double-logarithmic system, the viscosity is a linear function of the shear rate. This part of the flow curve can be often well described by the power law model (Eq. (1.104)).

The second area of the Newtonian behavior of polymers occurs at very high shear rates. The slope of the flow curve is again constant, and the tangent of the slope angle α_2 determines the viscosity limit η_∞ at the shear rate tending to infinity. It is referred to as the **infinite shear rate viscosity:**

$$\eta_\infty = \eta(\dot{\gamma} \to \infty) \tag{1.109}$$

As a simplification, it can be assumed that the first area of Newtonian behavior of polymers occurs at shear rates lower than 10^{-1}–10^{1} s^{-1}. In contrast, the second Newtonian area appears at shear rates of 10^{5}–10^{6} s^{-1}. At intermediate shear rates, polymers behave like non-Newtonian pseudoplastic fluids whose flow curves can be roughly described by the power law model. It is worth adding that polymer processing is usually carried out under non-Newtonian flow conditions, i.e. at shear rates of 10^{2}–10^{4} s^{-1}.

Figure 1.15 to Figure 1.22 show examples of viscosity curves as a function of shear rate and temperature for basic thermoplastics [42], semi-crystalline polymers (LDPE, HDPE, PP, PA), and amorphous ones (PS, PVC, PMMA, PC).

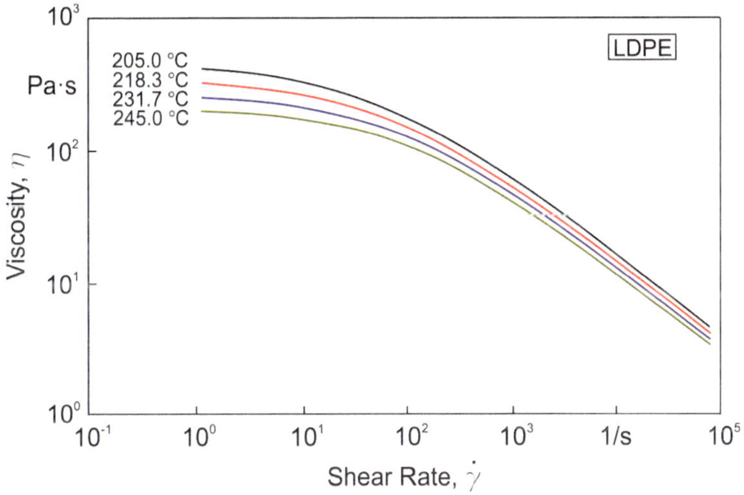

Figure 1.15 Viscosity curves of low density polyethylene (LDPE) Lupolen 1800 S (Basell Polyolefins Europe), adapted from Moldflow [42]

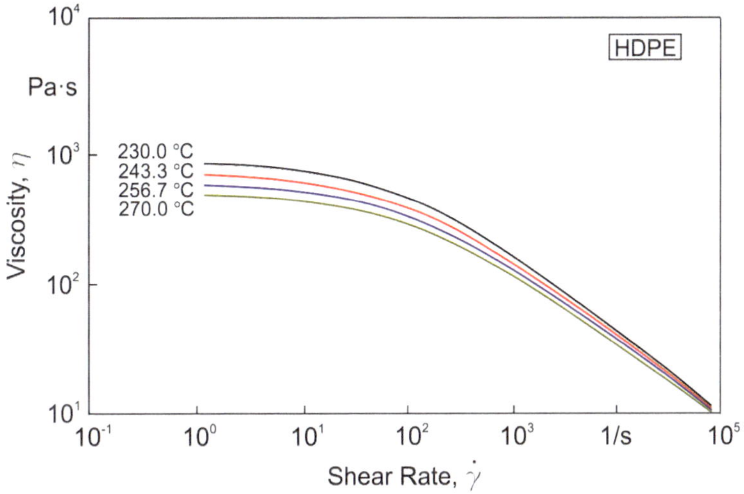

Figure 1.16 Viscosity curves of high density polyethylene (HDPE) Rigidex HD 6070 EA (BP Chemicals), adapted from Moldflow [42]

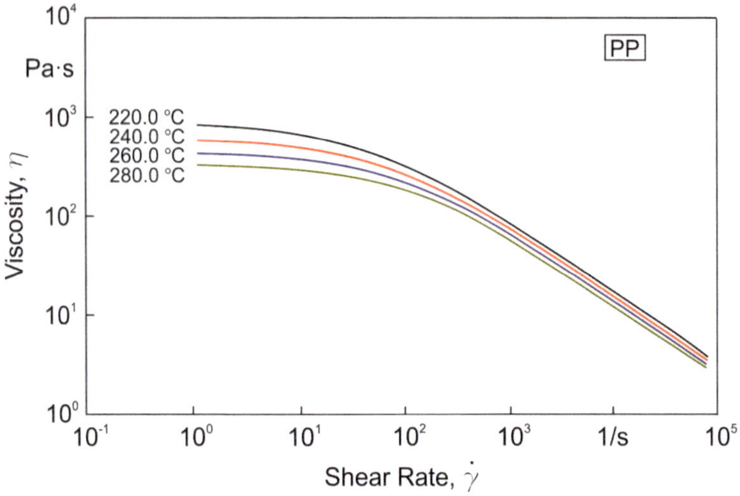

Figure 1.17 Viscosity curves of polypropylene (PP) Moplen EP240P (Basell Polyolefins Europe), adapted from Moldflow [42]

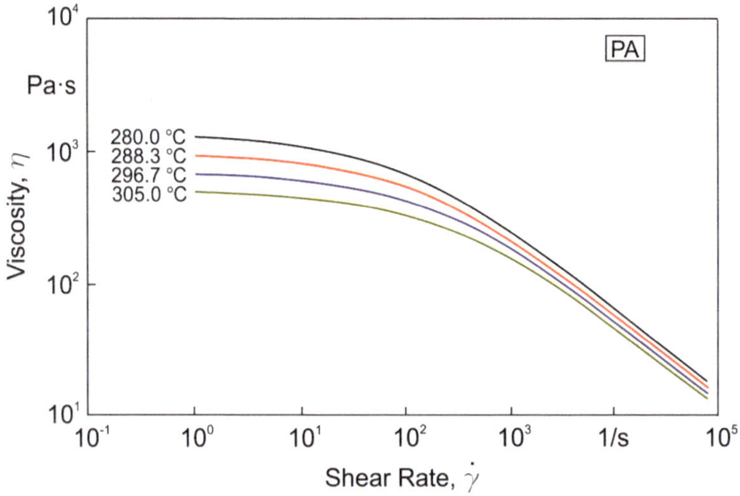

Figure 1.18 Viscosity curves of polyamide (PA 66) Zytel 70G33 (DuPont), adapted from Moldflow [42]

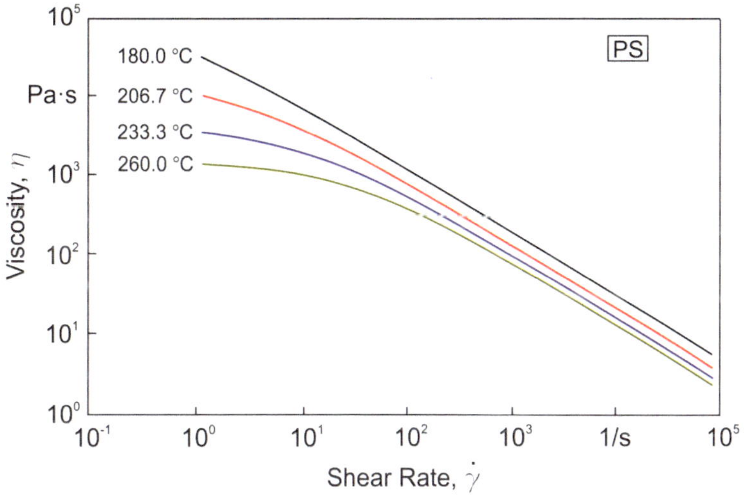

Figure 1.19 Viscosity curves of polystyrene (PS) Polystyrol 158 K (BASF), adapted from Moldflow [42]

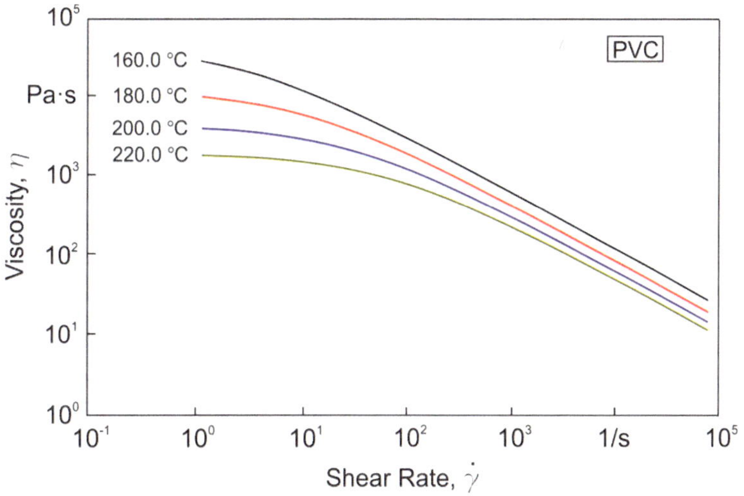

Figure 1.20 Viscosity curves of polyvinyl chloride (PVC) Vestolit 3233 (Begra), adapted from Moldflow [42]

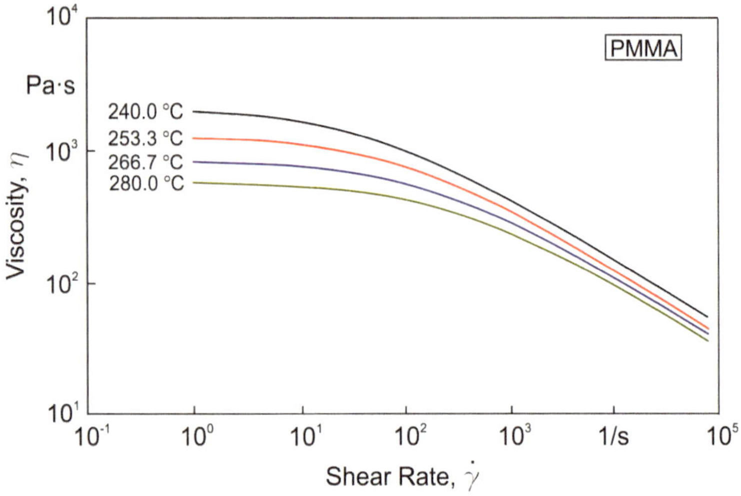

Figure 1.21 Viscosity curves of polymethyl methacrylate (PMMA) CP51 (Ineos), adapted from Moldflow [42]

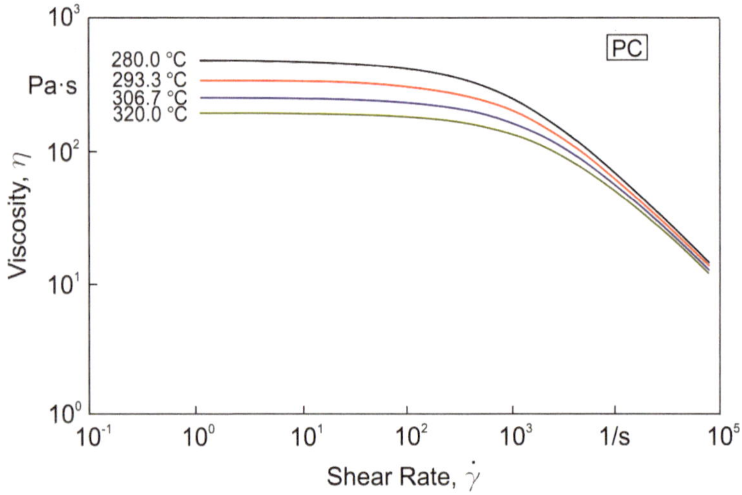

Figure 1.22 Viscosity curves of polycarbonate (PC) Makrolon AL2447 (Bayer), adapted from Moldflow [42]

1.2.3.1 Effect of Temperature on Viscosity

The viscosity of polymers decreases with increasing temperature, which can be explained by the increase in the mobility of macromolecules and the free volume between them.

For most polymers (Figure 1.15 to Figure 1.22) it is characteristic that

- the effect of temperature on the viscosity is much higher at lower shear rates than at higher shear rates, and it is especially important for values close to the zero shear rate viscosity η_0, i.e. for $\dot{\gamma} \to 0$,
- viscosity curves at different temperatures have a similar shape and are only shifted relative to each other.

The shift of viscosity curves is usually characterized by the parameter a_T. It determines the position of the viscosity curve at the current temperature relative to the viscosity curve at the reference temperature. This parameter can be expressed by

$$\lg a_T = \lg\left(\frac{\eta(T)}{\eta(T_r)}\right) \tag{1.110}$$

where a_T is the shift factor; $\eta(T)$ is the viscosity at a temperature T; $\eta(T_r)$ is the viscosity at a reference temperature T_r.

The dependence of viscosity on the temperature can be approximated by three different exponential formulas.

For semi-crystalline polymers, the **Arrhenius law** is used, which is expressed in terms of the formula

$$\eta(T) = \eta_r \exp\left[\frac{E}{R}\left(\frac{1}{T} - \frac{1}{T_r}\right)\right] \tag{1.111}$$

where $\eta(T)$ is the viscosity of polymer at a temperature T, Pa·s; T is the temperature, K; η_r is the viscosity at a reference temperature T_r, Pa·s; T_r is the reference temperature, K; E is the activation energy of polymer, J/mol; R is the gas constant, $R = 8.314$ J/(mol·K).

Since the activation energy of polymers depends on the temperature, Eq. (1.111) is a certain approximation for a given value E, at a temperature of about 50 K. From this equation, the viscosity at the temperature T can be determined when the viscosity at the reference temperature T_r is known, i.e. $\eta_r = \eta(T_r)$.

In the case of amorphous materials, the **WLF equation** (Williams, Landel, and Ferry) is used:

$$\eta(T) = \eta_r \exp\left[\frac{-C_1(T - T_r)}{C_2 + (T - T_r)}\right] \tag{1.112}$$

where $\eta(T)$ is the polymer viscosity at a temperature T, Pa·s; T is the temperature, °C; η_r is the viscosity at a reference temperature T_r, Pa·s; T_r is the reference temperature, °C; C_1, C_2 are the material constants dependent on the reference temperature T_r.

Eq. (1.112) is valid when the temperature is within the range between the glass transition temperature T_g and T_g + 100 °C. For a large group of polymers, the parameters C_1, C_2 have constant values, $C_1 = 8.86$ and $C_2 = 101.6$, when the reference temperature T_r is the so-called standard temperature $T_s = T_g$ + 50 °C. Then, the WLF equation takes the form

$$\eta(T) = \eta_r \exp\left[\frac{-8.86\left(T - T_g - 50\right)}{101.6 + \left(T - T_g - 50\right)}\right] \tag{1.113}$$

From Eq. (1.113), the viscosity of polymer at the temperature T can be determined, i.e. $\eta(T)$, when the viscosity of this polymer at other temperature T_1 is known, i.e. $\eta(T_1)$, according to the formula

$$\eta(T) = \eta(T_1)\exp\left[\frac{8.86\left(T_1 - T_g - 50\right)}{101.6 + \left(T_1 - T_g - 50\right)} - \frac{8.86\left(T - T_g - 50\right)}{101.6 + \left(T - T_g - 50\right)}\right] \tag{1.114}$$

The third equation is an empirical rule, unlike the Arrhenius and WLF equations, which have some theoretical justification, and expresses the effect of temperature on the viscosity as follows:

$$\eta(T) = \eta_r \exp\left[-b\left(T - T_r\right)\right] \tag{1.115}$$

where b is the material constant dependent on the temperature.

The viscosity dependence on the temperature for different polymers is varied. In general, it can be assumed that it is stronger for amorphous polymers than for partially crystalline polymers. For example, the viscosity of polyvinyl chloride (PVC) (Figure 1.20) and polymethyl methacrylate (PMMA) (Figure 1.21), which are amorphous polymers, is strongly dependent on the temperature, whereas the viscosity of polyethylene (PE) (Figure 1.15 and Figure 1.16) and polypropylene (PP) (Figure 1.17), as semi-crystalline polymers, is insensitive to temperature changes. In addition, the relationship between viscosity and temperature is stronger for values close to the glass transition temperature, and weakens as the temperature rises.

The sensitivity of polymer viscosity to the temperature change is expressed by the relative viscosity change per 1 °C, that is

$$\frac{1}{\eta}\frac{\partial \eta}{\partial T} \tag{1.116}$$

The sensitivity of polymer viscosity to temperature and the glass transition temperature T_g for the selected polymers are shown in Table 1.6.

Table 1.6 The Glass Transition Temperature T_g and the Sensitivity of Polymer Viscosity to Temperature for Selected Polymers

Polymer	T_g, °C	$\frac{1}{\eta}\frac{\partial \eta}{\partial T}$, °C^{-1}
Polystyrene	100	0.10
Polyvinyl chloride	80	0.20
Polycarbonate	150	0.05
Low density polyethylene	-100	0.03
High density polyethylene	-100	0.02
Polypropylene	-20	0.02
Polyamide 6.6	60	0.02

1.2.3.2 Effect of Pressure on Viscosity

The pressure affects the viscosity of polymers inversely to the temperature, i.e. as the pressure increases, the viscosity also increases. This is due to the decrease in the free volume and mobility of macromolecules.

The viscosity dependence on the pressure can be represented by an exponential formula similar to Eq. (1.115), that is

$$\eta(p) = \eta_r \exp\left[b_p\left(p - p_r\right)\right] \tag{1.117}$$

where $\eta(p)$ is the viscosity at a pressure p; p is the pressure; η_r is the viscosity at a reference pressure p_r; p_r is the reference pressure; b_p is the material constant.

Based on this equation, the polymer viscosity at the pressure p can be determined when the viscosity at the reference pressure p_r is known, i.e. $\eta_r = \eta(p_r)$.

The influence of pressure on the viscosity of polymers is negligible when the pressure does not exceed 35 MPa, that is, for most processes (excluding injection molding). However, the influence of pressure becomes very significant when it is higher than 35 MPa. The viscosity dependence on the pressure for various polymers is different but similar to the viscosity dependence on the temperature.

The sensitivity of polymer viscosity to the pressure change is expressed by the relative viscosity change per 1 Pa:

$$\frac{1}{\eta}\frac{\partial \eta}{\partial p} \tag{1.118}$$

For many polymers, the ratio of the temperature sensitivity of viscosity to the pressure sensitivity of viscosity is approximately constant and equal to

$$\frac{\left(\frac{1}{\eta}\frac{\partial\eta}{\partial p}\right)}{\left(\frac{1}{\eta}\frac{\partial\eta}{\partial T}\right)} \approx -5e^{-7},\ {}^\circ\mathrm{C/Pa} \tag{1.119}$$

1.2.3.3 Effect of Molecular Weight on Viscosity

The viscosity of polymers also depends on their molecular weight and its distribution.

With increase of the molecular weight in the limit zero shear rate viscosity of polymers η_0 is increased, which is depicted Figure 1.23. For polymers with a molecular weight less than some critical weight M_c, this dependence is linear. On the other hand, for polymers with a molecular weight higher than this critical weight, the dependence is exponential and it is expressed by a formula that has some theoretical justification:

$$\eta_0 = KM_w^\alpha \tag{1.120}$$

where η_0 is the zero viscosity (i.e. for $\dot{\gamma} \to 0$); M_w is the weight average molecular mass; K, α are the material constants.

Exponent α is approximately constant for most polymers and is $\alpha = 3.4$.

The influence of the molecular weight distribution on the viscosity of polymers is generally known (Figure 1.24). Namely, the broad molecular weight distribution results in a decrease in the Newtonian range of the flow curve and in a greater shear thinning.

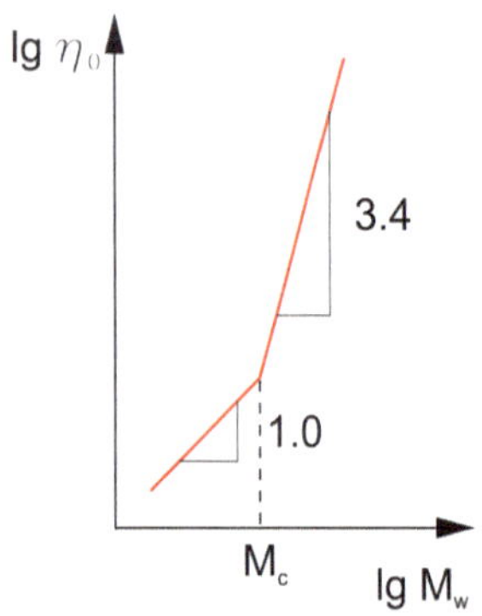

Figure 1.23
Dependence of the zero viscosity η_0 of polymers on the molecular weight M_w: M_c – the critical weight

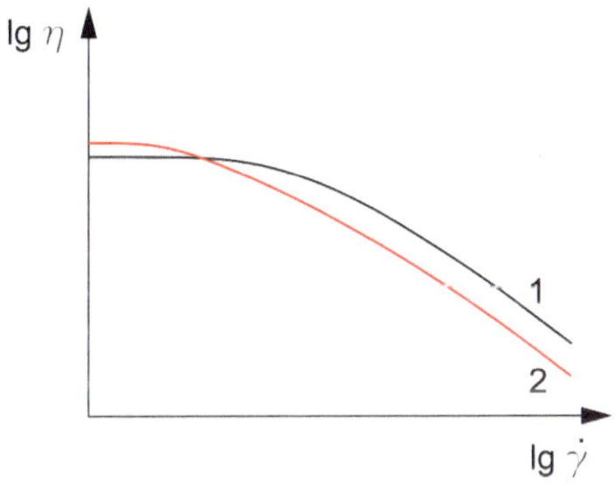

Figure 1.24
An effect of distribution of the molecular weight of polymers on the viscosity of polymers: 1 - narrow molecular weight distribution, 2 - broad molecular weight distribution

1.2.3.4 Effect of Time of Deformation on Viscosity

The viscosity of some, relatively rare, polymeric materials also depends on the time of deformation. They are called rheologically unstable, thixotropic or anti-thixotropic, fluids.

The **phenomenon of thixotropy** is that under the conditions of isothermal flow of the fluid, which previously had a longer rest, the shear stress decreases reversibly over time (at a constant shear rate). On the other hand, the **phenomenon of antithixotropy** is characterized by the fact that in the same conditions the shear stress increases reversibly over time. Thus, the viscosity of thixotropic fluids decreases with time, while the viscosity of antithixotropic fluids increases.

The characteristic hysteresis loop (Figure 1.25(a)) for the thixotropic system can be obtained by measuring the dependence of the shear stress on the shear rate. This is done by continuously increasing the shear rate from zero to a certain maximum value and, immediately upon reaching it, reducing the shear rate to zero.

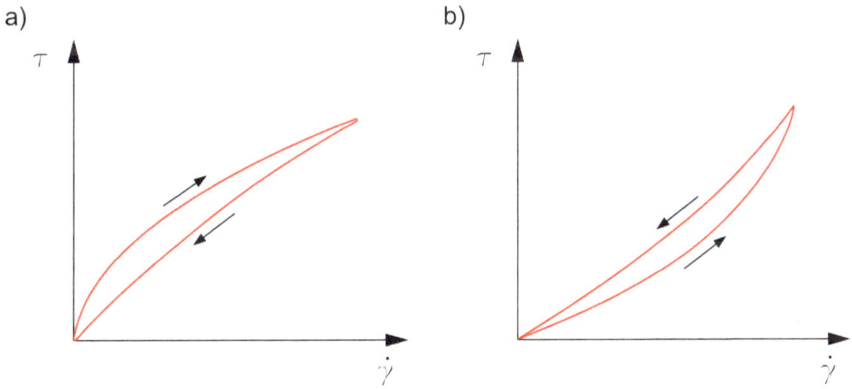

Figure 1.25 Hysteresis loop of flow curves: a) thixotropic fluid, b) antithixotropic fluid

As can be demonstrated by similar measurements for the antithixotropic systems, hysteresis of the flow curve will also occur, but its course will be different (Figure 1.25(b)).

The phenomenon of thixotropy is related to the shear thinning of polymers, whereas the phenomenon of antithixotropy is related to the shear thickening. The difference is that in the case of rheologically unstable fluids (thixotropic and antithixotropic), the rheological equilibrium at a given shear rate is obtained very slowly, so that it cannot be tracked using the available instruments. This difference is not qualitative, but only quantitative.

1.2.3.5 Effect of Crosslinking on Viscosity

The viscosity of crosslinking (curable) polymers depends on the degree of crosslinking, which is determined by the duration of the crosslinking reaction, so that

$$\eta = \eta\left(\dot{\gamma}, T, t\right) \tag{1.121}$$

where η is the viscosity; $\dot{\gamma}$ is the shear rate; T is the temperature; t is the time.

With an increase of crosslinking time, the viscosity increases continuously until the polymer reaches a solid state. The viscosity changes as a function of time and temperature can be generally represented by the following equation:

$$\ln \eta = C_1 + \frac{E_\eta}{RT} + C_2 t \exp\left(\frac{E_k}{RT}\right) \tag{1.122}$$

where E_η is the activation energy of flow of uncured polymer; E_k is the kinetic activation energy; R is the gas constant; t is the time; T is the temperature; C_1, C_2 are the equation parameters.

1.2.4 Rheological Models of Viscous Fluids

1.2.4.1 Newtonian Fluid

Generalization of Newton's Law

Newton's law, expressed by Eq. (1.100), represents the linear relationship between the shear stress τ_{yx} and the velocity gradient dv_x/dy and refers to one-dimensional flow, a simple shear flow (Figure 1.9), for which the velocity field has the form

$$v_x = \dot{\gamma} y, \quad v_y = 0, \quad v_z = 0 \tag{1.123}$$

where $\dot{\gamma} = dv_x/dy$ is the shear rate.

In order to extend the scope of this law to multidimensional flows, for which the velocity field in the most general case of three-dimensional flow takes the form

$$v_x = v_x(x,y,z), \quad v_y = v_y(x,y,z), \quad v_z = v_z(x,y,z) \tag{1.124}$$

the relationship in Eq. (1.100) must be presented as a tensor dependence. Then, Newton's law (for an incompressible fluids) is expressed by

$$\underset{\sim}{\boldsymbol{\tau}} = \mu 2 \underset{\sim}{\boldsymbol{D}} \tag{1.125}$$

$$\underset{\sim}{\boldsymbol{\sigma}} = -p\underset{\sim}{\mathbf{1}} + \mu 2 \underset{\sim}{\boldsymbol{D}} \tag{1.126}$$

where $\underset{\sim}{\boldsymbol{\sigma}}$ is the total stress tensor; $\underset{\sim}{\boldsymbol{\tau}}$ is the extra-stress tensor; p is the hydrostatic pressure; $\underset{\sim}{\mathbf{1}}$ is the tensor unity defined by the Eqs. (1.7) to (1.9); $\underset{\sim}{\boldsymbol{D}}$ is the rate deformation tensor (Eq. (1.20)).

Based on these equations, it is possible to determine the components of the extra-stress tensor $\underset{\sim}{\boldsymbol{\tau}}$, namely

$$\tau_{xx} = 2\mu \frac{\partial v_x}{\partial x} \tag{1.127}$$

$$\tau_{yy} = 2\mu \frac{\partial v_y}{\partial y} \tag{1.128}$$

$$\tau_{zz} = 2\mu \frac{\partial v_z}{\partial z} \tag{1.129}$$

$$\tau_{xy} = \tau_{yx} = \mu \left(\frac{\partial v_x}{\partial y} + \frac{\partial v_y}{\partial x} \right) \tag{1.130}$$

$$\tau_{xz} = \tau_{zx} = \mu \left(\frac{\partial v_x}{\partial z} + \frac{\partial v_z}{\partial x} \right) \tag{1.131}$$

$$\tau_{yz} = \tau_{zy} = \mu \left(\frac{\partial v_y}{\partial z} + \frac{\partial v_z}{\partial y} \right) \tag{1.132}$$

and components of the total stress tensor $\underset{\sim}{\boldsymbol{\sigma}}$, namely

$$\sigma_{xx} = -p + 2\mu \frac{\partial v_x}{\partial x} \tag{1.133}$$

$$\sigma_{yy} = -p + 2\mu \frac{\partial v_y}{\partial y} \tag{1.134}$$

$$\sigma_{zz} = -p + 2\mu \frac{\partial v_z}{\partial z} \tag{1.135}$$

$$\sigma_{xy} = \sigma_{yx} = \mu \left(\frac{\partial v_x}{\partial y} + \frac{\partial v_y}{\partial x} \right) \tag{1.136}$$

$$\sigma_{xz} = \sigma_{zx} = \mu\left(\frac{\partial v_x}{\partial z} + \frac{\partial v_z}{\partial x}\right) \tag{1.137}$$

$$\sigma_{yz} = \sigma_{zy} = \mu\left(\frac{\partial v_y}{\partial z} + \frac{\partial v_z}{\partial y}\right) \tag{1.138}$$

Similarly, the components of the extra-stress tensor of Newtonian fluid in the cylindrical system can be represented (Table 1.7).

It can be seen that under the simple shear conditions (Figure 1.9) described by the velocity field (Eq. (1.123)), the Newton equation (1.126) is reduced to the formula

$$\sigma_{ij} = -p\begin{vmatrix}1 & 0 & 0\\0 & 1 & 0\\0 & 0 & 1\end{vmatrix} + \mu\begin{vmatrix}0 & \dot{\gamma} & 0\\\dot{\gamma} & 0 & 0\\0 & 0 & 0\end{vmatrix} \tag{1.139}$$

where $\dot{\gamma} = \partial v_x/\partial y$.

As a result, we have

$$\tau_{xy} = \tau_{yx} = \mu\frac{\partial v_x}{\partial y} \tag{1.140}$$

which corresponds to the Newton equation (1.100).

In addition, under the conditions of this flow, the only normal component of the stress tensor $\underset{\sim}{\sigma}$ is the hydrostatic pressure p, so that the difference in normal stresses defined by Eqs. (1.11) and (1.12) is zero, i.e. $N_1 = N_2 = 0$.

Table 1.7 Components of the Extra-Stress Tensor $\underset{\sim}{\tau}$ of an Incompressible Fluid in Different Coordinate Systems

Cartesian system *xyz*		
$\tau_{xx} = 2\mu\frac{\partial v_x}{\partial x}$	$\tau_{xy} = \tau_{yx}$	$\tau_{xz} = \tau_{zx}$
$\tau_{yx} = \mu\left(\frac{\partial v_x}{\partial y} + \frac{\partial v_y}{\partial x}\right)$	$\tau_{yy} = 2\mu\frac{\partial v_y}{\partial y}$	$\tau_{yz} = \tau_{zy}$
$\tau_{zx} = \mu\left(\frac{\partial v_x}{\partial z} + \frac{\partial v_z}{\partial x}\right)$	$\tau_{zy} = \mu\left(\frac{\partial v_y}{\partial z} + \frac{\partial v_z}{\partial y}\right)$	$\tau_{zz} = 2\mu\frac{\partial v_z}{\partial z}$
Cylindrical system *rφz*		
$\tau_{rr} = 2\mu\frac{\partial v_r}{\partial r}$	$\tau_{r\varphi} = \tau_{\varphi r}$	$\tau_{rz} = \tau_{zr}$
$\tau_{\varphi r} = \mu\left[r\frac{\partial}{\partial r}\left(\frac{v_\varphi}{r}\right) + \frac{1}{r}\frac{\partial v_r}{\partial \varphi}\right]$	$\tau_{\varphi\varphi} = 2\mu\left(\frac{1}{r}\frac{\partial v_\varphi}{\partial \varphi} + \frac{v_r}{r}\right)$	$\tau_{\varphi z} = \tau_{z\varphi}$
$\tau_{zr} = \mu\left(\frac{\partial v_r}{\partial z} + \frac{\partial v_z}{\partial r}\right)$	$\tau_{z\varphi} = \mu\left(\frac{\partial v_\varphi}{\partial z} + \frac{1}{r}\frac{\partial v_z}{\partial \varphi}\right)$	$\tau_{zz} = 2\mu\frac{\partial v_z}{\partial z}$

Newton's equation of Eq. (1.125) or Eq. (1.126) is a constitutive equation that provides in the general case six independent equations. These equations allow you to close the system of equations of continuum mechanics and make them solvable (see Section 1.1.4). Table 1.8 and Table 1.9 show the equation of momentum (formulated in this case by Navier and Stokes) and the energy equation in the rectangular (Cartesian) and cylindrical system for the incompressible Newtonian fluid.

Table 1.8 Equation of Momentum (Navier-Stokes) for an Incompressible Fluid in Different Coordinate Systems

Cartesian system *xyz*

$$\rho\left(\frac{\partial v_x}{\partial t}+v_x\frac{\partial v_x}{\partial x}+v_y\frac{\partial v_x}{\partial y}+v_z\frac{\partial v_x}{\partial z}\right)=-\frac{\partial p}{\partial x}+\mu\left(\frac{\partial^2 v_x}{\partial x^2}+\frac{\partial^2 v_x}{\partial y^2}+\frac{\partial^2 v_x}{\partial z^2}\right)+\rho g_x$$

$$\rho\left(\frac{\partial v_y}{\partial t}+v_x\frac{\partial v_y}{\partial x}+v_y\frac{\partial v_y}{\partial y}+v_z\frac{\partial v_y}{\partial z}\right)=-\frac{\partial p}{\partial y}+\mu\left(\frac{\partial^2 v_y}{\partial x^2}+\frac{\partial^2 v_y}{\partial y^2}+\frac{\partial^2 v_y}{\partial z^2}\right)+\rho g_y$$

$$\rho\left(\frac{\partial v_z}{\partial t}+v_x\frac{\partial v_z}{\partial x}+v_y\frac{\partial v_z}{\partial y}+v_z\frac{\partial v_z}{\partial z}\right)=-\frac{\partial p}{\partial z}+\mu\left(\frac{\partial^2 v_z}{\partial x^2}+\frac{\partial^2 v_z}{\partial y^2}+\frac{\partial^2 v_z}{\partial z^2}\right)+\rho g_z$$

Cylindrical system $r\varphi z$

$$\rho\left(\frac{\partial v_r}{\partial t}+v_r\frac{\partial v_r}{\partial r}+\frac{v_\varphi}{r}\frac{\partial v_r}{\partial \varphi}-\frac{v_\varphi^2}{r}+v_z\frac{\partial v_r}{\partial z}\right)=$$

$$=-\frac{\partial p}{\partial r}+\mu\left[\frac{\partial}{\partial r}\left(\frac{1}{r}\frac{\partial}{\partial r}(rv_r)\right)+\frac{1}{r^2}\frac{\partial^2 v_r}{\partial \varphi^2}-\frac{2}{r^2}\frac{\partial v_\varphi}{\partial \varphi}+\frac{\partial^2 v_r}{\partial z^2}\right]+\rho g_r$$

$$\rho\left(\frac{\partial v_\varphi}{\partial t}+v_r\frac{\partial v_\varphi}{\partial r}+\frac{v_\varphi}{r}\frac{\partial v_\varphi}{\partial \varphi}-\frac{v_r v_\varphi}{r}+v_z\frac{\partial v_\varphi}{\partial z}\right)=$$

$$=-\frac{1}{r}\frac{\partial p}{\partial \varphi}+\mu\left[\frac{\partial}{\partial r}\left(\frac{1}{r}\frac{\partial}{\partial r}(rv_\varphi)\right)+\frac{1}{r^2}\frac{\partial^2 v_\varphi}{\partial \varphi^2}+\frac{2}{r^2}\frac{\partial v_r}{\partial \varphi}+\frac{\partial^2 v_\varphi}{\partial z^2}\right]+\rho g_\varphi$$

$$\rho\left(\frac{\partial v_z}{\partial t}+v_r\frac{\partial v_z}{\partial r}+\frac{v_\varphi}{r}\frac{\partial v_z}{\partial \varphi}+v_z\frac{\partial v_z}{\partial z}\right)=$$

$$=-\frac{\partial p}{\partial z}+\mu\left[\frac{1}{r}\frac{\partial}{\partial r}\left(r\frac{\partial v_z}{\partial r}\right)+\frac{1}{r^2}\frac{\partial^2 v_z}{\partial \varphi^2}+\frac{\partial^2 v_z}{\partial z^2}\right]+\rho g_z$$

Table 1.9 Equation of Energy for an Incompressible Fluid in Different Coordinate Systems

Cartesian system xyz
$\rho c_p\left(\frac{\partial T}{\partial t}+v_x\frac{\partial T}{\partial x}+v_y\frac{\partial T}{\partial y}+v_z\frac{\partial T}{\partial z}\right)=k\left(\frac{\partial^2 T}{\partial x^2}+\frac{\partial^2 T}{\partial y^2}+\frac{\partial^2 T}{\partial z^2}\right)+$ $+2\mu\left\{\left(\frac{\partial v_x}{\partial x}\right)^2+\left(\frac{\partial v_y}{\partial y}\right)^2+\left(\frac{\partial v_z}{\partial z}\right)^2\right\}+$ $+\mu\left\{\left(\frac{\partial v_x}{\partial y}+\frac{\partial v_y}{\partial x}\right)^2+\left(\frac{\partial v_x}{\partial z}+\frac{\partial v_z}{\partial x}\right)^2+\left(\frac{\partial v_y}{\partial z}+\frac{\partial v_z}{\partial y}\right)^2\right\}$
Cylindrical system $r\varphi z$
$\rho c_p\left(\frac{\partial T}{\partial t}+v_r\frac{\partial T}{\partial r}+\frac{v_\varphi}{r}\frac{\partial T}{\partial \varphi}+v_z\frac{\partial T}{\partial z}\right)=k\left[\frac{1}{r}\frac{\partial}{\partial r}\left(r\frac{\partial T}{\partial r}\right)+\frac{1}{r^2}\frac{\partial^2 T}{\partial \varphi^2}+\frac{\partial^2 T}{\partial z^2}\right]+$ $+2\mu\left\{\left(\frac{\partial v_r}{\partial r}\right)^2+\left[\frac{1}{r}\left(\frac{\partial v_\varphi}{\partial \varphi}+v_r\right)\right]^2+\left(\frac{\partial v_z}{\partial z}\right)^2\right\}+$ $+\mu\left\{\left[r\frac{\partial}{\partial r}\left(\frac{v_\varphi}{r}\right)+\frac{1}{r}\frac{\partial v_r}{\partial \varphi}\right]^2+\left(\frac{\partial v_z}{\partial r}+\frac{\partial v_r}{\partial z}\right)^2+\left(\frac{1}{r}\frac{\partial v_z}{\partial \varphi}+\frac{\partial v_\varphi}{\partial z}\right)^2\right\}$

Elongational Viscosity

The concept of viscosity introduced by Newton in 1687 refers to the shear flow; it is called shear viscosity or transverse viscosity. According to Eq. (1.100), it is defined as the ratio of shear stress to shear rate:

$$\mu=\frac{\tau}{\dot{\gamma}}=\frac{\tau_{yx}}{\mathrm{d}v_x/\mathrm{d}y} \tag{1.141}$$

As shear flows are common in polymer processing, e. g. in extrusion and injection molding, the shear viscosity is of fundamental importance, but does not exhaust all possibilities for describing the behavior of polymers during processing. There are many processes in which stretching plays a dominant role, such as fiber formation, blow molding, thermoforming, etc.

The concept of viscosity in relation to elongational flows was introduced by Trouton in 1906. He determined the measure of material resistance during such flows as the ratio of normal stress to strain rate under stretching.

In order to define the elongational viscosity, a fluid element of length L and cross-section A, which is uniformly stretched by force F with a velocity U (Figure 1.26), is considered. The ratio of force F to area A expresses the normal stress $\sigma_n=\sigma_{xx}$, while the ratio of velocity U to the element length L is equal to the velocity gradient

v_x in the elongation direction x, i. e. dv_x/dx. It can be shown that the velocity gradient is equivalent to the velocity of deformation of the fluid element caused by the normal stress σ_{xx}. Hence, this gradient is often called the elongation rate.

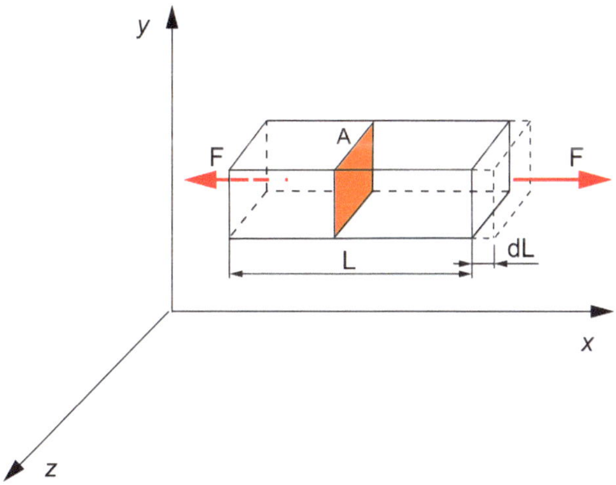

Figure 1.26 Uniaxial elongation (stretching)

As shown in Figure 1.26, the fluid element of length L after the time dt is extended by the length dL. Thus, the deformation will be

$$d\varepsilon = \frac{dL}{L} \tag{1.142}$$

Since $dL = U dt$, we obtain the expression for deformation rate, i. e. the **elongation rate:**

$$\frac{d\varepsilon}{dt} = \dot{\varepsilon} = \frac{dL}{L}\frac{1}{dt} = \frac{U}{L} = \frac{dv_x}{dx} \tag{1.143}$$

Ultimately, according to the definition of Trouton, one can write

$$\eta_e = \frac{\sigma_n}{\dot{\varepsilon}} = \frac{\sigma_{xx}}{dv_x / dx} \tag{1.144}$$

where η_e is the viscosity at elongation.

The elongational viscosity is also called the longitudinal viscosity.

To find a relationship between shear viscosity and elongational viscosity, uniaxial elongational flow of a Newtonian fluid (Eq. (1.126)) was investigated (Figure 1.26). The velocity field of such flow is expressed as

$$v_x = \dot{\varepsilon}_1 x, \quad v_y = \dot{\varepsilon}_2 y, \quad v_z = \dot{\varepsilon}_3 z \tag{1.145}$$

where

$$\dot{\varepsilon}_1 = \frac{\partial v_x}{\partial x}, \quad \dot{\varepsilon}_2 = \frac{\partial v_y}{\partial y}, \quad \dot{\varepsilon}_3 = \frac{\partial v_z}{\partial z} \tag{1.146}$$

At such conditions, the components of the rate of deformation tensor $\underset{\sim}{\boldsymbol{D}}$ (Eq. (1.20)) are equal to

$$D_{ij} = \begin{bmatrix} \dot{\varepsilon}_1 & 0 & 0 \\ 0 & \dot{\varepsilon}_2 & 0 \\ 0 & 0 & \dot{\varepsilon}_3 \end{bmatrix} \tag{1.147}$$

Since uniaxial elongation is a symmetric flow, $v_y = v_z$, and therefore $\dot{\varepsilon}_2 = \dot{\varepsilon}_3$. Then for incompressible fluids, based on the equation of continuity (Eq. (1.37)), we have

$$\dot{\varepsilon}_1 = -2\dot{\varepsilon}_2 = -2\dot{\varepsilon}_3 = \dot{\varepsilon} \tag{1.148}$$

where $\dot{\varepsilon}$ is the elongation rate.

As a result, the tensor of deformation rate expresses as

$$D_{ij} = \begin{bmatrix} \dot{\varepsilon} & 0 & 0 \\ 0 & -\dot{\varepsilon}/2 & 0 \\ 0 & 0 & -\dot{\varepsilon}/2 \end{bmatrix} \tag{1.149}$$

Thus, under conditions of uniaxial elongation (stretching), Newton's equation (1.126) is reduced to

$$\sigma_{ij} = -p \begin{bmatrix} 1 & 0 & 0 \\ 0 & 1 & 0 \\ 0 & 0 & 1 \end{bmatrix} + \mu \begin{bmatrix} 2\dot{\varepsilon} & 0 & 0 \\ 0 & -\dot{\varepsilon} & 0 \\ 0 & 0 & -\dot{\varepsilon} \end{bmatrix} \tag{1.150}$$

where $\dot{\varepsilon} = \partial v_x / \partial x$.

The normal stress tensor components are equal, respectively, to

$$\sigma_{xx} = -p + 2\mu\dot{\varepsilon} \tag{1.151}$$

$$\sigma_{yy} = \sigma_{zz} = -p - \mu\dot{\varepsilon} \tag{1.152}$$

Taking into account that in this flow $\sigma_{yy} = \sigma_{zz} = 0$, it leads to

$$\sigma_{xx} = 3\mu\dot{\varepsilon} \tag{1.153}$$

Based on Eq. (1.153) and Eq. (1.144) defining the elongational viscosity, the important formula is obtained:

$$\eta_e = 3\mu \tag{1.154}$$

Thus, for Newtonian fluids, the viscosity η_e at uniaxial stretching (i.e. longitudinal viscosity) is three times higher than the viscosity μ at shearing (i.e. transverse viscosity). It is called the **Trouton's viscosity.**

Equation (1.154) is valid for Newtonian fluids only, so in the case of polymers it is valid for very low shear and elongation rates. At higher deformation rates, the longitudinal viscosity may vary in different ways, depending on the polymer and its structure. Typically, in the case of polymers, it is from 10 to 100 times higher than the shear (transverse) viscosity. A typical course of changes in longitudinal viscosity as a function of the elongation (stretching) rate is shown in Figure 1.27.

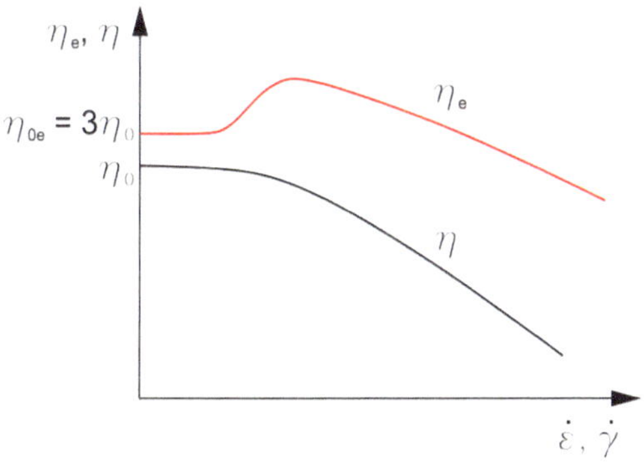

Figure 1.27
Longitudinal viscosity (at elongation) η_e and transverse viscosity (at shearing) η_e for polymers as a function of the rate of deformation: $\dot{\gamma}$ - shear rate, $\dot{\varepsilon}$ - elongation rate, η_{0e} - zero elongational viscosity, η_0 - zero shear viscosity

1.2.4.2 Generalized Newtonian Fluids

Polymers are mostly viscoelastic fluids, but many practical polymer processing issues can be solved considering them as purely viscous fluids, although of course non-Newtonian and usually thinned with shearing.

Non-Newtonian fluids (purely viscous) whose viscosity is a function of shear rate are called **generalized Newtonian fluids.** The constitutive equation of such fluids, assuming their incompressibility and limitation to shear flows, takes the form

$$\underset{\sim}{\tau} = \eta\left(\mathrm{II}_{2\underset{\sim}{D}}\right)2\underset{\sim}{\boldsymbol{D}} \tag{1.155}$$

or

$$\underset{\sim}{\sigma} = -p\underset{\sim}{\mathbf{1}} + \eta\left(\mathrm{II}_{2\underset{\sim}{D}}\right)2\underset{\sim}{\boldsymbol{D}} \tag{1.156}$$

where $\underset{\sim}{\sigma}$ is the total stress tensor; $\underset{\sim}{\tau}$ is the extra-stress tensor; p is the hydrostatic pressure; $\underset{\sim}{\mathbf{1}}$ is the tensor unity; η is the viscosity; $\mathrm{II}_{2\underset{\sim}{D}} = 2\underset{\sim}{\boldsymbol{D}} : 2\underset{\sim}{\boldsymbol{D}}$ is the second invariant of the rate of deformation tensor 2 $\underset{\sim}{\boldsymbol{D}}$; $\underset{\sim}{\boldsymbol{D}}$ is the rate of deformation tensor (Eq. (1.20)).

The dependence of viscosity on the shear rate is a function of the second invariant of the rate of deformation tensor 2 $\underset{\sim}{\boldsymbol{D}}$, i.e. $\eta = \eta\left(\mathrm{II}_{2\underset{\sim}{D}}\right)$. The expressions for this invariant in the rectangular (Cartesian) and cylindrical coordinate systems are shown in Table 1.10.

Table 1.10 The Second Invariant of the Rate of Deformation Tensor $2\underset{\sim}{\boldsymbol{D}}$ in Different Coordinate Systems

Cartesian system *xyz*
$\frac{1}{2}\mathrm{II}_{2\underset{\sim}{D}} = 2\left[\left(\frac{\partial v_x}{\partial x}\right)^2 + \left(\frac{\partial v_y}{\partial y}\right)^2 + \left(\frac{\partial v_z}{\partial z}\right)^2\right] + \left(\frac{\partial v_x}{\partial y} + \frac{\partial v_y}{\partial x}\right)^2 + \left(\frac{\partial v_x}{\partial z} + \frac{\partial v_z}{\partial x}\right)^2 + \left(\frac{\partial v_y}{\partial z} + \frac{\partial v_z}{\partial y}\right)^2$
Cylindrical system $r\varphi z$
$\frac{1}{2}\mathrm{II}_{2\underset{\sim}{D}} = 2\left[\left(\frac{\partial v_r}{\partial r}\right)^2 + \left(\frac{1}{r}\frac{\partial v_\varphi}{\partial \varphi} + \frac{v_r}{r}\right)^2 + \left(\frac{\partial v_z}{\partial z}\right)^2\right] + \left[r\frac{\partial}{\partial r}\left(\frac{v_\varphi}{r}\right) + \frac{1}{r}\frac{\partial v_r}{\partial \varphi}\right]^2 +$ $+\left(\frac{\partial v_z}{\partial r} + \frac{\partial v_r}{\partial z}\right)^2 + \left(\frac{1}{r}\frac{\partial v_z}{\partial \varphi} + \frac{\partial v_\varphi}{\partial z}\right)^2$

Usually, instead of the function $\eta = \eta\left(\mathrm{II}_{2\underset{\sim}{D}}\right)$, the function called **generalized shear rate** $\dot{\gamma}_D$ is used, i.e. $\eta = \eta(\dot{\gamma}_D)$. It is defined as

$$\dot{\gamma}_D = \sqrt{\frac{1}{2}\mathrm{II}_{2\underset{\sim}{D}}} \tag{1.157}$$

Based on Table 1.10, it can be seen that at simple shear conditions (Eq. (1.123)) for which only two components of the rate of deformation tensor are different from zero, i.e. $D_{xy} = D_{yx} = (\partial v_x/\partial y)/2$, the generalized shear rate $\dot{\gamma}_D$ is simply equal to the shear rate $\dot{\gamma}$.

There are many empirical equations that define the functions $\eta = \eta\left(\mathrm{II}_{2\underset{\sim}{D}}\right)$ or $\eta = \eta(\dot{\gamma}_D)$. They are called the rheological models of fluids.

Some rheological models that have practical application in the modeling of polymer processing are presented below. The functions of viscosity and shear stress are presented in the dependence on the shear rate $\dot{\gamma}$, noting that in the general case of three-dimensional flow they must be replaced by the generalized shear rate $\dot{\gamma}_D$.

The Power Law Model of Ostwald and de Waele

One of the basic empirical models is the power law model of Ostwald and de Waele. The basis for this model was the fact that for many polymers, in a wide range of shear rates, in a double logarithmic coordinate system there is a linear dependence of the viscosity on the shear rate. Based on this observation, the following equation was proposed:

$$\eta\left(\dot{\gamma}\right) = m\dot{\gamma}^{n-1} \tag{1.158}$$

where m, n are the model parameters; m is the consistency coefficient, $(\mathrm{N}\cdot\mathrm{s}^n)/\mathrm{m}^2$; n is the power law exponent (dimensionless).

After substituting Eq. (1.158) into Eq. (1.155), we have

$$\underset{\sim}{\tau} = m\dot{\gamma}^{n-1} 2\underset{\sim}{\boldsymbol{D}} \tag{1.159}$$

The effect of temperature on viscosity in this model is taken into account using Eq. (1.115) by

$$m(T) = m_r \exp\left[-b\left(T - T_r\right)\right] \tag{1.160}$$

where m_r is the consistency coefficient at the reference temperature T_r.

The power law model of Ostwald-de Waele is the simplest (two parameters) rheological model of generalized Newtonian fluid. Thanks to its simplicity, it is useful in practice because it enables the analytical solution of almost all theoretically developed problems for Newtonian fluids.

An important disadvantage of this model is its inconsistency to the results of rheological experiments at very low and very high shear rates. Namely,

$$\eta \to \infty \qquad \text{for } \dot{\gamma} \to 0$$

$$\eta \to 0 \qquad \text{for } \dot{\gamma} \to \infty$$

Viscosity increases to infinity at shear rates close to zero, creating serious computational problems in tasks where such close-to-zero shearing regions occur, e.g. in the cylindrical channel axis. These problems are solved using the reduced model, according to which

$$\eta\left(\dot{\gamma}\right) = m\dot{\gamma}^{n-1} \quad \text{for} \quad \dot{\gamma} > \dot{\gamma}_0 \tag{1.161}$$

and

$$\eta\left(\dot{\gamma}\right) = m_0 \quad \text{for} \quad \dot{\gamma} \le \dot{\gamma}_0 \tag{1.162}$$

where $m_0 = \eta(\dot{\gamma}_0)$ is the zero viscosity defined by Eq. (1.108); $\dot{\gamma}_0$ is the limit zero shear rate below which the parameter m_0 is used.

Parameters of the power law for polymers are presented in Table 1.11.

Table 1.11 Parameters of the Power Law Model for Common Polymers [33]

Polymer	T, °C	m, Pa·s^n	n
Polystyrene	170	$2.8 \cdot 10^4$	0.28
Polyvinyl chloride	180	$1.7 \cdot 10^4$	0.26
Polycarbonate	300	$6.0 \cdot 10^2$	0.98
Low density polyethylene	160	$6.0 \cdot 10^3$	0.39
High density polyethylene	180	$2.0 \cdot 10^4$	0.41
Polypropylene	200	$7.5 \cdot 10^3$	0.38
Polyamide 6.6	290	$6.0 \cdot 10^2$	0.66

The Bird-Carreau-Yasuda Model

Very good compatibility with rheological tests, also at very low and very high shear rates, is provided by the Bird-Carreau-Yasuda model, according to which

$$\frac{\eta - \eta_\infty}{\eta_0 - \eta_\infty} = \left[1 + \left(\lambda \dot{\gamma}\right)^a\right]^{(n-1)/a} \tag{1.163}$$

where η_0 is the zero shear rate viscosity; η_∞ is the infinite shear rate viscosity; λ is the time constant; n is the power law exponent; a is the dimensionless parameter defining the transition from the Newtonian range to the non-Newtonian range of the viscosity (described by the power law model).

A simplified version of Eq. (1.163) is often used. Bearing in mind that in the original Bird-Carreau model $a = 2$ and knowing that in the case of molten polymers we can assume that $\eta_\infty = 0$, this equation reduces to

$$\eta = \eta_0 \left[1 + \left(\lambda \dot{\gamma}\right)^2\right]^{(n-1)/2} \tag{1.164}$$

The effect of temperature on viscosity in this model is determined using Eq. (1.115) by

$$\eta_0\left(T\right) = \eta_{0r} \exp\left[-b\left(T - T_r\right)\right] \tag{1.165}$$

where η_{0r} is the zero shear rate viscosity at the reference temperature T_r.

Equation (1.163) has been modified to the form that is used in commercial polymer databases. The effect of temperature on viscosity is taken into account using the Arrhenius equation (1.111) for partially crystalline materials and the WLF equation (1.114) for amorphous polymers:

$$\eta = \frac{K_1 a_T}{\left(1 + K_2 \dot{\gamma} a_T\right)^{K_3}} \tag{1.166}$$

where a_T is the parameter that determines the location of the viscosity curve at the temperature T relative to the viscosity curve at the reference temperature T_r.

For partially crystalline polymers, the parameter a_T is defined as

$$a_T = \exp\left[\frac{E}{R}\left(\frac{1}{T} - \frac{1}{T_r}\right)\right] \tag{1.167}$$

while for amorphous polymers

$$a_T = \exp\left[\frac{8.86\left(K_4 - K_5\right)}{101.6 + K_4 - K_5} - \frac{8.86\left(T - K_5\right)}{101.6 + T - K_5}\right] \tag{1.168}$$

The Bird-Carreau-Yasuda model corresponds well to rheological studies but it is not applicable to solving analytical problems of fluid mechanics. However, it is excellent for computer numerical calculations. Parameters of Eqs. (1.166) to (1.168) for common polymers are presented in Table 1.12.

Table 1.12 Parameters of the Carreau-WLF and Carreau-Arrhenius Models [33]

Polymer	K_1 Pa·s	K_2 s	K_3	K_4 °C	K_5 °C	T_r °C	E J/mol
Polystyrene	1777	0.06400	0.73	200	123	-	-
Polyvinyl chloride	1786	0.05400	0.73	185	88	-	-
Polycarbonate	305	0.00046	0.48	320	153	-	-
Low density polyethylene	317	0.01500	0.61	-	-	189	43694
High density polyethylene	24 198	1.38000	0.60	-	-	200	22272
Polypropylene	1386	0.09100	0.68	-	-	220	427 198
Polyamide 6.6	44	0.00059	0.40	-	-	300	123058

The Klein Model

The model proposed by Klein captures in one equation the viscosity dependence on shear rate and temperature. However, like the Bird-Carreau-Yasuda model, it does not provide analytical solutions, but it is well suited for computer calculations. This model is expressed by

$$\lg \eta = A_0 + A_1 \lg \dot{\gamma} + A_{11} \lg^2 \dot{\gamma} + A_{12} T \lg \dot{\gamma} + A_2 T + A_{22} T^2 \tag{1.169}$$

where A_0, A_1, A_{11}, A_{12}, A_2, A_{22} are the model parameters.

The Muenstedt Model

The Muenstedt model is close to the Klein model and is expressed by

$$\begin{aligned} \lg \eta = \lg a_T + A_0 + A_1 \lg\left(a_T \dot{\gamma}\right) + A_2 \left[\lg\left(a_T \dot{\gamma}\right)\right]^2 + \\ + A_3 \left[\lg\left(a_T \dot{\gamma}\right)\right]^3 + A_4 \left[\lg\left(a_T \dot{\gamma}\right)\right]^4 \end{aligned} \tag{1.170}$$

where A_0, A_1, A_2, A_3, A_4 are the model parameters; a_T is the parameter that determines the location of the viscosity curve at the temperature T relative to the viscosity curve at the reference temperature T_r.

The Universal Model of Vinogradov and Malkin

The universal model of Vinogradov and Malkin is valid for many polymers to a good approximation. This is the single-parameter model given by

$$\eta\left(\dot{\gamma}, T\right) = \frac{\eta_0\left(T\right)}{1 + A_1\left[\dot{\gamma}\eta_0\left(T\right)\right]^a + A_2\left[\dot{\gamma}\eta_0\left(T\right)\right]^{2a}} \tag{1.171}$$

where $\eta(\dot{\gamma},T)$ is the viscosity corresponding to the shear rate $\dot{\gamma}$ at the temperature T; $\eta_0(T)$ is the zero shear rate viscosity ($\dot{\gamma} \to 0$) at the temperature T and $A_1 = 1.386 \cdot 10^{-2}$, $A_2 = 1.462 \cdot 10^{-3}$, $a = 0.355$.

The only parameter of this model is the zero shear rate viscosity that is dependent on the temperature (and pressure). So, this parameter defines the viscosity curve for a given polymer at a given temperature (and pressure). The other parameters of the model A_1, A_2, a are fixed (universal) and do not depend on the polymer.

The universal Vinogradov-Malkin model makes it possible to approximate the polymer viscosity curves based on the single value of this viscosity only (e.g. calculated on the basis of the MFI index). The procedure for such an application is described in Section 2.3.2.

The Bingham Model

Viscoplastic fluids that have a yield stress also belong to the generalized Newtonian fluids.

The Bingham model, which describes the linear relationship between shear stress and shear rate (Eq. (1.105)), is the most important in this group.

The apparent viscosity of a Bingham fluid, like any non-Newtonian fluid, is not constant and is expressed as

$$\begin{aligned} &\eta = \infty \quad \text{for} \quad \tau < \tau_0 \\ &\eta = \mu_0 + \frac{\tau_0}{\dot{\gamma}} \quad \text{for} \quad \tau \geq \tau_0 \end{aligned} \tag{1.172}$$

It results from Eq. (1.172) that the apparent viscosity decreases with increasing shear rate. In the case of a limit where the shear rate tends to infinity ($\dot{\gamma} \to \infty$), the apparent viscosity tends to the plastic viscosity ($\eta \to \mu_0$). Thus, the rheological parameter μ_0, i.e. the plastic viscosity, can be interpreted as the apparent viscosity at a shear rate tending to infinity. Then the Bingham equation (1.105) can be represented as

$$\tau = \tau_0 + \eta_\infty \dot{\gamma} \quad \text{for} \quad \tau \geq \tau_0 \tag{1.173}$$

where η_∞ is the infinite shear rate viscosity.

It should be noted that in the case of flow of viscoplastic fluids with uneven stress distribution, only part of the fluid (located in the area of stress greater than the yield stress) will flow, while the remaining part (located in the stress area less than the yield stress) will remain motionless.

The Herschel-Bulkley Model

The Herschel-Bulkley model is an example of a viscoplastic model that describes the nonlinear relationship between shear stress and shear rate, which refers to the power law model (Eq. (1.104)) and is expressed by

$$\begin{aligned} &\tau = \tau_0 + \mu_{0m}\dot{\gamma}^n \quad \text{for } \tau \geq \tau_0 \\ &\dot{\gamma} = 0 \quad \text{for } \tau < \tau_0 \end{aligned} \qquad (1.174)$$

where τ_0 is the yield stress; μ_{0m} is the consistency coefficient; n is the flow exponent.

The apparent viscosity of the Herschel-Bulkley fluid is as follows:

$$\begin{aligned} &\eta = \infty \quad \text{for } \tau < \tau_0 \\ &\eta = \frac{\tau_0}{\dot{\gamma}} + \mu_{0m}\dot{\gamma}^{n-1} \quad \text{for } \tau \geq \tau_0 \end{aligned} \qquad (1.175)$$

1.2.5 Multiphase Systems of Polymers

A distinct class of polymers is comprised by the polymer multiphase systems, for example composites, mixtures, and dispersion systems with low dispersion concentrations, i. e. suspensions, emulsions, and foams.

Multiphase systems are materials that have more than one phase. These can be mixtures of solids, liquids, and gases. The **phase** means a physically homogeneous part of the system separated by the phase boundary from the rest. In fluid mechanics, a continuous phase and a dispersed phase can be distinguished. The continuous phase is a liquid (or gas) and the dispersed phase can be a substance of any state, i. e. solid, liquid, or gas.

Multiphase systems of polymers can be broadly defined as dispersion systems of different type and size, and different degree of concentration. One can distinguish between systems with high and low concentration of dispersion. The first group includes composites and mixtures of polymers, while the second group includes slurries, emulsions, and foams. **Composites** are dispersion systems of solid particles, and **mixtures** are dispersion systems of liquid drops. **Suspensions** are dispersions of solid particles in a liquid, **emulsions** are dispersions of liquid drops, and **foams** are dispersions of gas.

The presence of dispersed phase in the continuous phase usually results in a significant change in velocity, pressure, and temperature fields during the flow and significantly influences the viscosity of this phase.

Viscosity of dispersion systems can be characterized by the following parameters:

- the relative or reduced viscosity

$$\eta_r = \frac{\eta}{\eta_m} \quad (1.176)$$

where η_r is the relative or reduced viscosity; η is the viscosity of dispersion system; η_m is the viscosity of the continuous phase (matrix),

- the specific viscosity

$$\eta_{sp} = \frac{\eta - \eta_m}{\eta_m} = \eta_r - 1 \quad (1.177)$$

where η_{sp} is the specific viscosity,

- the intrinsic viscosity

$$[\eta] = \lim_{\Phi \to 0} \left(\frac{\eta_{sp}}{\Phi} \right) \quad (1.178)$$

where $[\eta]$ is the intrinsic viscosity; Φ is the degree of concentration of the dispersed phase in the continuous phase,

- the viscosity ratio

$$\lambda = \frac{\eta_d}{\eta_m} \quad (1.179)$$

where λ is the viscosity ratio; η_d is the viscosity of the dispersed phase (dispersion); η_m is the viscosity of the continuous phase.

The rheology of multiphase systems is very complex, and the degree of complexity usually increases with increase of concentration of dispersion in the system.

With regard to the dispersion systems of solid particles, a fundamental relationship, although of limited significance, is the law of Einstein, which describes the viscosity of a system with very low concentration of spherical solid particles dispersed in a Newtonian liquid:

$$\eta = \eta_m (1 + 2.5\Phi) \quad (1.180)$$

where η is the viscosity of the dispersion system; η_m is the viscosity of the continuous phase; Φ is the degree of concentration of the dispersed phase in the continuous phase.

For more complex systems (e.g. composites), many different solutions are proposed, the most common of which is the power equation in the form

$$\eta = \eta_m (1 + a_1\Phi + a_2\Phi^2 + a_3\Phi^3 + ...) \quad (1.181)$$

where a_1, a_2, a_3 are the polynomial coefficients.

With reference to dispersion systems of liquid drops (e.g. mixtures of polymers), the fundamental relationship, although of limited significance, is the law of Taylor, which describes the viscosity of dispersions with low concentration of spherical drops of Newtonian liquid dispersed in a Newtonian liquid:

$$\eta = \eta_m \left[1 + \frac{1+2.5\lambda}{1+\lambda}\Phi\right] \tag{1.182}$$

where λ is the viscosity ratio.

This equation for $\lambda \to \infty$, i.e. for the dispersion of solid particles, is reduced to the Einstein equation (1.180), and for $\lambda \to 0$, i.e. for the gas dispersion, takes the form

$$\eta = \eta_m \left(1 + \Phi\right) \tag{1.183}$$

Sometimes, with respect to the mixture of polymers, a simple additive logarithmic rule is used, according to which the viscosity of mixture is determined by

$$\lg \eta = \sum_i w_i \lg \eta_i \tag{1.184}$$

where w_i is the weight fraction of the component i of the mixture; η_i is the viscosity of the component i.

This rule, however, is only valid for thermodynamically miscible polymers, i.e. those that do not show a multiphase structure. Therefore, it is not reasonable to apply to multiphase systems.

The rheology of multiphase polymer systems is so complex that the search for reasonable models of general significance does not make much sense. Therefore, the most sensible way of solving the flow problems of such systems is to determine the rheological properties of the currently considered system, based on appropriate rheological measurements.

■ 1.3 Viscoelasticity

1.3.1 The Concept of Viscoelasticity

Viscoelasticity means simultaneous occurrence of viscous and elastic properties. The viscous properties are related to the occurrence of irreversible deformations, whose degree, under a certain force, increases continuously over time. **Viscosity** is a measure of the resistance of a material to flow. The elastic properties are related to the occurrence of elastic (reversible) deformations, which disappear as soon as the force ceases. **Elasticity** is a measure of the ability of a material to immediately recover.

The specific behavior of the real material, i.e. more elastic or more viscous, depends on a time, and more precisely, on the relationship of the deformation time of a material to the certain characteristic natural time of this material. In general, when deformation is faster, the material is more elastic, and when deformation is slower, the material is more viscous. This is due to the molecular structure of the material.

The characteristic time of particular materials spans a wide range; it ranges from 10^{-13} to 10^{13} s. For example, the characteristic time of water is 10^{-12} s. In such a short characteristic time, practically every real deformation is relatively slow and therefore the water behaves like a purely viscous liquid. In contrast to water, the characteristic time of glass is very long: it is about 100 years. Thus, deformations are in practice very fast in this case and therefore the glass is elastic. Materials whose characteristic time is of the same order as the time of phenomena (deformations) occurring in reality, i.e. of the order of seconds, minutes, or hours, exhibit viscoelastic properties. Such materials include polymers, whose characteristic time is from 10^{-2} to 10^{2} s. Thus, the ancient statement that **"everything flows"** ("panta rhei") is justified, and only the time scale in which we observe the phenomenon determines whether or not this flow can be observed. The measure of the viscoelastic behavior of a material in a given process is the ratio of the characteristic time of the material to the characteristic time of the process. This defines the so-called **Deborah number:**

$$De = \frac{\lambda}{t_p} \tag{1.185}$$

where De is the Deborah number; λ is the characteristic material time; t_p is the characteristic process time.

A typical polymer with the characteristic time of 1 s ($\lambda = 1$ s) will behave in a different way, depending on how long the process lasts. With a very long characteristic time of the process ($t_p \to \infty$), the Deborah number will tend to zero ($De \to 0$) and the material will behave like a viscous liquid. At a very short time of the process ($t_p \to 0$), the Deborah number will be very large ($De \to \infty$) and the same material will behave like an elastic solid. The characteristic time of many flows in polymer processing is from 0.1 to 1 s, which means that for $\lambda = 1$ s, the Deborah number $De = 1$–10. Thus, the polymers exhibit both viscous and elastic properties, and are viscoelastic in processing practice.

1.3.2 Characteristic Phenomena of Viscoelasticity

Viscoelastic properties of polymers are revealed and can be observed in many characteristic phenomena, e.g.

- the Weissenberg effect,
- the Barus effect,
- time effects (static and dynamic).

The **Weissenberg effect** consists in unusual formation of the free surface of a liquid in the Couette flow that is the shear flow between two coaxial cylinders, one of which rotates (Figure 1.28). During such a flow, in the case of molten polymers, the characteristic climbing of the free surface at the rotating inner cylinder is observed (Figure 1.28(a)). This phenomenon also occurs, for example, when mixing paints or varnishes. On the other hand, it does not occur in the case of Newtonian fluids, when the free surface, by neglecting the inertia force, remains flat (Figure 1.28(b)).

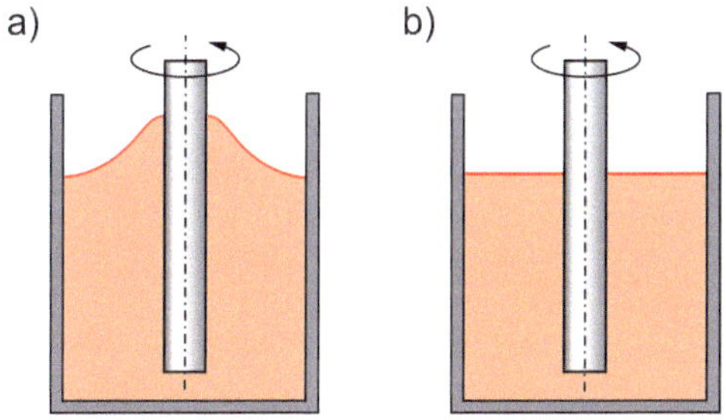

Figure 1.28
Weissenberg effect in the Couette flow:
a) molten polymer (viscoelastic),
b) Newtonian liquid

The Weissenberg effect is the result of generation of additional stresses during the shear flow, i.e. normal stresses. This is evident in Figure 1.29, which shows the shear flow between two parallel disks and, accompanying this flow, unequal formation of the liquid columns in the manometric tubes attached to one of these disks.

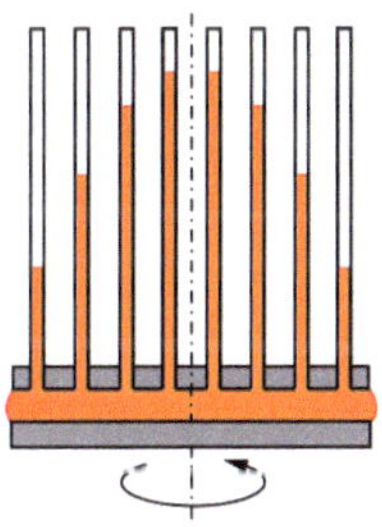

Figure 1.29
Weissenberg effect in the shear flow between two parallel disks

The **Barus effect** is an expansion of the stream of liquid flowing out of a capillary (Figure 1.30). It is often called extrudate swell or die swell, and is usually characterized by the ratio of the diameter of the stream (extrudate) d to the diameter of the capillary D, that is, the so-called **degree of swelling** $B = d/D$.

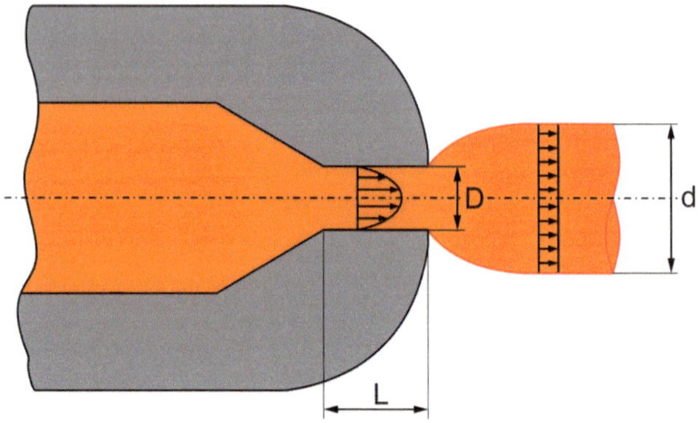

Figure 1.30 Barus effect: d - diameter of polymer stream (extrudate), D - capillary diameter, L - capillary length

Extrudate swelling is characteristic for viscoelastic materials, but also to a certain extent occurs in the case of Newtonian liquids. Then, it depends on the Reynolds number (Figure 1.31). For a Reynolds number close to zero, the Newtonian swelling is approximately 13% ($B = 1.13$). Then, it decreases with increase of the Reynolds number, and at $Re = 16$ the swelling degree is equal to unity, $B = 1.0$. In the case of Newtonian fluids, further increasing the Reynolds number results in asymptotically approaching 87% ($B = 0.87$).

In the case of polymers the extrudate swell is very large. The swelling degree is usually from 1.2 to 2.5, although sometimes it may be higher. It is characteristic that it depends largely on the flow rate of the material and the geometry of the capillary, more precisely on the ratio of the capillary length L to its diameter D (L/D).

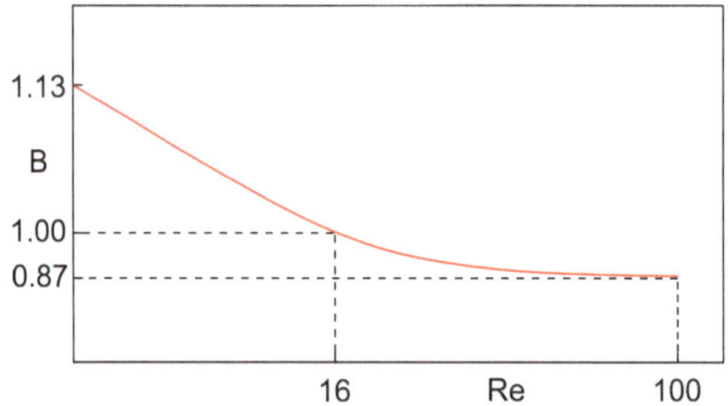

Figure 1.31 Dependence of swelling degree B on the Reynolds number Re, for the flow of Newtonian liquid through a cylindrical channel

The Barus effect, similarly to the Weissenberg effect, is strongly related to the occurrence of non-zero differences in normal stresses in the shear flow, N_1 and N_2, i. e. it is the normal stress effect.

Time effects are related to the occurrence of time dependence of stress and strain in materials. They can appear under static and dynamic conditions. The **static time effects** include changes in time of the material stress at constant strain (stress relaxation), or the material strain at constant stress (strain creeping). The **dynamic time effects** arise from the dynamic (oscillatory) loads (or strains) of the material and are based on the time shifting between the load and strain.

1.3.3 Linear Viscoelasticity

1.3.3.1 The Concept of Linear Viscoelasticity

The basic method of describing the viscoelasticity is a linear combination of viscous and elastic properties. As a result, linear relationships of stresses and deformations (or their time derivatives) are obtained. The coefficients of these linear relationships, i. e. modulus of elasticity and viscosity, are constant, i. e. not dependent on deformation.

A condition of the correct description of real viscoelastic systems based on linear relations of stresses and deformations is to limit these stresses and deformations to small values. The basic testing methods in which such conditions can be provided are

- stress relaxation test, i. e. measurements of the change of stress in the material after the step change of strain (or strain rate),
- strain creeping tests, i. e. measurements of the change of strain after the step change of stress,
- dynamic tests that record changes in stress (or strain) caused by dynamic (sinusoidal) changes in strain (or stress).

Figure 1.32 shows the course of stress relaxation for three types of material: elastic, viscous, and viscoelastic. As one can see, in the case of a step increase in deformation performed in time Δt (Figure 1.32(a)), the elastic material does not exhibit stress relaxation (Figure 1.32(b)), the viscous material relaxes immediately (upon reaching the deformation) to zero (Figure 1.32(c)), while the viscoelastic material relaxes to zero with some delay (Figure 1.32(d)).

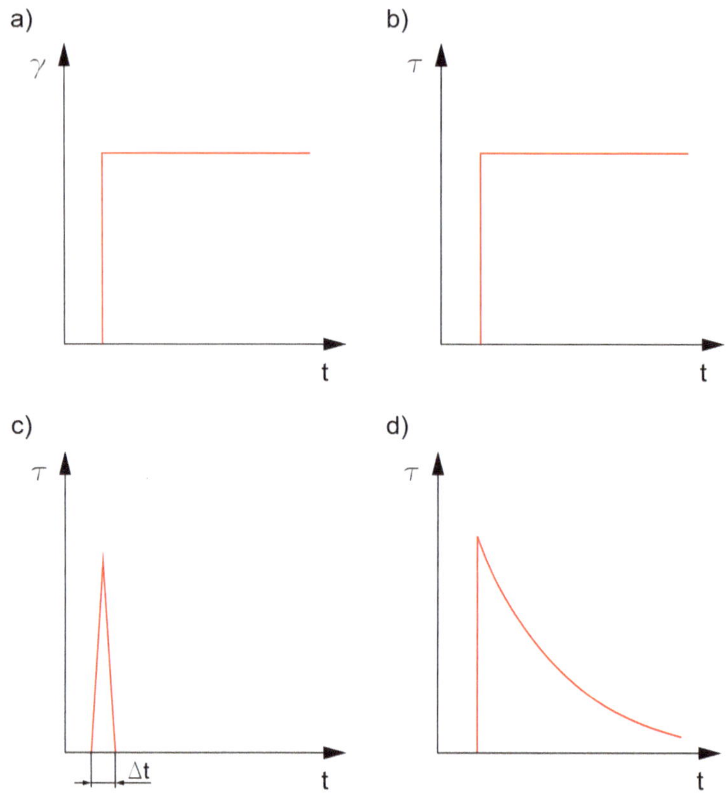

Figure 1.32 Relaxation of stress τ after a step increase in strain γ performed in time Δt: a) step increase in strain, b) response in stress for elastic Hookean solids, c) response for viscous Newtonian liquids, d) response for viscoelastic fluids

Stress relaxation is often described by the **relaxation stress modulus,** defined by the ratio of the relaxation stress $\tau(t)$ to the constant strain γ applied, according to the formula

$$G(t) = \frac{\tau(t)}{\gamma} \tag{1.186}$$

where $G(t)$ is the modulus of relaxation stress; $\tau(t)$ is the shear stress; γ is the strain; t is the time.

Figure 1.33 shows the typical plots of stress for molten polymers and the plots of the relaxation modulus for different strains. In the case of small deformations, smaller than unity (for $\gamma < 1$), the stress relaxation modulus is independent of deformation and represents one curve. Thus, Eq. (1.186), which represents the linear dependence of strain relaxation on deformation, defines the **linear viscoelasticity**.

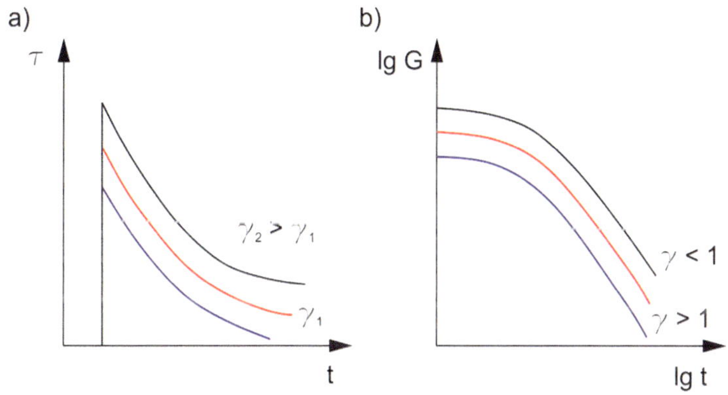

Figure 1.33 Stress relaxation and relaxation modulus for different values of strain γ: a) relaxation of stress τ, b) relaxation modulus G

In the case of larger deformations (for $\gamma > 1$), the relaxation modulus is no longer independent of deformation and it is then determined by

$$G(t,\gamma) = \frac{\tau(t,\gamma)}{\gamma} \tag{1.187}$$

It represents the nonlinear dependence of stress relaxation on deformation and defines the nonlinear viscoelasticity.

In the range of linear viscoelasticity, due to the small deformations, the type of deformation (shear, stretching) is not relevant, so that the considerations presented are of general character.

1.3.3.2 Mechanical Rheological Models

Models of Perfect Bodies

Some mechanical models are introduced to describe the rheological properties of materials. They reflect the behavior of these materials under certain load and deformation. The models of perfect bodies, whose rheological properties are precisely defined, are of paramount importance. They represent the basic types of material deformation:

- elastic deformation (reversible),
- plastic deformation (irreversible),
- deformation during a flow (irreversible).

The perfect bodies are the

- perfect elastic Hookean body,
- perfect plastic St. Venant body,
- perfect viscous Newtonian fluid.

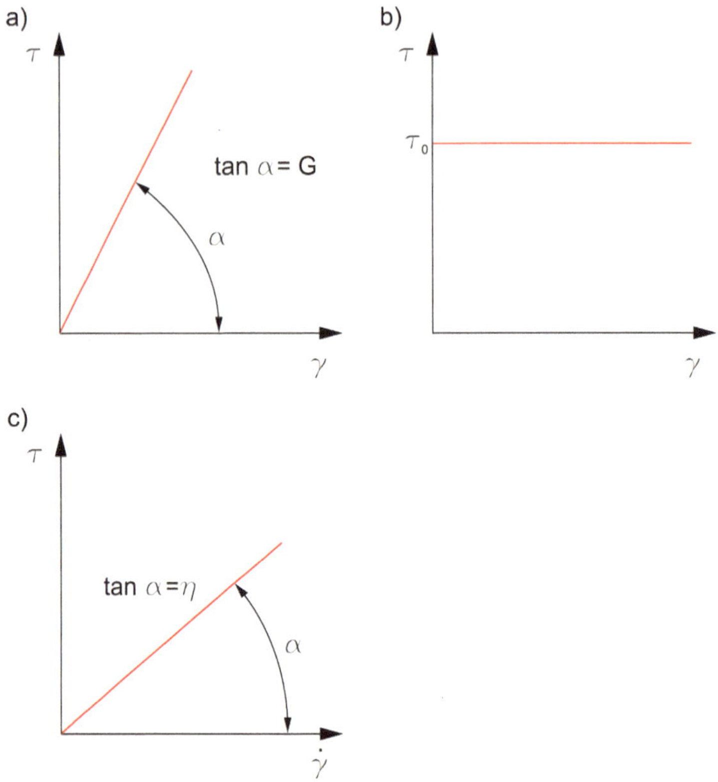

Figure 1.34 Rheological plots: a) Hookean elastic body, b) St. Venant plastic body, c) Newtonian viscous fluid

In the case of simple shearing, the rheological properties of these bodies (Figure 1.34) are described by the following relationships:

$$\tau = G\gamma \tag{1.188}$$

$$\tau = \tau_0 \tag{1.189}$$

$$\tau = \eta\dot{\gamma} \tag{1.190}$$

where τ is the shear stress; G is the shear modulus; τ_0 is the yield stress; η is the viscosity; γ is the strain; $\dot{\gamma}$ is the shear rate.

The mechanical models of perfect bodies reflect their rheological properties in accordance with Eqs. (1.188) to (1.190). There are some analogies between the elongation of the model (or the elongation rate) and the deformation of the body (or the strain rate), and between the force acting on the model and the stress in the body.

The mechanical model of a perfectly elastic body is a helical spring, as shown in Figure 1.35(a). The elongation of this spring is proportional to the applied force F, and this force immediately generates the appropriate elongation, while its removal immediately returns the spring to its original state. Such behavior is thus a complete analogy of the perfectly elastic body.

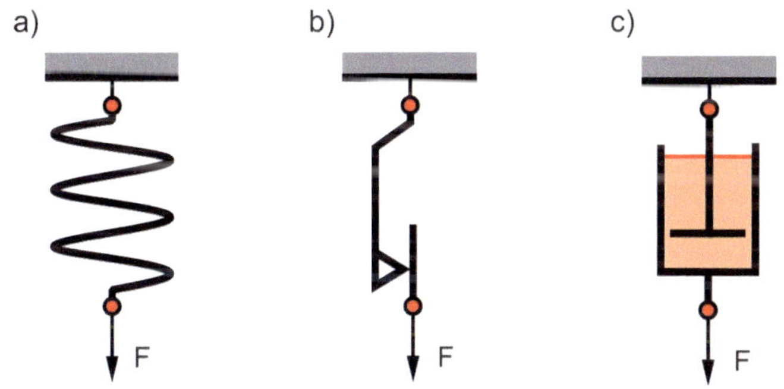

Figure 1.35 Mechanical models: a) Hookean elastic body, b) St. Venant plastic body, c) Newtonian viscous fluid

The mechanical model of a perfectly plastic body is a slider, shown in Figure 1.35(b). The elongation of this system is not possible until the force F applied to the slider does not exceed the limit value equal to the static friction force. Only after exceeding this value, the elongation will occur under the constant force. Removal of this force will not return the slider to its original position. This is an analogy of the perfectly plastic body.

The mechanical model of a perfect fluid is a dashpot, shown in Figure 1.35(c). This is a piston with holes moving in a cylinder filled with viscous liquid. In this case, the applied force F will produce a suitable constant rate of elongation. Reduction of this force will not return the piston to its original position. This is an analogy of the perfectly viscous fluid.

Mechanical models of bodies with complex rheological properties are obtained by a suitable combination of the models of the perfect bodies. Such models may vary by the type and number of elements used for their construction and how they are combined. Connections of the model elements can be serial or parallel. In the serial connection, the deformation of the elements is summed up and the stresses are the same. In the parallel connection, the deformation is the same, and the stresses are added together. The basic models of linear viscoelasticity are series and parallel combinations of Hookean and Newtonian models.

The Maxwell Model

The Maxwell model is a serial connection of the spring with a constant modulus of elasticity G and the dashpot of constant viscosity η (Figure 1.36).

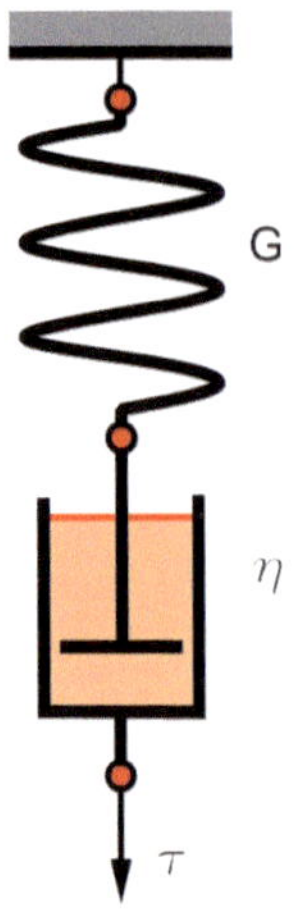

Figure 1.36
The Maxwell mechanical model: τ - the stress in the Maxwell system

The total deformation of such a model γ is equal to the sum of the deformation of the spring γ_1 and the deformation of the dashpot γ_2. In contrast, the stress induced in these elements is the same, i.e.

$$\gamma = \gamma_1 + \gamma_2 \tag{1.191}$$

$$\tau = \tau_1 = \tau_2 \tag{1.192}$$

where τ is the total stress in a Maxwell system; τ_1 is the stress in the spring; τ_2 is the stress in the dashpot.

The rate of deformation of the Maxwell model, after differentiation of Eq. (1.191) with respect to time, is expressed in the formula

$$\frac{d\gamma}{dt} = \frac{d\gamma_1}{dt} + \frac{d\gamma_2}{dt} \tag{1.193}$$

The behavior of the spring is described by

$$\tau = G\gamma_1 \tag{1.194}$$

which, after differentiation of this equation with redpect to time, expresses as

$$\frac{d\tau}{dt} = G\frac{d\gamma_1}{dt} \tag{1.195}$$

The behavior of the dashpot is described by

$$\tau = \eta\frac{d\gamma_2}{dt} \tag{1.196}$$

After introducing Eqs. (1.195) and (1.196) into Eq. (1.193), we receive

$$\frac{\mathrm{d}\gamma}{\mathrm{d}t}=\frac{1}{G}\frac{\mathrm{d}\tau}{\mathrm{d}t}+\frac{\tau}{\eta} \tag{1.197}$$

and, after transformations, we obtain the **Maxwell equation:**

$$\tau+\frac{\eta}{G}\frac{\mathrm{d}\tau}{\mathrm{d}t}=\eta\frac{\mathrm{d}\gamma}{\mathrm{d}t} \tag{1.198}$$

where the ratio of viscosity η and elastic modulus G has the dimension of time, and is referred to as the **characteristic time** λ:

$$\lambda=\eta\,/\,G \tag{1.199}$$

Thus, Eq. (1.198) can be presented as

$$\tau+\lambda\dot{\tau}=\eta\dot{\gamma} \tag{1.200}$$

where $\dot{\tau}=\mathrm{d}\tau\,/\,\mathrm{d}t;\dot{\gamma}=\mathrm{d}\gamma\,/\,\mathrm{d}t$.

In the Maxwell model, the stress τ depends on the rate of deformation $\dot{\gamma}$, according to Eq. (1.198) or Eq. (1.200). So, it describes the rheological properties of a viscoelastic liquid. At constant stress, i.e. at $\dot{\tau}=0$, the Maxwell equation is reduced to the Newton equation (1.190).

The Kelvin-Voigt Model

The Kelvin-Voigt model (in short, Kelvin) is a parallel connection of the spring with a constant modulus of elasticity G and the dashpot of constant viscosity η (Figure 1.37).

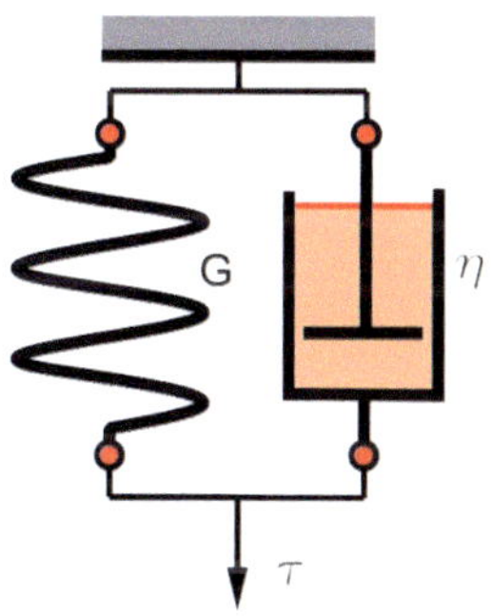

Figure 1.37
The Kelvin-Voigt mechanical model: τ - the stress in the Kelvin-Voigt system

The total stress in such a model γ is equal to the sum of the stress in the spring τ_1 and the stress in the dashpot τ_2. In contrast, the deformations of these elements are equal, i.e.

$$\tau = \tau_1 + \tau_2 \tag{1.201}$$

$$\gamma = \gamma_1 = \gamma_2 \tag{1.202}$$

where γ is the total deformation of the Kelvin system; γ_1 is the deformation of the spring; γ_2 is the deformation of the dashpot.

The behavior of the spring is described by

$$\tau_1 = G\gamma \tag{1.203}$$

and the behavior of the dashpot is determined by

$$\tau_2 = \eta \frac{d\gamma}{dt} \tag{1.204}$$

After introducing Eqs. (1.203) and (1.204) into Eq. (1.201), we receive the **Kelvin-Voigt equation:**

$$\tau = G\gamma + \eta \frac{d\gamma}{dt} \tag{1.205}$$

This can be presented as

$$\tau = G\gamma + \eta\dot{\gamma} \tag{1.206}$$

where $\dot{\gamma} = d\gamma / dt$.

In the Kelvin-Voigt model, the stress depends on the deformation γ, according to Eqs. (1.205) or (1.206). Thus, it describes the rheological properties of a viscoelastic body. At constant deformation, i. e. at $\dot{\gamma} = 0$, the Kelvin-Voigt equation reduces to the Hooke equation (1.188).

1.3.3.3 Time Effects of Viscoelasticity

Stress Relaxation

To investigate the relaxation phenomenon, it is necessary to trace the stress induced in the Maxwell fluid undergoing shearing, which is suddenly stopped. Then, the shear rate becomes zero, i. e. $d\gamma/dt = 0$ and the deformation is constant $\gamma = \gamma_0 =$ const. As a result, the Maxwell equation (1.198) takes the form

$$\tau + \lambda \frac{d\tau}{dt} = 0 \tag{1.207}$$

This is a linear first-order differential equation that can be solved by separation the variables and integrating with the initial condition $\tau = \tau_0$ for $t = 0$. As a result, we obtain

$$\int_{\tau_0}^{\tau} \frac{d\tau}{\tau} = -\frac{1}{\lambda} \int_0^t dt \tag{1.208}$$

and then

$$\ln\left(\frac{\tau}{\tau_0}\right) = -\frac{t}{\lambda} \tag{1.209}$$

After transforming Eq. (1.209), the equation of **stress relaxation** of the Maxwell liquid is developed in the form

$$\tau(t) = \tau_0 \exp(-t/\lambda) \tag{1.210}$$

where $\tau_0 = G_0\gamma_0$ is the initial stress at $t = 0$; G_0 is the initial modulus; γ_0 is the constant deformation.

It follows from this equation that the stress decreases from τ_0 (for $t = 0$) to zero (for $t \to \infty$). Thus, the characteristic time λ characterizes a fading of the stress (i.e. relaxation) and it is therefore termed as the **relaxation time.** It is worth noting that for $t = \lambda$

$$\frac{\tau(t)}{\tau_0} = e^{-1} = 0.37$$

which means that λ represents the time after which the stress decreases $1/e$ times (i.e. 0.37).

The stress relaxation plot of the Maxwell liquid is presented in Figure 1.38.

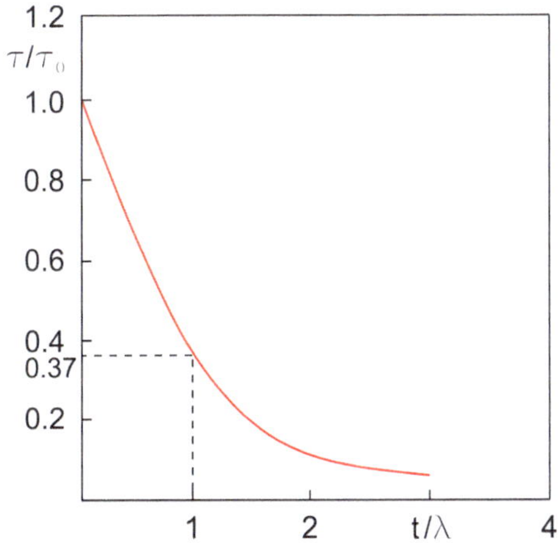

Figure 1.38
Stress relaxation curves for the Maxwell fluid: τ - stress, τ_0 - initial stress (for $t = 0$), t - time, λ - characteristic relaxation time

The stress relaxation modulus (Eq. (1.186)) will have the same course (over time) as the stress, that is

$$G(t) = \frac{\tau(t)}{\gamma_0} = G_0 \exp(-t/\lambda) \tag{1.211}$$

Such changes of this modulus correspond to the gradual relaxation of molten polymers (Figure 1.33).

Creep Deformation

To investigate the creeping phenomenon, it is necessary to trace the deformation of the Kelvin-Voigt body undergoing the stress, which is suddenly removed. Then, the stress becomes zero, i.e. $\tau = 0$ and the Kelvin-Voigt equation (1.205) takes the form

$$G\gamma + \eta\left(\frac{d\gamma}{dt}\right) = 0 \tag{1.212}$$

This is a linear first-order differential equation that can be solved by separation the variables and integrating with the initial condition $\gamma = \gamma_0$ for $t = 0$. As a result, we obtain

$$\int_{\gamma_0}^{\gamma} \frac{d\gamma}{\gamma} = -\frac{1}{\lambda}\int_0^t dt \tag{1.213}$$

and then

$$\ln\left(\frac{\gamma}{\gamma_0}\right) = -\frac{t}{\lambda} \tag{1.214}$$

After transforming Eq. (1.214), the equation of **creep deformation** (or retardation) of the Kelvin-Voigt body is developed in the form

$$\gamma(t) = \gamma_0 \exp(-t/\lambda) \tag{1.215}$$

where γ_0 is the initial deformation, for ($t = 0$).

It follows from this equation that the strain decreases from γ_0 (for $t = 0$) to zero (for $t \to \infty$). Thus, the characteristic time λ characterizes the delay of the strain relative to the change in stress (i.e. retardation or creep deformation) and it is therefore referred to as the **retardation time.**

Dynamic Time Effects

Dynamic time effects occur with dynamic (oscillating) loads (or deformations) of the material, and rely on the time shift between load (stress) and deformation.

The change of strain in the material subjected to the oscillating shear is expressed by the formula

$$\gamma = \gamma_0 \sin(\omega t) \tag{1.216}$$

where $\gamma = \gamma(t)$ is the strain (time varying); γ_0 is the strain amplitude; ω is the frequency of the strain oscillation; t is the time.

The shear rate in this case, after differentiation of Eq. (1.216) with respect to time, will be equal to

$$\dot{\gamma} = \gamma_0 \omega \cos(\omega t) = \dot{\gamma}_0 \cos(\omega t) \tag{1.217}$$

where $\dot{\gamma}_0 = \gamma_0 \omega$ is the amplitude of shear rate.

In the case of a perfectly elastic body for which

$$\tau = G\gamma \tag{1.218}$$

the stress will vary according to the equation

$$\tau = G\gamma_0 \sin(\omega t) = \tau_0 \sin(\omega t) \tag{1.219}$$

where $\tau_0 = G\gamma_0$ is the stress amplitude (of the elastic body).

In this case, the stress will be in phase with the strain (Figure 1.39).

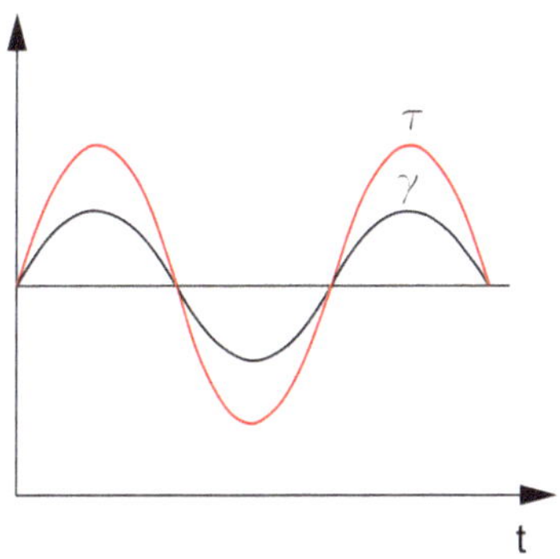

Figure 1.39
Sinusoidally oscillating strain and stress for the perfectly elastic body

In the case of a perfectly viscous liquid, for which

$$\tau = \eta \dot{\gamma} \tag{1.220}$$

the stress will vary, taking into account Eq. (1.217), according to the formula

$$\tau = \eta \gamma_0 \omega \cos(\omega t) = \eta \dot{\gamma}_0 \cos(\omega t) = \tau_0 \sin(\omega t + \pi / 2) \tag{1.221}$$

where $\tau_0 = \eta \dot{\gamma}_0 = \eta \gamma_0 \omega$ is the stress amplitude (of the viscous liquid).

Thus, the stress will not be in phase with the strain, but will be shifted by a phase angle of $\pi/2$ with respect to the strain wave (Figure 1.40). However, it will be in phase with the shear rate defined by Eq. (1.217).

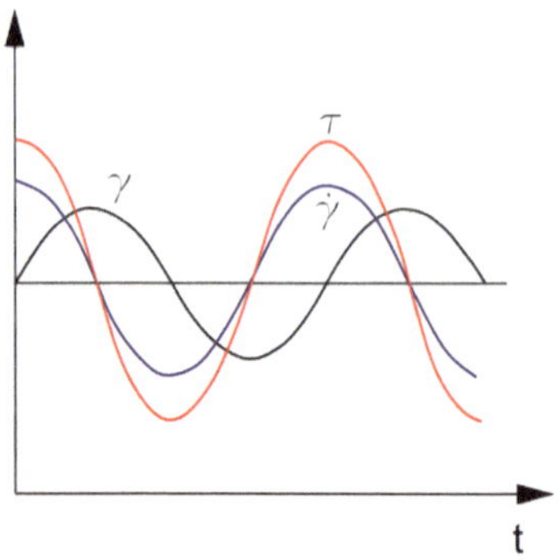

Figure 1.40
Sinusoidally oscillating strain, stress, and shear rate for a perfectly viscous fluid

In the case of viscoelastic material, the stress will be shifted with respect to the strain wave by a phase angle δ (Figure 1.41), according to the formula

$$\tau = \tau_0 \sin\left(\omega t + \delta\right) \tag{1.222}$$

where δ is the phase angle, contained within limits $0 < \delta < \pi/2$.

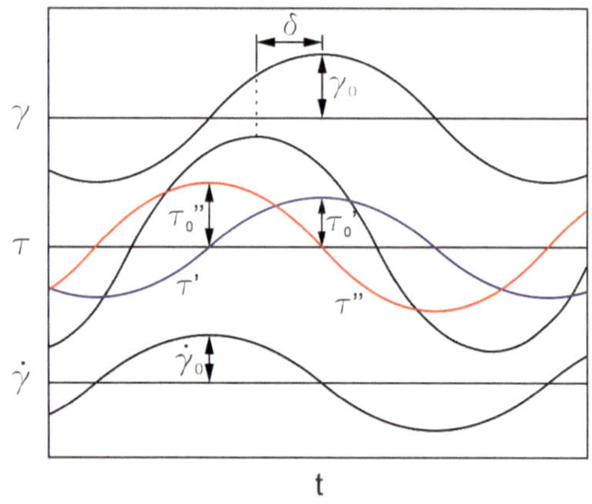

Figure 1.41 Oscillating strain, stress, and shear rate for a viscoelastic liquid; γ - strain, γ_0 - amplitude of strain oscillation, τ - stress, τ' - stress component in phase with strain wave, τ'' - stress component $\pi/2$ out of phase with strain wave, τ_0' - amplitude of oscillation of elastic component of stress, τ_0'' - amplitude of oscillation of viscous component of stress, $\dot{\gamma}$ - shear rate, $\dot{\gamma}_0$ - amplitude of oscillation of shear rate, δ - phase angle, t - time

Note that at the limit values of the angle δ (i.e. for $\delta = 0$ and $\delta = \pi/2$), Eq. (1.222), describing the behavior of a viscoelastic material, is reduced to Eqs. (1.219) and (1.221), which describe a perfectly elastic and perfectly viscous material.

In order to distinguish in the viscoelastic response of a material the elastic stress component (in phase with strain) and the viscous stress component ($\pi/2$ out of phase with strain), the oscillating stress τ is decomposed into two components with the same oscillation frequency. One of them τ' will be in phase with the oscillating strain γ, while the other τ'' will be shifted by a phase angle $\pi/2$ (Figure 1.41). Then one can write that

$$\tau = \tau' + \tau'' = \tau_0' \sin(\omega t) + \tau_0'' \cos(\omega t) \tag{1.223}$$

where τ_0' is the amplitude of oscillation of the elastic stress component; τ_0'' is the amplitude of oscillation of the viscous stress component.

When the modulus of elastic stress component is defined as the elastic modulus G' (storage), and the modulus of viscous stress component as the viscous modulus (loss) G'', according to the formulas

$$G' = \frac{\tau_0'}{\gamma_0} \tag{1.224}$$

$$G'' = \frac{\tau_0''}{\gamma_0} \tag{1.225}$$

where γ_0 is the amplitude of oscillating stress, Eq. (1.223) will have the form

$$\tau = \gamma_0 \left(G' \sin(\omega t) + G'' \cos(\omega t)\right) \tag{1.226}$$

Then, Eq. (1.222) can be trigonometrically expanded to

$$\tau = \tau_0 \sin(\omega t + \delta) = \tau_0 \left(\sin(\omega t)\cos\delta + \cos(\omega t)\sin\delta\right) \tag{1.227}$$

and the ratio of the amplitude of oscillating stress τ_0 to the amplitude of oscillating strain γ_0 can be defined as G_d:

$$G_d = \frac{\tau_0}{\gamma_0} \tag{1.228}$$

where $G_d = G_d(\omega)$.

Then, using Eqs. (1.226) to (1.228) we obtain

$$G' = G_d \cos\delta \tag{1.229}$$

$$G'' = G_d \sin\delta \tag{1.230}$$

It should be noted that for a perfectly elastic body subjected to oscillating shear for which the stress varies according to Eq.(1.219), one can obtain, by comparing this equation with Eq. (1.226), $G' = G$, $G'' = 0$, and $\delta = 0$. On the other hand, for a perfectly viscous body for which the stress varies according to Eq. (1.221), we obtain $G' = 0$, $G'' = \eta\omega$, and $\delta = \pi/2$. Thus, the **storage modulus** G' represents the ability of the material to accumulate energy while the **loss modulus** G'' represents the ability of the material to dissipate energy. A measure of the ability of the material to dissipate energy (damping) is the ratio of the loss modulus to the storage modulus, called the **damping factor.** It is equal, according to Eqs. (1.229) and (1.230), to the tangent of the phase angle δ:

$$\tan\delta = \frac{G''}{G'} \tag{1.231}$$

The moduli G' and G'' are often presented as the real and imaginary components of the complex modulus G^*, i.e.

$$G^* \equiv G' + iG'' \tag{1.232}$$

although the modulus G'' is a real, and not imaginary, measure of a dissipated energy.

The modulus (absolute value) of the complex number G^* expresses as

$$\left|G^*\right| = \sqrt{(G')^2 + (G'')^2} \tag{1.233}$$

and it is equal, using Eqs. (1.228) to (1.230), to the ratio of the amplitude of stress and the amplitude of strain:

$$\left|G^*\right| = G_d = \frac{\tau_0}{\gamma_0} \tag{1.234}$$

Similarly to the presented analysis of the dynamic shear as a function of sinusoidally oscillating strain, the dynamic shear analysis can be performed as a function of the sinusoidally oscillating shear rate determined by Eq. (1.217).

When the oscillating stress τ is decomposed according to Eq. (1.223), the component τ'' will be in phase with the oscillating shear rate $\dot{\gamma}$, and the component τ' will be $\pi/2$ out of phase with respect to that rate.

After defining, by analogy to Eqs. (1.224) and (1.225), the viscous component of stress as the dynamic viscosity η', and the elastic component of stress as the elastic viscosity η'', according to the formula

$$\eta' = \frac{\tau_0''}{\dot{\gamma}_0} \tag{1.235}$$

$$\eta'' = \frac{\tau_0'}{\dot{\gamma}_0} \tag{1.236}$$

where $\dot{\gamma}_0$ is the amplitude of oscillating shear rate, Eq. (1.223) will have the form

$$\tau = \dot{\gamma}_0 \left(\eta' \cos(\omega t) + \eta'' \sin(\omega t)\right) \tag{1.237}$$

Then, after defining the ratio of the amplitude of stress τ_0 to the amplitude of shear rate $\dot{\gamma}_0$ as η_d

$$\eta_d = \frac{\tau_0}{\dot{\gamma}_0} \tag{1.238}$$

where $\eta_d = \eta_d(\omega)$, and using Eqs. (1.226) to (1.230) and (1.237), we obtain the following formulas:

$$\eta' = \eta_d \sin\delta = \frac{G''}{\omega} \tag{1.239}$$

$$\eta'' = \eta_d \cos\delta = \frac{G'}{\omega} \tag{1.240}$$

The viscosities η' and η'' are often presented as the real and imaginary components of the complex viscosity η^*, i.e.

$$\eta^* \equiv \eta' - i\eta'' \tag{1.241}$$

The modulus (absolute value) of the complex number η^* expresses as

$$\left|\eta^*\right| = \sqrt{(\eta')^2 + (\eta'')^2} \tag{1.242}$$

and it is equal, using Eqs. (1.238) to (1.240), to the ratio of the amplitude of stress and the amplitude of shear rate:

$$\left|\eta^*\right| = \eta_d = \frac{\tau_0}{\dot{\gamma}_0} \tag{1.243}$$

The presented moduli and viscosities allow the rheological characteristics of viscoelastic materials to be determined on the basis of dynamic measurements.

For example, it can be demonstrated [8] that for the Maxwell liquid (Eq. (1.200)) the following dependencies exist:

$$G' = \frac{\eta\lambda\omega^2}{1 + (\omega\lambda)^2} \tag{1.244}$$

$$G'' = \frac{\eta\omega}{1 + (\omega\lambda)^2} \tag{1.245}$$

and

$$\eta' = \frac{\eta}{1+(\omega\lambda)^2} \tag{1.246}$$

$$\eta'' = \frac{\eta\lambda\omega}{1+(\omega\lambda)^2} \tag{1.247}$$

$$\tan\delta = \frac{G''}{G'} = \frac{1}{\lambda\omega} \tag{1.248}$$

which were derived on the basis of Eqs. (1.244), (1.245), (1.231), (1.239), and (1.240).

Then, for example, Eq. (1.248) makes it possible to determine the relaxation time λ on the basis of the measurement of the modules G' and G''. While Eq. (1.246) makes it possible to determine the viscosity η by measuring the dynamic viscosity η' for the oscillation frequency close to zero (for $\omega \to 0$, $\eta' \to \eta$).

The important rule defining the relation between the static and dynamic measurements is the **empirical rule of Cox-Merz.** According to this rule, the viscosity η determined in the static conditions as a function of the shear rate $\dot{\gamma}$, i.e. $\eta = \eta(\dot{\gamma})$, is equal to the complex viscosity η^* determined as a function of the oscillation frequency ω, i.e. $\eta^* = \eta^*(\omega)$, when the shear rate $\dot{\gamma}$ is equal to the oscillation frequency ω, that is

$$\eta(\dot{\gamma}) = \left|\eta^*(\omega)\right| \quad \text{for} \quad \dot{\gamma} = \omega \tag{1.249}$$

The plots of moduli and viscosities as a function of oscillation frequency, typical for molten polymers, are depicted in Figure 1.42 and Figure 1.43.

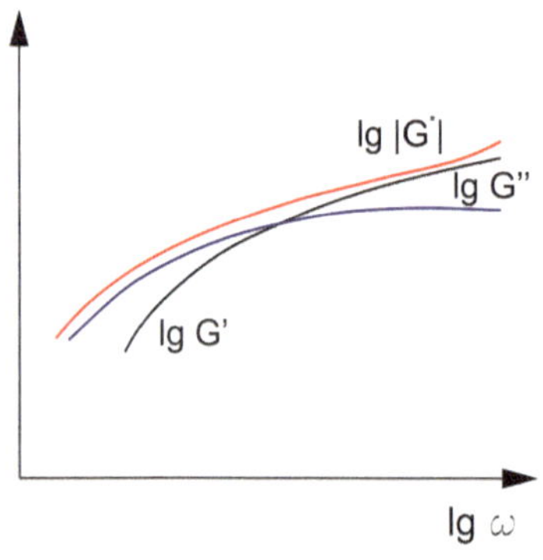

Figure 1.42
Typical moduli for molten polymers as a function of oscillation frequency ω: G^* - complex modulus, G' - storage modulus, G'' - loss modulus

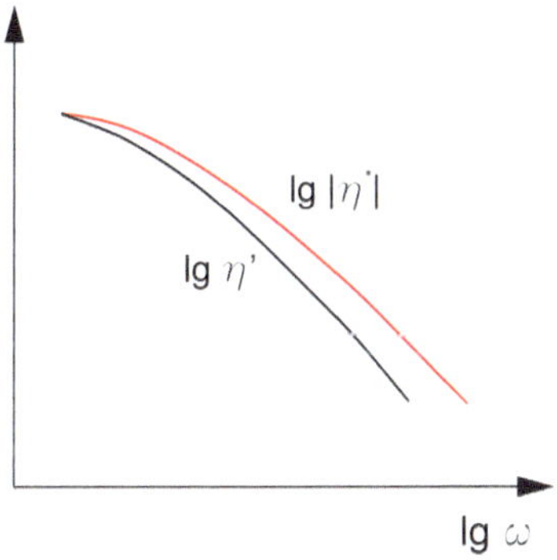

Figure 1.43
Typical viscosities for molten polymers as a function of oscillation frequency ω: η^* - complex viscosity, η' - dynamic viscosity

1.3.3.4 The General Differential Model of Linear Viscoelasticity

Description of the behavior of real viscoelastic materials is usually not possible with single Maxwell or Kelvin-Voigt elements. Therefore, various combinations of these elements (and Hookean and Newtonian elements) are used, thus leading to a general linear differential equation that defines the relationships between stress and strain as well as their time derivatives. The coefficients of these linear relationships are constant, i.e. independent of deformation. The general differential model of linear viscoelasticity has the following form:

$$\sum_i a_i \frac{d^i \tau}{dt^i} = \sum_j b_j \frac{d^j \gamma}{dt^j} \tag{1.250}$$

where $d^i\tau/dt^i$ is the time derivative i of the stress; $d^j\gamma/dt^j$ is the time derivative j of the strain; a_i, b_j are the coefficients of equation, and

$$\frac{d^0 \tau}{dt^0} = \tau, \quad \frac{d^0 \gamma}{dt^0} = \gamma$$

The previously discussed models of Hooke, Newton, Maxwell, and Kelvin-Voigt are the special cases of the general viscoelasticity model (Eq. (1.250)).

The **Hooke model** corresponds to Eq. (1.250), where the only non-zero parameters are the coefficients for $i = 0$ and $j = 0$, i.e. $a_0 = 1$ and $b_0 = G$. Then, Eq. (1.250) is reduced to the Hooke equation (1.188).

The **Newton model** is a special case of the general equation in which the coefficients corresponding to $i = 0$ and $j = 1$ are non-zero, i.e. $a_0 = 1$ and $b_1 = \eta$. Then, Eq. (1.250) is reduced to the Newton equation (1.190).

The **Maxwell model** corresponds to the case in which the coefficients for $i = 0$, $i = 1$, and $j = 1$ are non-zero, namely $a_0 = 1$, $a_1 = \eta/G$, and $b_1 = \eta$. In this case, the general equation is reduced to the Maxwell equation (1.198).

The **Kelvin-Voigt model** is a special case of the general equation where the non-zero coefficients $i = 0, j = 0$, and $j = 1$ are equal to $a_0 = 1$, $b_0 = G$, and $b_1 = \eta$. Then, Eq. (1.250) is transformed into the Kelvin-Voigt equation (1.205).

The presented simple concept of mechanical models leads to description of the behavior of viscoelastic materials by means of differential equations, i.e. to rheological models of the differential type. There is also another, equivalent form of description of such behavior by means of integral equations, which will not be presented here.

1.3.4 Nonlinear Viscoelasticity

1.3.4.1 The Concept of Nonlinear Viscoelasticity

Linear elasticity makes it possible to describe time effects occurring in polymeric materials, i.e., stress relaxation, creep deformation, and dynamic (oscillating) loads (or deformations). However, it does not make it possible to describe characteristic viscoelastic phenomena such as the Weissenberg effect or the Barus effect. Such phenomena can be explained by nonlinear viscoelasticity.

The range of linear viscoelasticity is limited to small deformations in accordance with Eq. (1.186). In the case of large deformations that occur during the processing of polymers, nonlinear effects of viscoelasticity are observed, e.g. according to Eq. (1.187).

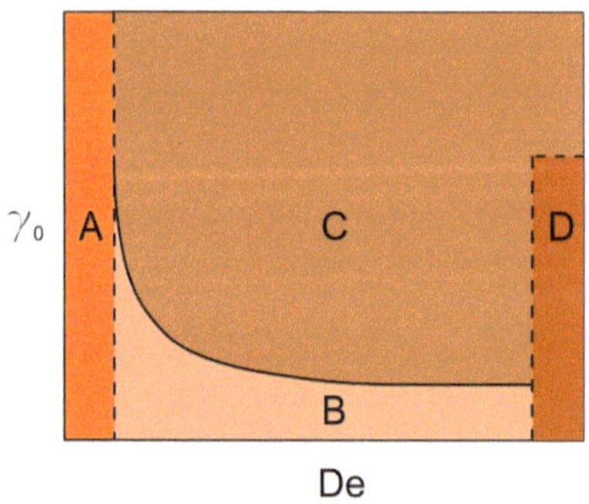

Figure 1.44
Range of various rheological material behavior depending on deformation γ_0 and Deborah number *De*: A - viscous (Newtonian), B - linear viscoelastic, C - nonlinear viscoelastic, D - elastic

The effects of linear or nonlinear viscoelasticity depend not only on the size of deformation but also on the duration of deformation; precisely from the relation of this time to the characteristic time of the material, i.e. from the Deborah number defined by Eq. (1.185).

A range of various rheological behaviors, i.e. viscous, linear viscoelastic, nonlinear viscoelastic, and elastic, as a function of deformation and Deborah number, is shown in Figure 1.44.

The basic phenomena of nonlinear viscoelasticity are

- differences in normal stresses,
- shear thinning,
- extensional thickening.

Nonlinear shear thinning and extensional thickening are also observed in nonlinear purely viscous fluids (Figure 1.27). On the other hand, the differences in normal stresses are the sole nonlinear characteristic of elastic materials.

1.3.4.2 Normal Stress Differences

The concept of differences of normal stresses is given by Eqs. (1.11) and (1.12). At the conditions of simple shear (Figure 1.9), where the velocity field is described by Eq. (1.123), the stress tensor defined by Eq. (1.9) has the form

$$\sigma_{ij} = -p\begin{bmatrix} 1 & 0 & 0 \\ 0 & 1 & 0 \\ 0 & 0 & 1 \end{bmatrix} + \begin{bmatrix} \tau_{xx} & \tau_{xy} & 0 \\ \tau_{yx} & \tau_{yy} & 0 \\ 0 & 0 & \tau_{zz} \end{bmatrix} \tag{1.251}$$

Thus, in the general case, the state of stress is defined by three independent functions, the so-called **viscometric functions:**

- **shear stress function**

$$\tau(\dot{\gamma}) = \tau_{xy} = \tau_{yx} \tag{1.252}$$

- **first normal stress difference function**

$$N_1(\dot{\gamma}) = \sigma_{xx} - \sigma_{yy} = \tau_{xx} - \tau_{yy} \tag{1.253}$$

- **second normal stress difference function**

$$N_2(\dot{\gamma}) = \sigma_{yy} - \sigma_{zz} = \tau_{yy} - \tau_{zz} \tag{1.254}$$

These are the functions of shear rate, and can be roughly presented for low shear rates as

$$\tau(\dot{\gamma}) = \eta\dot{\gamma} \tag{1.255}$$

$$N_1(\dot{\gamma}) = \psi_1\dot{\gamma}^2 \tag{1.256}$$

$$N_2(\dot{\gamma}) = \psi_2\dot{\gamma}^2 \tag{1.257}$$

where η is the viscosity; ψ_1 is the coefficient of the first normal stress difference; ψ_2 is the coefficient of the second normal stress difference.

In this case, the coefficients η, ψ_1, ψ_2 are not dependent on the shear rate.

For higher shear rates, the coefficients of viscometric functions are dependent on the shear rate, and then

$$\tau(\dot{\gamma}) = \eta(\dot{\gamma})\dot{\gamma} \tag{1.258}$$

$$N_1(\dot{\gamma}) = \psi_1(\dot{\gamma})\dot{\gamma}^2 \tag{1.259}$$

$$N_2(\dot{\gamma}) = \psi_2(\dot{\gamma})\dot{\gamma}^2 \tag{1.260}$$

Typical plots of viscometric functions of molten polymers are presented in Figure 1.45, while typical plots of their coefficients η and ψ_1 are depicted in Figure 1.46.

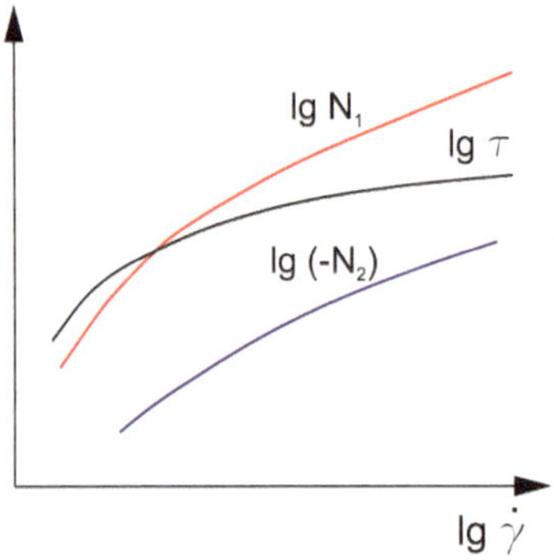

Figure 1.45
Viscometric functions for molten polymers: τ - shear stress function, N_1 and N_2 - first and second difference of normal stresses

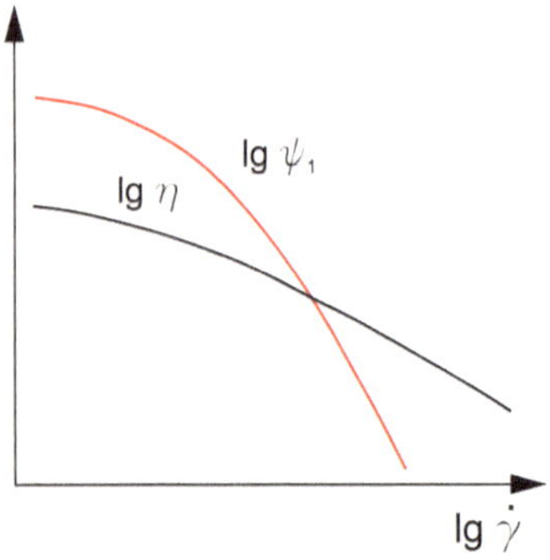

Figure 1.46
Coefficients of viscometric functions for molten polymers, viscosity η and first normal stress coefficient ψ_1 as a function of shear rate $\dot{\gamma}$

The first normal stress difference is usually much higher than the second normal stress difference, which is most often negative. Typically, N_1/N_2 is from -0.1 to -0.2, so that the second difference in normal stresses is often omitted in practical applications.

The first difference in normal stresses, as evidenced by an experience, appears in the materials exhibiting elasticity only, so it is the basic feature that distinguishes viscoelastic and viscous materials.

The ratio of the first normal stress difference N_1 to the shear stress τ is a measure of the viscoelastic behavior of the material and defines the Weissenberg number as

$$We = \frac{N_1}{2\tau} \tag{1.261}$$

If $We > 0.5$, the material is very elastic.

The above considerations are valid not only for the simple shear flow, but also for the whole class of the so-called **viscometric flows.** Simplifying, these are the flows that have only one component of velocity, and it is a function of only one coordinate which does not coincide with the direction of flow. Examples of viscous flows are Poiseuille flow and Couette flow. The following convention applies to the definition of viscometric functions:

- 1 – the flow direction,
- 2 – the direction along which the velocity varies,
- 3 – the neutral direction.

1.3.4.3 Normal Stress Effects

Weissenberg Effect

The Weissenberg effect is the primary effect of normal stresses. To explain this phenomenon, the Couette flow analysis may be performed [39], in which this effect is observed (Figure 1.47).

The **Couette flow** is a viscous flow whose velocity field (in the cylindrical coordinate system r, φ, z) is expressed as

$$v_r = 0, \quad v_\varphi = v_\varphi(r), \quad v_z = 0 \tag{1.262}$$

and the stress tensor (Eq. (1.6)) has the form

$$\sigma_{ij} = \begin{bmatrix} \sigma_{rr} & \sigma_{r\varphi} & 0 \\ \sigma_{\varphi r} & \sigma_{\varphi\varphi} & 0 \\ 0 & 0 & \sigma_{zz} \end{bmatrix} \tag{1.263}$$

where φ is the flow direction; r is the direction along which the velocity varies; z is the neutral direction.

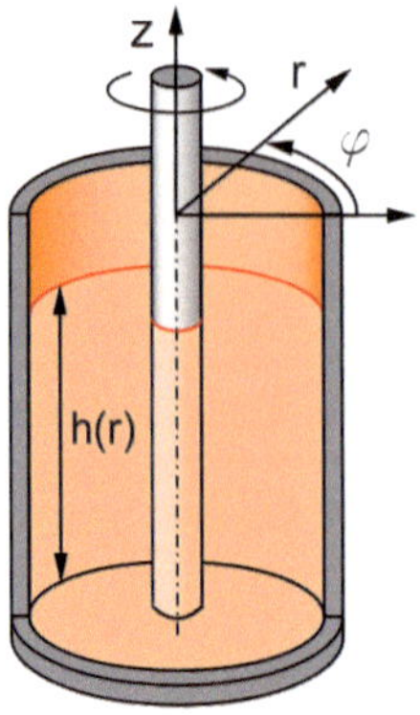

Figure 1.47
Scheme of Weissenberg effect in the Couette flow, r, φ, z - coordinates of cylindrical system, $h(r)$ - height of the liquid column

According to the assumed convention (see Section 1.3.4.2), the viscometric functions have the form

$$\sigma = \sigma_{r\varphi} \tag{1.264}$$

$$N_1 = \sigma_{\varphi\varphi} - \sigma_{rr} \tag{1.265}$$

$$N_2 = \sigma_{rr} - \sigma_{zz} \tag{1.266}$$

and

$$N_1 = \sigma_{\varphi\varphi} - \sigma_{rr} > 0 \tag{1.267}$$

$$N_2 = \sigma_{rr} - \sigma_{zz} \approx 0 \tag{1.268}$$

Let us consider the equation of momentum in the direction of the radius r (Table 1.3). At the assumed flow conditions (neglecting the inertia and gravity forces), it reduces to

$$\frac{\sigma_{rr}}{r} + \frac{\partial \sigma_{rr}}{\partial r} - \frac{\sigma_{\varphi\varphi}}{r} = 0 \tag{1.269}$$

and, taking into account Eq. (1.267), to the relationship

$$\frac{\partial \sigma_{rr}}{\partial r} = \frac{\sigma_{\varphi\varphi} - \sigma_{rr}}{r} > 0 \tag{1.270}$$

which, using Eq. (1.268), can be written as

$$\frac{\partial \sigma_{zz}}{\partial r} = \frac{\partial \sigma_{rr}}{\partial r} > 0 \tag{1.271}$$

The normal stress σ_{zz} must balance with the pressure of the liquid column of height $h(r)$, i.e.

$$\sigma_{zz} = -\rho g h(r) \tag{1.272}$$

where ρ is the density; g is the acceleration due to gravity.

It follows from Eq. (1.272) that $\partial h(r)/\partial r$ is negative, since $\partial \sigma_{zz}/\partial r$ is positive according to Eq. (1.271). This explains the rise of the free surface near the inner rotating cylinder.

Barus Effect

The Barus effect is very complex, and can also be observed for Newtonian fluids. In this case, extrudate swell can be explained by the change in the velocity profile (Figure 1.48). The change of the velocity component v_z along the z-axis means that $\partial v_z/\partial z \neq 0$, and this implies the appearance of the radial component v_r which forms the free surface of the stream.

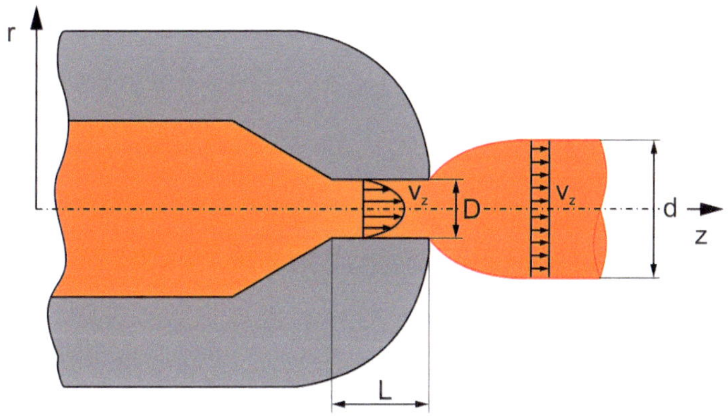

Figure 1.48 Barus effect - velocity profile: D - capillary diameter, d - extrudate diameter, v_z - velocity, r, z - coordinates of cylindrical system

The velocity component v_r results from the continuity equation (Table 1.2), which can be presented (omitting the velocity component v_φ) as

$$\frac{\partial v_r}{\partial r} + \frac{v_r}{r} = -\frac{\partial v_z}{\partial z} \neq 0 \tag{1.273}$$

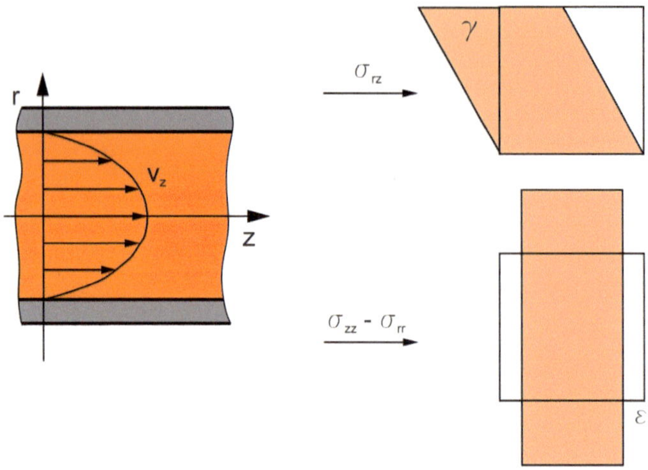

Figure 1.49 Stress and deformation in Barus effect: γ - shear strain, ε - extensional strain, v_z - velocity, σ_{rz} - shear stress, $\sigma_{zz} - \sigma_{rr}$ - normal stress difference, r, z - coordinates of cylindrical system, adapted from [39]

In order to analyze the effects of normal stresses in the swelling extrudate, the Poiseuille flow in the capillary (Figure 1.49) may be considered [39]. This is the viscous flow whose velocity field (in the cylindrical coordinate system r, φ, z) is expressed as

$$v_r = 0, \quad v_\varphi = 0, \quad v_z = v_z(r) \tag{1.274}$$

and the stress tensor (Eq. (1.6)) has the following form

$$\sigma_{ij} = \begin{bmatrix} \sigma_{rr} & 0 & \sigma_{rz} \\ 0 & \sigma_{\varphi\varphi} & 0 \\ \sigma_{rz} & 0 & \sigma_{zz} \end{bmatrix} \tag{1.275}$$

where φ is the flow direction; r is the direction along which the velocity varies; z is the neutral direction.

According to the assumed convention (see Section 1.3.4.2), the viscometric functions have the form

$$\sigma = \sigma_{rz} \tag{1.276}$$

$$N_1 = \sigma_{zz} - \sigma_{rr} \tag{1.277}$$

$$N_2 = \sigma_{rr} - \sigma_{\varphi\varphi} \tag{1.278}$$

and

$$N_1 = \sigma_{zz} - \sigma_{rr} > 0 \tag{1.279}$$

$$N_2 = \sigma_{rr} - \sigma_{\varphi\varphi} \approx 0 \tag{1.280}$$

The material flowing out of the capillary does not meet the constraints and freely deforms (reversibly) respectively to the stresses occurring in the capillary flow which are specified by Eq. (1.275). The components of this deformation are

$$\begin{bmatrix} \varepsilon & 0 & -\gamma \\ 0 & \varepsilon & 0 \\ -\gamma & 0 & -2\varepsilon \end{bmatrix} \tag{1.281}$$

where γ is the reversible shear strain caused by the shear stress σ_{rz}; ε is the reversible extensional strain caused by the normal stress difference $\sigma_{zz} - \sigma_{rr}$, i.e. the contraction in the flow direction and expansion in the radial direction.

The analysis of Figure 1.49 shows that the shear stress does not affect the expansion of the stream. This results from the first difference in normal stresses.

Tanner [14] obtained the relationship binding the swelling degree of the extrudate with the first difference of normal stresses that has the form

$$B = \left[1 + \left(\frac{N_{1w}}{2\sqrt{2}\sigma_w}\right)^2\right]^{1/6} \tag{1.282}$$

and, after taking into account the Newtonian expansion

$$B = 0.13 + \left[1 + \left(\frac{N_{1w}}{2\sqrt{2}\sigma_w}\right)^2\right]^{1/6} \tag{1.283}$$

where B is the extrudate swelling degree; N_{1w} is the first normal stress difference; σ_w is the shear stress at the capillary wall.

The extrudate swell phenomenon is due to the overlap of various factors that will not be discussed here.

1.3.5 Rheological Models of Viscoelastic Liquids

Most polymers are viscoelastic fluids. However, many practical problems of polymer processing can be addressed by considering them as purely viscous liquids, of course non-Newtonian and usually shear thinning, i.e. as generalized Newtonian liquids (see Section 1.2.4.2). However, there are also many problems that need to take into account viscoelastic properties of polymers, e.g. profile extrusion, blow molding, fiber formation, etc.

The viscoelastic rheological models formulated on the basis of the mechanical models of perfect bodies (see Section 1.3.3.2) are valid only in the case of small deformations, i.e. in the range of linear viscoelasticity. The serious problems in polymer processing are related to the presence of large deformations, so their solutions require the use of nonlinear viscoelastic models. Such models should describe the basic features of nonlinear viscoelasticity, i.e. differences in normal stresses, shear thinning, and extensional thickening, as well as time effects. Nonlinear phenomena of shear thinning and extensional thickening are also present in nonlinear purely viscous materials (Figure 1.27), but the occurrence of normal stress variations is an exclusive feature of nonlinear viscoelastic materials.

There are many viscoelastic rheological models, which can be broadly divided into differential and integral models. Because of the complexity of the problems of constitutive equations in the range of nonlinear viscoelasticity, only the basic models of the differential type are presented here.

The problems of constitutive equations (principles of their formulation and exhaustive review) are presented in the fundamental book of Bird, Armstrong, and Hassager [15]. It can be stated that, despite a large number of different constitutive equations, there is no equation that would describe the viscoelastic properties of polymers (i.e. the effect of normal stresses, shear thinning, extensional thickening, time effects) in full and under all processing conditions (e.g. shear rate). The constitutive equation usually describes only some of the viscoelastic properties under limited processing conditions. That is why it is important to choose the right equation for the problem in question. An exhaustive assessment of the suitability of particular constitutive equations in the modeling of real materials is included in the book of Tanner [14].

1.3.5.1 The Rivlin-Ericksen Model of Second Order

The simplest constitutive equation that captures the normal stress effect is the Rivlin-Ericksen model of the second order.

This equation can be expressed as

$$\underset{\sim}{\tau} = \eta_0 2\underset{\sim}{\boldsymbol{D}} - \psi_{1.0} \overset{\nabla}{\underset{\sim}{\boldsymbol{D}}} + \psi_{2.0} \left(2\underset{\sim}{\boldsymbol{D}}\right)^2 \tag{1.284}$$

or

$$\underset{\sim}{\sigma} = -p\underset{\sim}{\mathbf{1}} + \eta_0 2\underset{\sim}{\boldsymbol{D}} - \psi_{1.0} \overset{\nabla}{\underset{\sim}{\boldsymbol{D}}} + \psi_{2.0} \left(2\underset{\sim}{\boldsymbol{D}}\right)^2 \tag{1.285}$$

where $\underset{\sim}{\sigma}$ is the total stress tensor; $\underset{\sim}{\tau}$ is the extra-stress tensor; p is the hydrostatic pressure; $\underset{\sim}{\mathbf{1}}$ is the unity tensor; η_0 is the zero viscosity (i.e. for the shear rate close to zero, $\dot{\gamma} \to 0$); $\psi_{1.0}$ is the zero coefficient of the first normal stress difference (for $\dot{\gamma} \to 0$); $\psi_{2.0}$ is the zero coefficient of the second normal stress difference (for $\dot{\gamma} \to 0$); $\underset{\sim}{\boldsymbol{D}}$ is the rate of deformation tensor (Eq. (1.20)).

In this equation, there is a **convective derivative** (of Oldroyd) of the rate of deformation tensor $\overset{\nabla}{\underset{\sim}{D}}$, which is defined as

$$\overset{\nabla}{\underset{\sim}{D}} = \dot{\underset{\sim}{D}} - \underset{\sim}{L}^{T}\underset{\sim}{D} - \underset{\sim}{D}\underset{\sim}{L} \tag{1.286}$$

where $\dot{\underset{\sim}{D}}$ is the total (substantial) derivative of the rate of deformation tensor $\underset{\sim}{D}$; $\underset{\sim}{L}$ is the velocity gradient tensor of the velocity field $\underset{\sim}{v}$, defined by Eq. (1.15); $\underset{\sim}{L}^{T}$ is the transposed velocity gradient tensor.

In the Cartesian system, Eq. (1.286) can be expressed in the index notation as

$$\frac{\Delta}{\Delta t} D_{ij} = \frac{D}{Dt} D_{ij} - D_{kj}\frac{\partial v_i}{\partial x_k} - D_{ik}\frac{\partial v_j}{\partial x_k} \tag{1.287}$$

where $\Delta(D_{ij})/\Delta t = \overset{\nabla}{\underset{\sim}{D}}$ is the convective derivative; $D(D_{ij})/Dt = \dot{\underset{\sim}{D}}$ is the substantial derivative.

The convective derivative is a constitutive derivative. It satisfies the **principle of objectivity,** i.e. the principle of invariance with respect to the reference system. This principle expresses the fact of independence of constitutive equations, i.e. the mechanical properties of the material, from the motion of the observer which may be associated with any reference system.

The total stress tensor $\dot{\underset{\sim}{D}}$ expresses as

$$\dot{\underset{\sim}{D}} = \frac{D}{Dt}\underset{\sim}{D} = \frac{\partial}{\partial t}\underset{\sim}{D} + \left(\underset{\sim}{v}\cdot\text{grad}\right)\underset{\sim}{D} \tag{1.288}$$

and, in the index notation, in the Cartesian coordinate system as

$$\frac{D}{Dt} D_{ij} = \frac{\partial}{\partial t} D_{ij} + v_k \frac{\partial}{\partial x_k} D_{ij} \tag{1.289}$$

This derivative with respect to any scalar function, e.g. Eq. (1.35), vector function, e.g. Eq. (1.59), or tensor function, e.g. Eq. (1.287), is the sum of the local derivative $\partial/\partial t$ of this function and the convective derivative $\left(\underset{\sim}{v}\cdot\text{grad}\right)$, where the symbol $\left(\underset{\sim}{v}\cdot\text{grad}\right)$ is an operator that can be formally considered as the scalar product of the velocity and the gradient (grad) operator. The total derivative D/Dt determines the changes occurring in the moving fluid element, the local derivative $\partial/\partial t$ determines the changes occurring with time in a fixed point of space, i.e. local changes, and the convective derivative $\left(\underset{\sim}{v}\cdot\text{grad}\right)$ describes the changes related to the displacement of the fluid element to another location, i.e. the convection changes. The total derivative is objective in relation to scalar and vector functions, but it is not objective to tensor functions. Thus, it is not a constitutive derivative.

It is important to note that the second order equation is limited to the steady shear flows at low shear rates, where it is reasonable to assume constant values of parameters η_0, $\psi_{1.0}$, $\psi_{2.0}$.

1.3.5.2 The Criminale-Ericksen-Filbey Model

The Criminale-Ericksen-Filbey model (CEF model) is an extension of the concept of the second order model represented by Eq. (1.284). The fixed parameters η_0, $\psi_{1.0}$, $\psi_{2.0}$ are replaced by the coefficients η, ψ_1, ψ_2, which depend on the shear rate $\dot{\gamma}$, and in general on the generalized shear rate $\dot{\gamma}_D$, defined by Eq. (1.157). Thus, this equation can be expressed as

$$\underset{\sim}{\boldsymbol{\tau}} = \eta\left(\dot{\gamma}_D\right)2\underset{\sim}{\boldsymbol{D}} - \psi_1\left(\dot{\gamma}_D\right)\underset{\sim}{\boldsymbol{D}} + \psi_2\left(\dot{\gamma}_D\right)\left(2\underset{\sim}{\boldsymbol{D}}\right)^2 \tag{1.290}$$

The CEF model has no limitation of the second order in respect to low shear rate, and it is valid for high shear rates, but it is also limited to the steady shear flows.

1.3.5.3 The Maxwell Convective Model

The Maxwell convective model results from the simplest concept of simultaneous consideration of time effects and rheological nonlinearity. It was formulated by replacing the total derivative of the stress $\underset{\sim}{\boldsymbol{\tau}}$ in the tensor version of the Maxwell equation (1.200), by the convective derivative of Oldroyd. This model can be expressed as

$$\underset{\sim}{\boldsymbol{\tau}} + \lambda \overset{\nabla}{\underset{\sim}{\boldsymbol{\tau}}} = \eta_0 2\underset{\sim}{\boldsymbol{D}} \tag{1.291}$$

where $\underset{\sim}{\boldsymbol{\tau}}$ is the extra-stress tensor; $\overset{\nabla}{\underset{\sim}{\boldsymbol{\tau}}}$ is the convective Oldroyd derivative of the tensor $\underset{\sim}{\boldsymbol{\tau}}$; λ is the material parameter of the relaxation time; η_0 is the zero viscosity.

The Maxwell convective model describes the steady shear flows well, but does not reveal the second normal stress difference. It is further limited to low shear rates, where it is reasonable to assume constant values of the relaxation time λ and viscosity η_0. A description of time effects is not satisfactory in this model because in this case there is only one parameter of relaxation time.

1.3.5.4 The White-Metzner Model

The White-Metzner model is a generalization of the Maxwell convective model represented by Eq. (1.291) where the model parameters, i.e. relaxation time and viscosity, are the functions of the generalized function of shear rate $\dot{\gamma}_D$, defined by Eq. (1.157). This model expresses as

$$\underset{\sim}{\boldsymbol{\tau}} + \lambda\left(\dot{\gamma}_D\right)\overset{\nabla}{\underset{\sim}{\boldsymbol{\tau}}} = \eta\left(\dot{\gamma}_D\right)2\underset{\sim}{\boldsymbol{D}} \tag{1.292}$$

The White-Metzner model has no limitation of the Maxwell convective model in respect to low shear rates, and it is also applicable for high shear rates. However, it shows the other limitation of this model, regarding the inability to disclose the second normal stress difference and the good description of time effects.

An extensive review of constitutive equations is given in References [14], [15], [20], [27].

References

1. Reiner, M., *Twelve Lectures on Theoretical Rheology* (1949) North Holland Publishing Co., Amsterdam
2. Eirich, F.R. (Ed.), *Rheology. Theory and Applications* v.1 (1956), v.2 (1958), v.3 (1960) Academic Press Inc., New York
3. Bird, R.B., Stewart, W.E., Lightfoot, E.N., *Transport Phenomena* (1960) John Wiley & Sons, New York
4. Wilkinson, W.L., *Non-Newtonian Fluids* (1960) Pergamon Press Ltd., London
5. Plajer, O., *Praktische Rheologie für Kunststoffschmelzen* (1970) Zechner and Hüthig, Zürich
6. Houwink, R., de Decker, H.K., *Elasticity, Plasticity and Structure of Matter* (1971) Cambridge University Press, Cambridge
7. Lenk, R.S., *Rheologie der Kunststoffe* (1971) Carl Hanser Verlag, Munich
8. Kembłowski, Z., *Rheometry of Non-Newtonian Fluids* (1973) WNT, Warsaw (in Polish)
9. Astarita, G., Marucci, G., *Principles of Non-Newtonian Fluid Mechanics* (1974) McGraw Hill, London
10. Han, Ch. D., *Rheology in Polymer Processing* (1976) Academic Press, New York
11. Ferry, J., *Viscoelastic Properties of Polymers* (1980) John Wiley & Sons, New York
12. Zahorski, S., *Mechanics of Viscoelastic Fluids* (1981) Martinus Nijhoff, The Hague, (1978) PWN, Warsaw (in Polish).
13. Pahl, M.H., *Praktische Rheologie der Kunststoffschmelzen und Lösungen* (1983) VDI Verlag GmbH, Düsseldorf
14. Tanner, R.I., *Engineering Rheology* (1985) Clarendon Press, Oxford
15. Bird, R.B., Armstrong, R.C., Hassager, O., *Dynamic of Polymeric Liquids* (1987) John Wiley & Sons, New York
16. White, J. L, *Principles of Polymer Engineering Rheology* (1990) John Wiley & Sons, New York
17. Collyer, A.A., Utracki, L.A., *Polymer Rheology and Processing* (1990) Elsevier, London
18. Ferguson, J., Kembłowski, Z., *Applied Fluid Rheology* (1991) Elsevier, London
19. Barnes, H.A., Hutton, J.F., Walters, K., *An Introduction to Rheology* (1993) Elsevier, Amsterdam
20. Macosko, C.W., *Rheology: Principles, Measurements and Applications* (1994) VCH Publishers, New York
21. Malkin, A.Y., *Rheology Fundamentals* (1994) ChemTec Publishing, Toronto
22. Gordon, G.V., Shaw, M., *Computer Programs for Rheologists* (1994) Hanser Publishers, Munich
23. Schramm, G., *A Practical Approach to Rheology and Rheometry* (1994) Haake GmbH, Karlsruhe
24. Dealy, J.M., Wissbrun, K.F., *Melt Rheology and its Role in Plastics Processing* (1995) Chapman & Hall, London
25. Rohn, Ch. L., *Analytical Polymer Rheology* (1995) Carl Hanser Verlag, Munich
26. Kou, S., *Transport Phenomena and Materials Processing* (1996) John Wiley & Sons, New York

27. Carreau, P.J., De Kee, D.C.R., Chhabra, R.P., *Rheology of Polymeric Systems* (1997) Hanser Publishers, Munich
28. Barnes, H.A., *A Handbook of Elementary Rheology* (2000), University of Wales, Cardiff
29. Morrison, F.A., *Understanding Rheology* (2001), Oxford University Press, New York
30. Han, Ch. D., *Rheology and Processing of Polymeric Materials*, v.1. *Polymer Rheology*, v.2. *Polymer Processing* (2007), Oxford University Press, Oxford
31. Chhabra, R.P., Richardson, J.F., *Non-Newtonian Flow and Applied Rheology* (2008) Elsevier, Amsterdam
32. Vlachopoulos, J., *Polymer Rheology and Extrusion* (2011) McMaster University, Hamilton
33. Osswald, T.A., Menges, G., *Materials Science of Polymers for Engineers* (2012) Carl Hanser Verlag, Munich
34. Malkin, A.Y., Isayev, A.I., *Rheology: Concepts, Methods, and Applications* (2012) ChemTec Publishing, Toronto
35. Shaw, M.T., *Introduction to Polymer Rheology* (2012) John Wiley & Sons, Hoboken
36. Osswald, T., Rudolph, N., *Polymer Rheology. Fundamentals and Applications* (2015) Carl Hanser Verlag, Munich
37. Vlachopoulos, J., *Fundamentals of Fluid Mechanics* (2016) Polydynamics, Hamilton
38. Münstedt, H., *Rheological and Morphological Properties of Dispersed Polymeric Materials* (2016) Carl Hanser Verlag, Munich
39. Agassant, J.-F., Avenas, P., Carreau, P.J., Vergnes, B., Vincent, M., *Polymer Processing: Principles and Modeling* (2017) Carl Hanser Verlag, Munich
40. Hopmann, Ch., Michaeli, W., *Extrusion Dies for Plastics and Rubber* (2016) Carl Hanser Verlag, Munich
41. Vlachopoulos, J., Polychronopoulos, N.D., *Understanding Rheology and Technology of Polymer Extrusion* (2019) Polydynamics Inc., Dundas
42. *www.autodesk.moldflow.com*

2 Rheometry

2.1 The Concept of Rheometry

Rheometry is the science concerning the measurement of rheological properties of materials, and thus on the measurements of stress and strain. These are made using rheometers.

The aim of rheometry is to determine the rheological characteristics of the material. The most important of these characteristics were presented in Chapter 1. These are the material functions of

- the stress relaxation modulus $G(t)$, defined by Eq. (1.186),
- dynamic moduli and viscosities, i.e. the complex modulus $G^*(\omega)$, the elastic (storage) modulus $G'(\omega)$, and the viscous (loss) modulus $G''(\omega)$, as well as the complex viscosity $\eta^*(\omega)$, the dynamic viscosity $\eta'(\omega)$, and the elastic viscosity $\eta''(\omega)$, defined by Eqs. (1.232), (1.224), (1.225), (1.241), (1.235), and (1.236),
- the shear viscosity $\eta(\dot{\gamma})$ and the coefficients of the first normal stress difference and the second normal stress difference $\psi_1(\dot{\gamma})$, $\psi_2(\dot{\gamma})$, defined by Eqs. (1.255) to (1.257),
- the elongational viscosity $\eta_e(\dot{\varepsilon})$, defined by Eq. (1.144).

The functions of shear viscosity and normal stress differences as well as the function of elongational viscosity are of fundamental importance in the rheological description of polymer processing.

Rheometric tests are usually carried out by measuring forces or moments that represent the state of stress in the material, and by measuring displacements or velocities representing deformations of the material. There is also a group of optical methods in which the optical properties of the material are measured during an experiment. Then, the effects of energy absorption, radiation scattering, or optical birefringence are used. These methods will not be discussed here.

2.2 Classification of Rheometric Methods

Rheometric methods can be divided into two groups, depending on the kinematic conditions of the measurement, i. e.

- methods implemented under shear conditions,
- methods implemented under stretching conditions.

As a result, rheometers can be divided into the shear type rheometers and the stretching type rheometers.

The shear type rheometers can be divided according to the type of flow into:

- rheometers in which shearing takes place as a result of drag flow,
- rheometers in which shearing takes place as a result of pressure flow.

The **drag flow** is the flow between two surfaces, one of which is moving and the other stationary. In contrast, the **pressure flow** is defined as the flow that takes place in a closed channel due to the pressure difference along this channel (i. e. pressure driven flow).

The basic drag flows are shown in Figure 2.1, whereas the pressure flows are depicted in Figure 2.2. Both these flows belong to the class of viscometric flows (see Section 1.3.4.2), i. e. the flows in which there is only one component of velocity, being a function of only one coordinate, not coincident with the flow direction. These flows are very important in rheometry because one can get exact (or very close to exact) solutions of the momentum equation. In addition, these flows are relatively simple in practical implementation.

According to the viscometric flows shown in Figure 2.1 and Figure 2.2, there are different types of rheometers. The most important are

- rotating rheometers
  - with coaxial cylinders (Figure 2.1(b)),
  - cone-and-plate rheometers (Figure 2.1(c)),
  - plate-and-plate rheometers (Figure 2.1(d)),
- pressure rheometers
  - capillary rheometers (Figure 2.2(a)),
  - slit rheometers (Figure 2.2(b)).

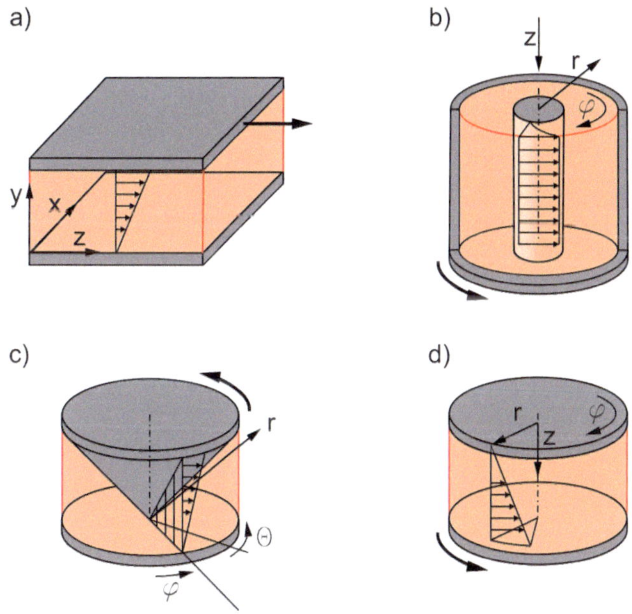

Figure 2.1 Drag flows: a) the flow between parallel (sliding) plates, b) the flow between coaxial cylinders (Couette flow), c) the flow between cone and plate, d) the flow between two plates (torsional flow)

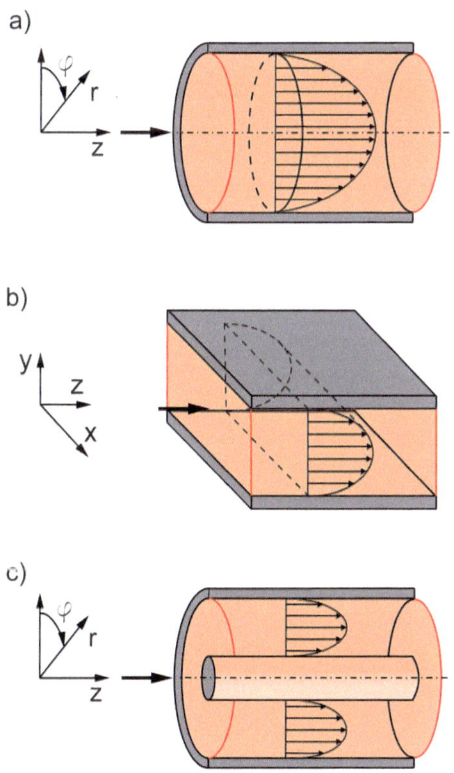

Figure 2.2
Pressure flows: a) the flow in a cylindrical channel (Poiseuille flow), b) the flow in a slit channel (between two parallel plates), c) the flow in an annular channel

The principle of rotating rheometers is to shear the material in the gap between two surfaces, one of which rotates and the other of which is stationary. These surfaces can be formed by two coaxial cylinders, a cone and a plate, or two parallel plates (Figure 2.1). The measurement consists in determining the relationship between the angular velocity of the rotating measuring element and the torque that is associated with this rotation. This makes it possible to determine the relationship between the shear rate and the shear stress, and thus makes it possible to determine the flow curve of the material to be tested.

The principle of operation of pressure rheometers consists in forcing the material to be tested through channels (cylinder or slit). The measurement consists in determining the dependence between the volume flow rate of the material and the pressure drop caused by the internal friction of the material. This dependence makes it possible to determine the relationship between the shear rate and the shear stress, and thus to determine the flow curve of the material to be tested.

The main difference between rotating and pressure rheometers is that in the case of rotating rheometers there is no time limit of measurement. Therefore, they enable the determination of the time and dynamic rheological characteristics. Using them, one can also determine differences in normal stresses. However, the limitation of these rheometers is the possibility of using them in the range of low shear rates, only.

Pressure rheometers have a simpler design and allow measurement over a wide range of shear rates, but only at higher shear rates. Using them, it is also possible to examine the phenomenon of extrudate swell and on this basis to assess the difference in normal stresses. Other, more accurate ways to measure the difference in normal stresses are technically difficult in this case.

There is also a group of eccentric rheometers that are based on the modification of viscometric flows. These modifications can be divided into two groups:

- modifications that involve the introduction of a small eccentricity between the axes of rotating elements, such as the Maxwell orthogonal rheometer with eccentric plates (Figure 2.3) or the Abbott and Walters rheometer with eccentric cylinders (Figure 2.4),
- modifications that rely on the mutual inclination of the axes of rotating elements, such as the Kepes balance rheometer with the axes of rotating hemispheres inclined to each other (Figure 2.5) or in the cone-and-plate rheometer with inclined axes of the cone and plate (Figure 2.6).

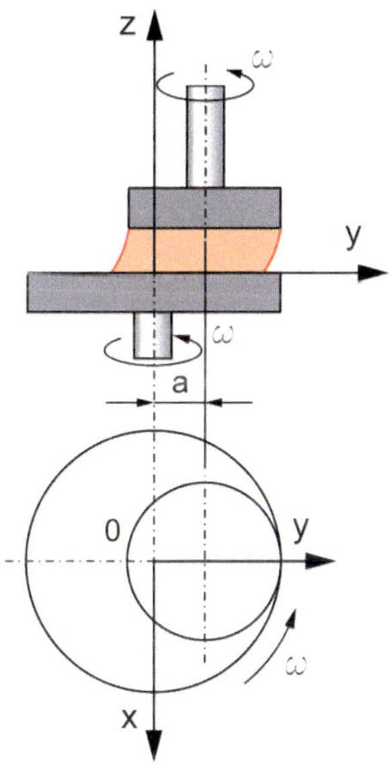

Figure 2.3 Maxwell orthogonal rheometer with eccentric plates

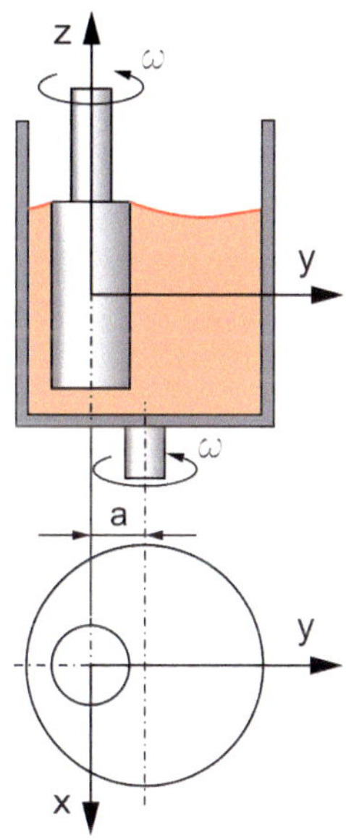

Figure 2.4 Abbott and Walters rheometer with eccentric cylinders

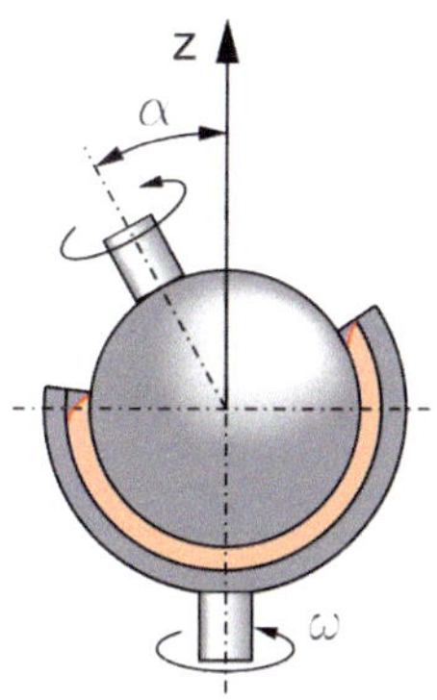

Figure 2.5 Kepes balancing rheometer with axes of rotating hemispheres inclined towards each other

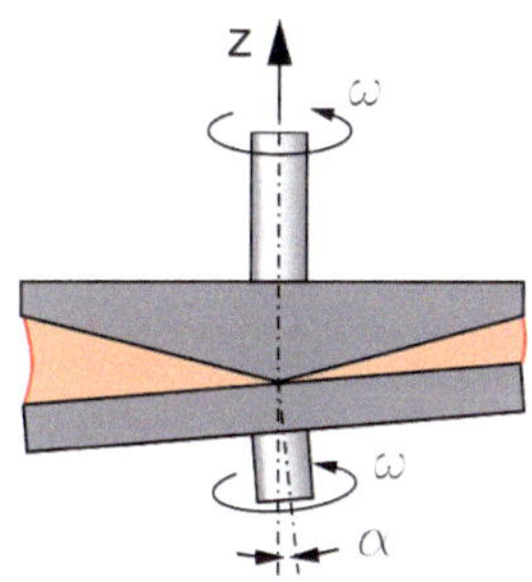

Figure 2.6 A cone-and-plate rheometer with inclined axes of cone-and-plate

Rheometers of this type make it possible to determine dynamic rheological characteristics of viscoelastic fluids; however, nowadays their practical application is not large.

Rheometers can also be divided, from the point of view of their construction, into rheometers operating on the principle of constant shear rate and rheometers operating on the principle of constant stress.

In rheometers operating on the constant shear rate principle (with controlled shear rate), the shear rate is an independent variable, and the stress is a dependent variable. In rotating rheometers, the rotational speed of the measuring element is set and the torque is measured, and in pressure rheometers, the piston speed is set and the load is measured.

In rheometers operating on the principle of constant stress (with controlled stress), the independent variable is the stress and the dependent variable is the shear rate. In rotating rheometers, the torque is set and the rotational speed of the measuring element is measured, and in pressure rheometers the load of the piston (or other element forcing the movement) is set and the flow rate of the material is measured.

Of fundamental importance in the rheometry of polymers are the capillary rheometers, in which there is a pressure flow, and the cone-and-plate rheometers, in which a drag flow is implemented. These rheometers complement each other, i.e. they enable determination of flow curves in the whole range of shear rates occurring in polymer processing practice (Figure 2.7). A cone-and-plate rheometer enables the determination of normal stress differences (capillary rheometer only in an approximate way), and the time and dynamic characteristics of the material.

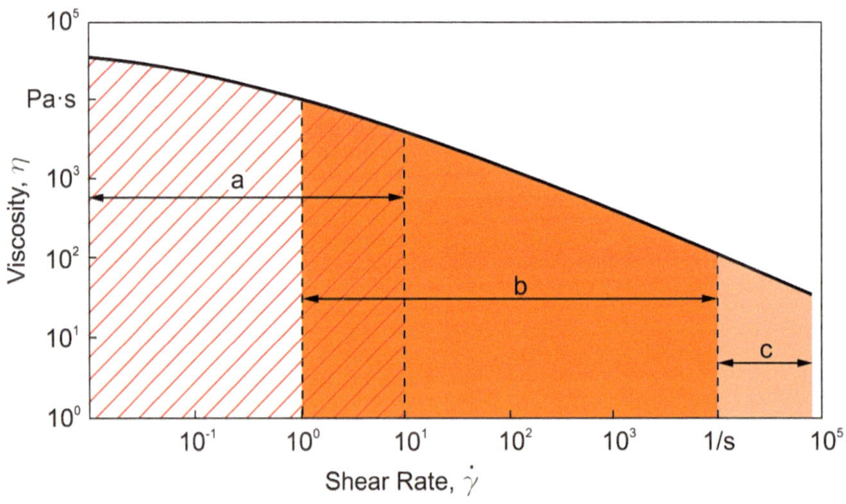

Figure 2.7 Typical range of use of capillary rheometers and cone-and-plate rheometers: a – cone-and-plate rheometer, b – capillary rheometer, c – measurement difficulties

In addition to the shear type rheometers, the tensile type rheometers that allow measurements of rheological properties under stretching conditions are distinguished. The importance of this type of research has increased in recent years. It is associated with an increase of awareness that tensile flows may play a significant and sometimes decisive role in many processes, such as fiber forming, blow molding, thermoforming, or convergent flows in extrusion dies and injection molds.

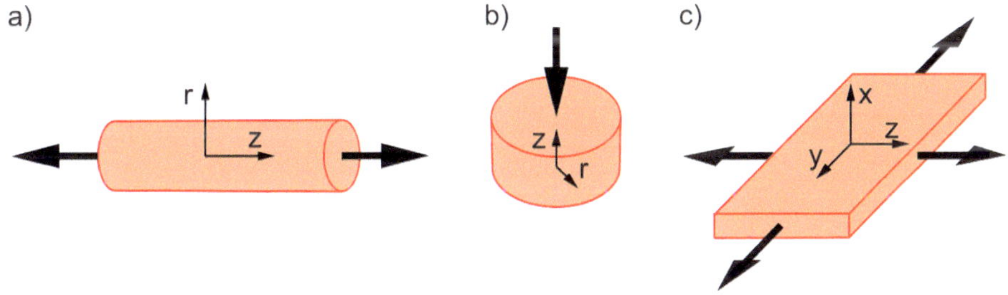

Figure 2.8 Basic tensile flows: a) simple stretching, b) compression, c) multi-axis stretching

Rheometric measurements under tensile conditions are difficult due to technical difficulties in the implementation of pure tensile flows. Particular problems are encountered with fluids of low viscosity. The more important methods for implementing tensile flows are shown in Figure 2.8. The method of straight stretching is of fundamental importance.

Rheometric methods enable determination of complex characteristics (functions) of polymeric materials, e.g. the viscosity function and the normal stress factors functions. However, they are expensive and time-consuming. Therefore, in industrial practice, the methods of simplified evaluation of rheological properties of polymers are often used that are based on the flow rate measurements. The **Melt Flow Index** (MFI) is of key importance. There are also known the methods for approximate determination of viscosity curves based on a single measuring point, e.g. MFI. Both methods can be classified as single-point rheometry methods.

2.3 Single-Point Methods

2.3.1 Melt Flow Index

Exact determination of the polymer viscosity curves, i.e. dependences of the viscosity on the shear rate, is expensive and time-consuming. Therefore, in industrial practice, most polymers are characterized by the melt flow index, only. It is defined by the mass flow rate of the material flowing out of the capillary under standard conditions. The scheme of the measuring device for its determination is shown in Figure 2.9.

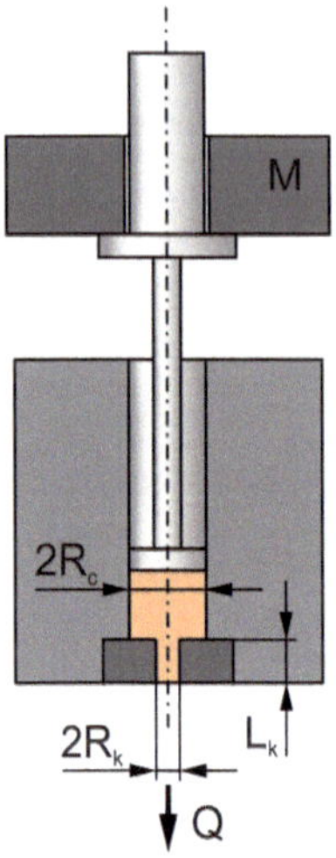

Figure 2.9
Melt flow indexer: M - piston load, R_c - cylinder radius, R_k - capillary radius, L_k - capillary length, Q - volume flow rate

The normalized measurement conditions (ASTM D1238 standard) concern the geometry of the capillary and the piston, the piston load as well as the time and temperature of measurement.

The device geometry and the measurement time are fixed, i.e.

- capillary diameter: $2R_k = 2.095$ mm,
- capillary length: $L_k = 8$ mm; the ratio of capillary length to capillary radius $L_k/R_k = 7.637$,
- piston diameter (cylinder): $2R_c = 9.55$ mm,
- measurement time: $t = 10$ min.

The piston load and the measurement temperature are variable depending on the type of material being tested. The most common conditions are as follows:

- piston load: $M = 2.16$ kg,
- measurement temperature: $T = 190$ °C.

The MFI measurement consists in extruding material samples under normalized conditions and weighing them. The measurement result is given in g/10 min.

The melt flow index is a measure of the material fluidity (the inverse of the material viscosity). It represents a single point on the viscosity curve of this material, determined at the measurement temperature (Figure 2.10). It does not define unambiguously the material to be tested and may be the same for materials of different viscosity characteristics.

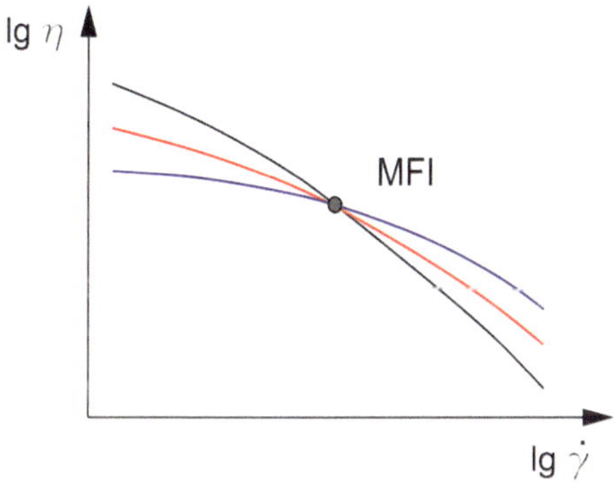

Figure 2.10
Viscosity interpretation of the melt flow index (MFI) in relation to three different viscosity curves

The point that represents the MFI on the viscosity curve can be determined from the formula

$$\eta_a = \frac{\tau_w}{\dot{\gamma}_a} \tag{2.1}$$

where η_a is the apparent viscosity; τ_w is the shear stress at the capillary wall; $\dot{\gamma}_a$ is the shear rate at the capillary wall (non-corrected).

Shear stress and shear rate are determined by Eqs. (2.14) and (2.39), respectively, which take the form

$$\tau_w = \frac{\Delta p R_k}{2L_k} = \frac{F/\pi R_c^2}{2L_k} R_k \tag{2.2}$$

$$\dot{\gamma}_a = \frac{4Q}{\pi R_k^3} \tag{2.3}$$

where Δp is the pressure drop along the capillary; R_k is the capillary radius; L_k is the capillary length; $F = Mg$ is the force acting on the piston; M is the piston load; g is the gravitational acceleration; R_c is the cylinder radius; $Q = (m/t)/\rho$ is the volume flow rate of the material; m is the sample weight; t is the measurement time; ρ is the material density.

The Melt Volume Rate (MVR) may be also used to characterize the flowability of polymers. It is defined by the volume flow rate of the material flowing out of the capillary under standard conditions. It is measured in the same way as the melt flow index, but converting the weight of the extruded strand to the volume using the polymer density. The measurement result is given in $cm^3/10$ min.

2.3.2 The Single-Point Method of Determination of Viscosity Curve

The basis for the approximate method of determining the viscosity curves based on the single rheometric measuring point, e.g. MFI, is the rheological model of Vinogradov and Malkin. It is universal and with good approximation applies to many materials (see Section 1.2.4.2). This is the single-parameter model represented by Eq. (1.171).

The procedure for determining the viscosity function is as follows:

1. Determination of the polymer viscosity $\eta\left(\dot{\gamma}^*, T^*\right)$ at the specific shear rate $\dot{\gamma}^*$ and temperature T^*, e.g. based on the MFI, according to Eqs. (2.1), (2.2), and (2.3).
2. Using this viscosity, determination of the zero viscosity $\eta_0\left(T^*\right)$ corresponding to the temperature T^*, according to Eq. (1.171), i.e. solving the equation

$$\eta\left(\dot{\gamma}^*, T^*\right)=\frac{\eta_0\left(T^*\right)}{1+A_1\left[\dot{\gamma}^* \eta_0\left(T^*\right)\right]^a+A_2\left[\dot{\gamma}^* \eta_0\left(T^*\right)\right]^{2a}} \tag{2.4}$$

3. The knowledge of the zero viscosity $\eta_0\left(T^*\right)$ allows, based on Eq. (1.171), the calculation of the viscosity at another shear rate $\dot{\gamma}$, i.e. to determine the viscosity curve at T^*, according to the formula

$$\eta\left(\dot{\gamma}, T^*\right)=\frac{\eta_0\left(T^*\right)}{1+A_1\left[\dot{\gamma} \eta_0\left(T^*\right)\right]^a+A_2\left[\dot{\gamma} \eta_0\left(T^*\right)\right]^{2a}} \tag{2.5}$$

4. Calculation of the zero viscosity $\eta_0(T)$, based on the WLF equation (1.114) with reference to another temperature T, according to the formula

$$\eta_0(T)=\eta_0\left(T^*\right)\exp\left[\frac{8.86\left(T^*-T_g-50\right)}{101.6+\left(T^*-T_g-50\right)}-\frac{8.86\left(T-T_g-50\right)}{101.6+\left(T-T_g-50\right)}\right] \tag{2.6}$$

   where T_g is the glass transition temperature.
5. Determination of the viscosity at different shear rates $\dot{\gamma}$ on the basis of Eq. (1.171), i.e. determination of the viscosity curve at the temperature T.
6. As a result of repeating steps from points 4 and 5 in relation to other temperatures, viscosity curves at different temperatures can be obtained.

On the basis of the presented procedure, a computer program can be developed, which allows quick determination of the viscosity curves based on the single measurement point, e.g. MFI.

2.4 Capillary Rheometers

2.4.1 Principle of Operation

The principle of operation of the capillary rheometer consists in forcing the test material through capillaries (cylindrical tubes). The measurement consists in determining the relationship between the volume flow rate of the material and the pressure drop caused by the internal friction of the material. The knowledge of the dependence of the volume flow rate and the pressure drop allows the determination of the relationship between the shear rate and the shear stress, and thus the determination of the flow curve of the material to be tested.

Capillary rheometers are of fundamental importance in rheometry. They allow determination of the material shear functions, whereby they are primarily used to determine the material function of shear (transverse) viscosity $\eta(\dot{\gamma})$. They have a relatively simple design and allow measurements over a wide range shear rates, but only at higher shear rates. Using them, one can also estimate the elongational (longitudinal) viscosity and the first difference in normal stresses.

2.4.2 Theoretical Basics

The flow in the capillary rheometer is shown in the cylindrical coordinate system r, φ, z in Figure 2.11. The basic assumptions of this flow are as follows:

- the flow is steady, laminar, and isothermal,
- the flow is fully developed,
- there is no slip on the capillary wall,
- gravity and inertia forces are neglected,
- the fluid is incompressible and its viscosity does not depend on the pressure,
- the velocity field has the form

$$v_r = 0, \quad v_\varphi = 0, \quad v_z = v_z(r) \tag{2.7}$$

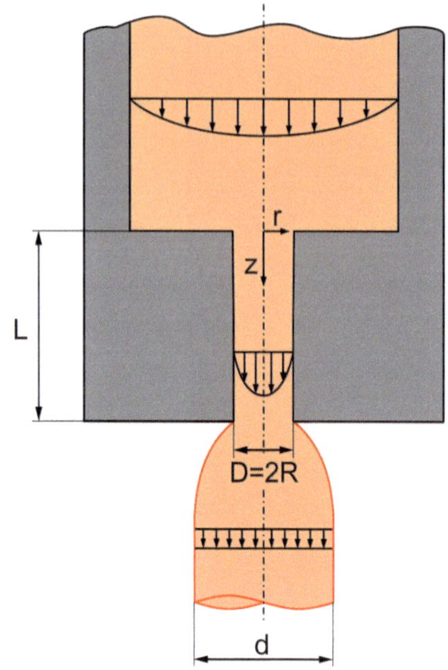

Figure 2.11
The flow in a capillary rheometer: *R* - capillary radius, *D* - capillary diameter, *L* - capillary length, *d* - extrudate diameter

Under these conditions, the equation of motion in the flow direction (Table 1.3) is expressed as

$$0=-\frac{\partial p}{\partial z}+\frac{1}{r}\frac{\partial}{\partial r}\left(r\tau_{rz}\right) \tag{2.8}$$

Since the pressure is the function of the *z*-coordinate only, and the stress τ_{rz} depends on the coordinate *r* only, the partial derivatives can be replaced with ordinary derivatives and Eq. (2.8) can be presented as

$$0=-\frac{\mathrm{d}p}{\mathrm{d}z}+\frac{1}{r}\frac{\mathrm{d}}{\mathrm{d}r}\left(r\tau_{rz}\right) \tag{2.9}$$

The pressure gradient in this flow is constant and equal to

$$-\frac{\mathrm{d}p}{\mathrm{d}z}=\frac{\Delta p}{L} \tag{2.10}$$

where $\Delta p = p_0 - p_L$ is the pressure drop along the capillary; p_0 is the pressure at the beginning of the channel; p_L is the pressure at the end of the channel; L is the capillary length.

Thus, Eq. (2.9) can be transformed into

$$\frac{\mathrm{d}}{\mathrm{d}r}\left(r\tau_{rz}\right)=-\frac{\Delta p}{L}r \tag{2.11}$$

and, after integration

$$\tau_{rz} = -\frac{\Delta p}{L}\frac{r}{2} + \frac{C_1}{r} \tag{2.12}$$

where the constant $C_1 = 0$ because the stress in the channel axis, i.e. $\tau_{rz}(r = 0)$, cannot be infinite.

Thus, the stress distribution is linear and is expressed by

$$\tau_{rz} = -\frac{\Delta p}{L}\frac{r}{2} \tag{2.13}$$

According to this formula, the shear stress τ_w at the capillary wall is equal to

$$\tau_w = \tau_{rz}(r = R) = -\frac{\Delta p R}{2L} \tag{2.14}$$

so the stress distribution can be represented as

$$\frac{\tau_{rz}}{\tau_w} = \frac{r}{R} \tag{2.15}$$

The minus sign in Eq. (2.14) means that the force, defined by the product of shear stress and surface, that is exerted by the channel wall on the fluid is directed against the motion of the fluid. In contrast, the force exerted by the fluid on the channel wall is directed according to the movement of the fluid, so that it has the opposite sign, i.e. positive. Accepting the appropriate stress sign is conventional and often the stress τ_w is assumed to be a positive quantity.

It follows from Eq. (2.15) that the distribution of stress in this flow is linear and independent of the constitutive equation of the material.

Further considerations will apply to Newtonian fluids. Then the stress component τ_{rz} will be equal (Table 1.7) to

$$\tau_{rz} = \mu \frac{\mathrm{d}v_z}{\mathrm{d}r} \tag{2.16}$$

where μ is the Newtonian viscosity.

After substituting Eq. (2.16) into Eq. (2.13), we have

$$\mu \frac{\mathrm{d}v_z}{\mathrm{d}r} = -\frac{\Delta p}{L}\frac{r}{2} \tag{2.17}$$

and then, after transformation

$$\frac{\mathrm{d}v_z}{\mathrm{d}r} = -\frac{1}{\mu}\frac{\Delta p}{L}\frac{r}{2} \tag{2.18}$$

and, after integration

$$v_z = -\frac{1}{\mu}\frac{\Delta p}{L}\frac{r^2}{4} + C_2 \tag{2.19}$$

where the constant C_2 can be determined from the condition of no slip on the capillary wall, i. e. for $v_z(r = R) = 0$, and is equal to

$$C_2 = -\frac{1}{\mu}\frac{\Delta p}{L}\frac{R^2}{4} \tag{2.20}$$

As a result, the parabolic velocity distribution is obtained:

$$v_z = \frac{R^2}{4\mu}\frac{\Delta p}{L}\left[1 - \left(\frac{r}{R}\right)^2\right] \tag{2.21}$$

where the maximum velocity v_{max} occurs for $r = 0$, i. e. in the channel axis, and it is expressed by

$$v_{max} = \frac{R^2}{4\mu}\frac{\Delta p}{L} \tag{2.22}$$

Distributions of shear stress (Eq. (2.15)), velocity (Eq. (2.21)), and shear rate (Eq. (2.18)) in the capillary channel are shown in Figure 2.12.

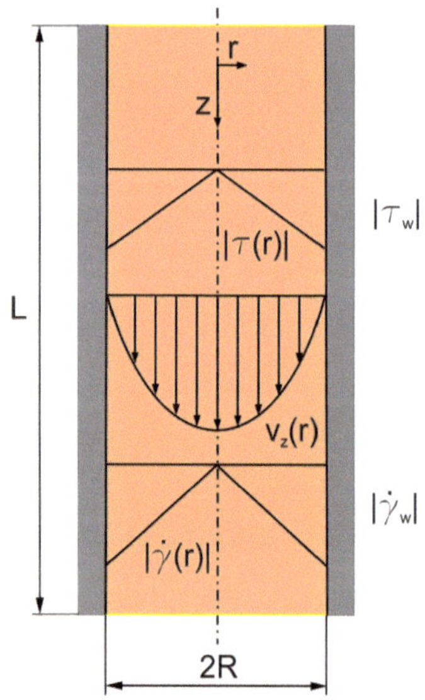

Figure 2.12
Distribution of the shear stress $\tau(r)$, velocity $v_z(r)$, and shear rate $\dot{\gamma}(r)$ in the Newtonian pressure flow in a cylindrical channel: τ_w - shear stress at the wall, $\dot{\gamma}_w$ - shear rate at the wall

Considering an elementary flow dQ through an annular channel with radius r and thickness dr, occurring at the local velocity $v_z(r)$, i. e.

$$dQ = v_z(r)2\pi r dr \tag{2.23}$$

the flow rate can be determined in relation to the entire cross-section of the channel as

$$Q = 2\pi \int_0^R v_z(r) r \mathrm{d}r \tag{2.24}$$

After substituting Eq. (2.21) into Eq. (2.24), we obtain

$$Q = 2\pi \int_0^R \frac{R^2}{4\mu} \frac{\Delta p}{L} \left[1 - \left(\frac{r}{R}\right)^2\right] r \mathrm{d}r \tag{2.25}$$

and, after integration, we have the formula determining the total volume flow rate, i.e. the **Hagen-Poiseuille law:**

$$Q = \frac{\pi R^4}{8\mu} \frac{\Delta p}{L} \tag{2.26}$$

On the basis of the Hagen-Poiseuille equation, the viscosity μ can be determined by measuring the pressure drop Δp and the flow rate Q, in the channel of length L and radius R, according to the formula

$$\mu = \frac{\pi R^4}{8Q} \frac{\Delta p}{L} \tag{2.27}$$

Eq. (2.27) can be transformed into

$$\mu = \frac{\dfrac{\Delta p R}{2L}}{\dfrac{4Q}{\pi R^3}} \tag{2.28}$$

The numerator in Eq. (2.28) expresses the shear stress τ_w at the channel wall, defined by Eq. (2.14), while the denominator determines the shear rate $\dot{\gamma}_w$ at the wall, expressed by Eq. (2.18) for $r = R$.

This can be demonstrated by inserting the dependence in Eq. (2.26) into Eq. (2.18). After transformation, we get

$$\dot{\gamma}_w = \frac{\mathrm{d}v_z}{\mathrm{d}r}(r = R) = -\frac{4Q}{\pi R^3} \tag{2.29}$$

where the shear rate $\dot{\gamma}_w$, just like the shear stress at the channel wall τ_w, is usually conventionally taken as a positive value, i.e. as

$$\dot{\gamma}_w = -\left(\frac{\mathrm{d}v_z}{\mathrm{d}r}\right) = \frac{4Q}{\pi R^3} \tag{2.30}$$

The Hagen-Poiseuille equation in the form of Eq. (2.27) or Eq. (2.28) makes it possible to determine the Newtonian viscosity; however, it does not allow the deter-minion of the non-Newtonian fluid flow curves, for which the viscosity is a function of the shear rate, i. e. $\eta = \eta(\dot{\gamma})$.

In order to determine the flow curve of non-Newtonian fluids, we need to have a relationship between shear stress and shear rate in the widest possible range of these. Since both the shear stress and the shear rate vary with the distance from the capillary axis, it is necessary that both of them are determined at the same point in the capillary. The most convenient point is the wall of the capillary.

The shear stress at the channel wall is independent of the constitutive equation of the material and is expressed as in the case of Newtonian fluid, i. e. according to Eq. (2.14). However, the shear rate at the wall in the case of non-Newtonian fluid is not equal to the shear rate of the Newtonian fluid at this point and cannot be determined by Eq. (2.30).

Determination of this shear rate basing on experimental tests is possible using the Rabinowitsch-Mooney equation. In order to derive this equation, the volume flow rate in the capillary should be analyzed.

After integrating Eq. (2.24) by parts we have

$$Q = \pi r^2 v_z \Bigg|_0^R - \int_0^R \pi r^2 \left(\frac{\mathrm{d}v_z}{\mathrm{d}r}\right) \mathrm{d}r \tag{2.31}$$

Since the assumption of no slip at the wall shows that $v_z(r = R) = 0$, the first expression on the right side of Eq. (2.31) equals zero and we have

$$Q = -\pi \int_0^R r^2 \left(\frac{\mathrm{d}v_z}{\mathrm{d}r}\right) \mathrm{d}r \tag{2.32}$$

It follows from Eq. (2.15) that

$$r = \frac{\tau_{rz}}{\tau_w} R \tag{2.33}$$

and

$$\mathrm{d}r = \frac{R}{\tau_w} \mathrm{d}\tau_{rz} \tag{2.34}$$

After substituting these relationships into Eq. (2.32) we have

$$\frac{Q\tau_w^{\;3}}{\pi R^3} = -\int_0^{\tau_w} \tau_{rz}^{\;2} \left(\frac{\mathrm{d}v_z}{\mathrm{d}r}\right) \mathrm{d}\tau_{rz} \tag{2.35}$$

By differentiating Eq. (2.35) with respect to τ_w and using the Leibnitz theorem, we have

$$\frac{3\tau_w^{\;2} Q}{\pi R^3} + \frac{\tau_w^{\;3}}{\pi R^3} \frac{\mathrm{d}Q}{\mathrm{d}\tau_w} = \tau_w^{\;2} \left(-\frac{\mathrm{d}v_z}{\mathrm{d}r}\right) \Bigg|_{r=R} \tag{2.36}$$

After transforming this relationship, the equation which allows the determination of the shear rate at the capillary wall is obtained, usually called the **Rabinowitsch-Mooney equation:**

$$\dot{\gamma}_w = \left(-\frac{dv_z}{dr}\right)_w = \frac{1}{\pi R^3}\left(3Q + \tau_w \frac{dQ}{d\tau_w}\right) \tag{2.37}$$

or

$$\dot{\gamma}_w = \left(-\frac{dv_z}{dr}\right)_w = \frac{4Q}{\pi R^3}\left(\frac{3}{4} + \frac{1}{4}\frac{d\ln Q}{d\ln \tau_w}\right) \tag{2.38}$$

It is worth noting that the part $4Q/\pi R^3$ in Eq. (2.38) expresses, according to Eq. (2.30), the shear rate of the Newtonian fluid at the capillary wall. In the case of non-Newtonian fluid, it is called a nominal or non-corrected (or apparent) shear rate and is marked as $\dot{\gamma}_a$ *(a - apparent):*

$$\dot{\gamma}_a = \frac{4Q}{\pi R^3} \tag{2.39}$$

The Rabinowitsch-Mooney equation (2.38) can also be presented in a slightly different form as

$$\dot{\gamma}_w = \frac{4Q}{\pi R^3}\left(\frac{3}{4} + \frac{1}{4}\frac{d\ln(4Q/\pi R^3)}{d\ln \tau_w}\right) \tag{2.40}$$

and after using Eq. (2.39)

$$\dot{\gamma}_w = \dot{\gamma}_a\left(\frac{3}{4} + \frac{1}{4}\frac{d\ln \dot{\gamma}_a}{d\ln \tau_w}\right) \tag{2.41}$$

Then, the derivative can be replaced with the symbol n', which is called the characteristic flow index and defined as

$$n' = \frac{d\ln \tau_w}{d\ln \dot{\gamma}_a} \tag{2.42}$$

and finally, the following relationship is obtained:

$$\dot{\gamma}_w = \dot{\gamma}_a\left(\frac{3n'+1}{4n'}\right) \tag{2.43}$$

The shear rate $\dot{\gamma}_w$ of the non-Newtonian fluid at the capillary wall is called the corrected shear rate or the true shear rate. The expression in brackets in Eq. (2.43) can be considered as the correction of the nominal shear rate $\dot{\gamma}_a$. This correction is often called the **Rabinowitsch correction.**

The characteristic flow index n', defined by Eq. (2.42), is equal to one for the Newtonian fluids since the stress does not depend on the shear rate in this case. On the other hand, for the non-Newtonian fluids this index is different from one and, of course, varies for different materials. For most polymers it is approximately constant in a large range of shear rates.

In each shear rate range, where $n' = const$, integration of Eq. (2.42) leads to the formula

$$\tau_w = m'\dot{\gamma}_a^{n'} \tag{2.44}$$

where m' is also constant.

If Eq. (2.43) is substituted into Eq. (2.44) we have

$$\tau_w = m'\left[\left(\frac{4n'}{3n'+1}\right)\dot{\gamma}_w\right]^{n'} \tag{2.45}$$

Comparing this formula with the power law of Ostwald-de Waele (1.104), which in this case will take the form

$$\tau_w = m\dot{\gamma}_w^n \tag{2.46}$$

it can be seen that Eq. (2.44) is a particular form of Eq. (2.46) with

$$n = n' \quad \text{and} \quad m = m'\left(\frac{4n'}{3n'+1}\right)^n \tag{2.47}$$

The flow index n' is a measure of the degree of non-Newtonian behavior of materials. The more n' differs from unity, the more non-Newtonian the fluid is. For a fluid thinning with shearing $n' < 1$, while for a shear thickening fluid $n' > 1$. This is the same as in the case of the flow exponent n in the power law (see Section 1.2.2). On the other hand, the parameter m' defines the consistency of the fluid, and the bigger it is, the more viscous the fluid is.

2.4.3 Errors of Capillary Rheometry

End Effects

In the previous considerations, it was assumed that the flow in the capillary is fully developed over the entire length, and the pressure drop is caused by the internal friction of the fluid only. Thus, the end-flow effects at the inlet and outlet of the capillary, which result in an additional pressure drop in the capillary, were omitted. However, the inlet effects are particularly important here. As a result of omitting these, calculation of the shear stress, defined by Eq. (2.14), is subjected to a certain error and provides higher values than would result from the assumptions made.

The inlet effect is associated with the formation of the velocity profile in the initial part of the capillary, in the so-called inlet section L_e (Figure 2.13). Additional energy is needed for this formation, which results in a higher pressure gradient in this area than in the area of the fully developed flow. As a result, the problem arises to determine, on the basis of experimental data, the value of the pressure drop caused by internal friction in the conditions of the fully developed velocity profile. The result of measurements is a total pressure drop resulting from the inlet effect and the developed flow.

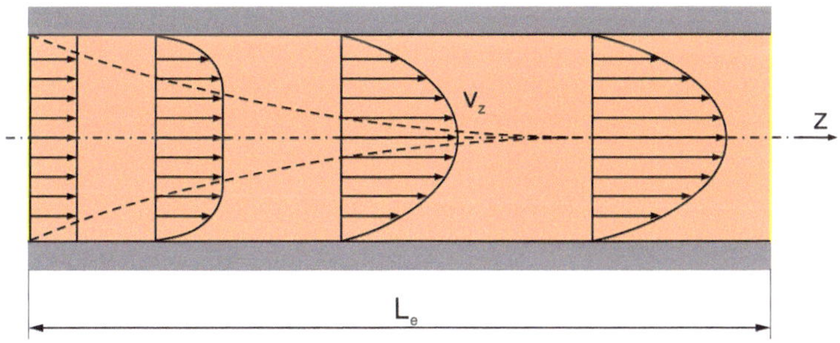

Figure 2.13 Formation of the velocity profile v_z in the inlet section L_e of a cylindrical channel

The pressure profile along the capillary is shown in Figure 2.14. It is seen that the actual pressure drop Δp_c, i. e. measured during the test, is the sum of the theoretical pressure drop Δp that occurs due to the flow in the capillary (under the conditions of the fully developed velocity profile), and the pressure drop Δp_e at the inlet of the capillary, i. e.

$$\Delta p_c = \Delta p + \Delta p_e \tag{2.48}$$

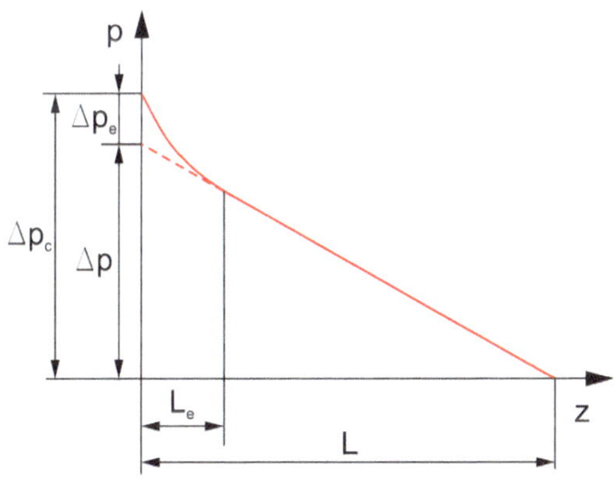

Figure 2.14 Pressure profile along the rheometer capillary: p - pressure, z - coordinate in the capillary length direction, Δp_c - actual pressure drop, Δp - theoretical pressure drop, Δp_e - pressure drop at the capillary inlet, L_e - inlet section, L - capillary length

The pressure drop at the inlet of the capillary (the so-called inlet loss) Δp_e depends on the flow rate of the fluid and the diameter of the capillary, but does not depend on the length of the capillary. It follows that the effect of the inlet effect can be eliminated by taking measurements for different length capillaries, but with the same diameter and the same flow rate. This method of eliminating the influence of the inlet effect is called the **Couette method.**

Let us consider the measurement according to this method for two capillaries of different length L_1 and L_2, assuming that $L_1 < L_2$. The pressure drop in these capillaries will be equal to (Figure 2.15)

$$\Delta p_{c1} = \Delta p_1 + \Delta p_e \tag{2.49}$$

$$\Delta p_{c2} = \Delta p_2 + \Delta p_e \tag{2.50}$$

where the inlet loss Δp_e will be equal in both cases as long as the shorter capillary has a length greater than the inlet section L_e, on which the fully developed velocity profile is formed, i.e. when the following condition is met:

$$L_1 > L_e \tag{2.51}$$

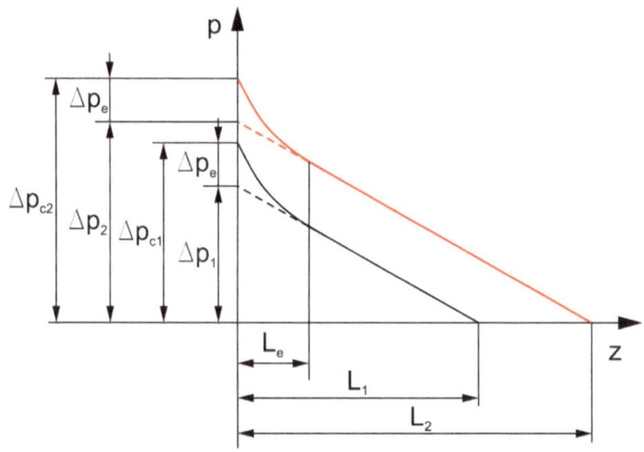

Figure 2.15 Pressure profile for capillaries of different length, the same diameter, and the same flow rate (Couette method): p - pressure, z - coordinate in the capillary length direction, Δp_{c1}, Δp_{c2} - actual pressure drops, Δp_1, Δp_2 - theoretical pressure drops, Δp_e - pressure drop at the capillary inlet, L_e - inlet section, L_1, L_2 - capillary length

From Eqs. (2.49) and (2.50) it follows that the difference in measured pressure drops ($\Delta p_{c2} - \Delta p_{c1}$) is equal to the difference in theoretical drops ($\Delta p_2 - \Delta p_1$), which depends on the length of the capillaries, only. Thus, the pressure drop Δp per unit length L of the capillary is equal to

$$\frac{\Delta p}{L} = \frac{\Delta p_2}{L_2} = \frac{\Delta p_1}{L_1} = \frac{\Delta p_{c2} - \Delta p_{c1}}{L_2 - L_1} \tag{2.52}$$

In order to check if the condition in Eq. (2.51) is met, the pressure drop Δp_3 (at the same flow rate) for the third capillary of length $L_3 > L_2$ should be measured. Eq. (2.52) will be true when

$$\frac{\Delta p_{c2} - \Delta p_{c1}}{L_2 - L_1} = \frac{\Delta p_{c3} - \Delta p_{c2}}{L_3 - L_2} = \frac{\Delta p_{c3} - \Delta p_{c1}}{L_3 - L_1} \tag{2.53}$$

In this way, the calculated pressure gradient is not burdened with an inlet effect error and can be the basis for calculating the shear stress τ_w according to Eq. (2.14).

Modification of the Couette method is the **Bagley method**, the essence of which is expressed in Figure 2.16. This method involves determining the pressure drop per unit length of the capillary under the conditions of the developed velocity profile. This is done by referring the actual pressure drop Δp_c to a certain hypothetical length of the capillary L_c, including an additional pressure drop over the inlet section L_e.

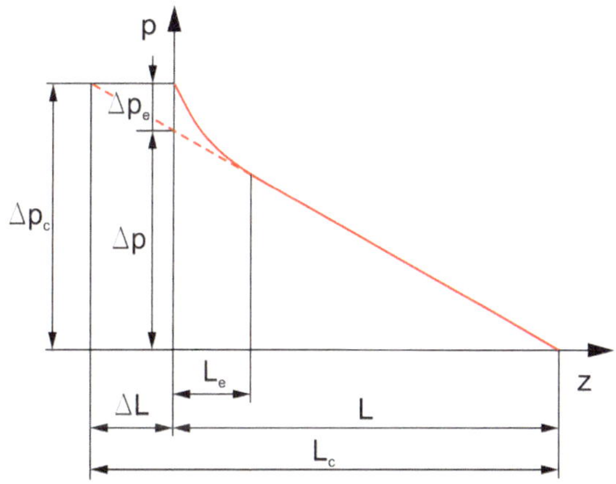

Figure 2.16 Schematic of Bagley method: p - pressure, z - coordinate in the capillary length direction, Δp_c - actual pressure drop, Δp - theoretical pressure drop, Δp_e - pressure drop at the capillary inlet, L_e - inlet section, L - capillary length, ΔL - additional length of the capillary, taking into account the additional pressure drop at the inlet section L_e, L_c - hypothetical length of the capillary, taking into account the additional pressure drop at the inlet section L_e

Thus, in this method the following relationship is valid:

$$\frac{\Delta p}{L} = \frac{\Delta p_c}{L_c} \tag{2.54}$$

where

$$L_c = L + \Delta L \tag{2.55}$$

where $\Delta L = eR$ is the additional (fictional) capillary length, taking into account the additional pressure drop at the inlet section L_e; e is the correction factor, i.e. Bagley correction; R - capillary radius.

The shear stress defined by Eq. (2.14) can be calculated using this method as

$$\tau_w = \frac{\Delta p_c R}{2(L + eR)} \tag{2.56}$$

The value of Bagley correction can be determined on the basis of measurements made for capillaries with different L/R ratio. The pressure drop Δp_c should be determined for each capillary giving a constant value of the non-corrected shear rate $\dot{\gamma}_a$ defined by Eq. (2.39).

The plot of the dependence

$$\Delta p_c = f\left(\frac{L}{R}\right) \tag{2.57}$$

should be a straight line, cutting the abscissa axis (L/R) at the value ($-e$) for $\Delta p_c = 0$ (Figure 2.17).

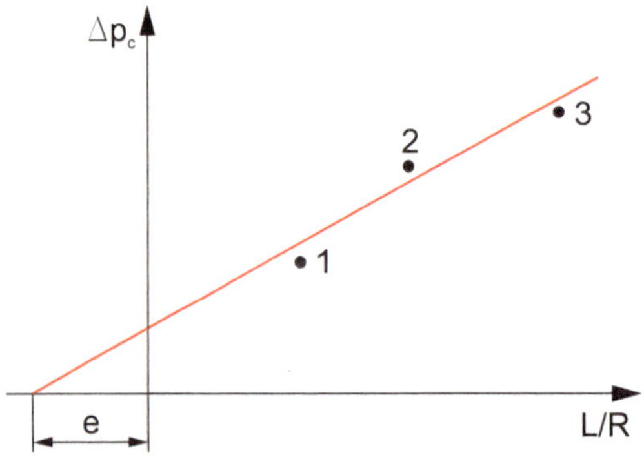

Figure 2.17
Method of determining the Bagley correction (e): 1, 2, 3 - measurement results Δp_c for capillaries of different ratio L/R

Similar plots can be made for other constant shear rates $\dot{\gamma}_a$, which allows one to create the dependency plot:

$$e = f(\dot{\gamma}_a) \tag{2.58}$$

which makes it possible to calculate the shear stress τ_w according to Eq. (2.56), for any shear rate $\dot{\gamma}_a$.

The influence of inlet losses on the measurement results decreases with an increase in the ratio of the capillary length L to its diameter D. Typically, this effect is considered negligible when $L/D > 60$.

Slip Effects

The considerations in Section 2.4.2 were carried out assuming a no slip condition on the capillary wall, i. e. it was assumed that the velocity of the fluid at the wall is equal to the velocity of the wall, i. e. zero. In fact, in the case of polymer flows, such slipping often occurs. It occurs after exceeding a certain critical value of the shear stress at the capillary wall (e. g. for polyethylene equal to about 0.1 MPa).

The dependence of the slip velocity V_s on the shear stress τ_w can be expressed by the formula

$$V_s = A\tau_w^b \tag{2.59}$$

where A and b are the material parameters (usually, $b = 2\text{–}4$).

Slip is an important phenomenon, but difficult to assess. However, it is relatively easy to determine the slip velocity using the **Mooney method.**

In order to present this method, let us consider the flow without slip and with slipping. In the first case, the flow Q in the capillary with radius R will be equal to

$$Q = \pi R^2 V \tag{2.60}$$

where V is the average flow velocity.

The shear rate $\dot{\gamma}_a$ will be expressed by Eq. (2.39), which after applying Eq. (2.60), will take the form

$$\dot{\gamma}_a = \frac{4V}{R} \tag{2.61}$$

If there is a slip, the shear rate will be expressed by

$$\dot{\gamma}_{as} = \frac{4\left(V - V_s\right)}{R} \tag{2.62}$$

where $(V - V_s)$ is the average velocity of the material flow relative to its velocity at the capillary wall.

After using Eq. (2.39), we have

$$\dot{\gamma}_{as} = \frac{4Q}{\pi R^3} - \frac{4V_s}{R} \tag{2.63}$$

The Mooney method is based on a series of measurements of non-corrected shear rate $\dot{\gamma}_a$, i. e. $4Q/\pi R^3$, for a capillary with different radius, but the same L/R ratio. This is done while maintaining the constant value of the shear stress τ_w, i. e. with a constant pressure drop Δp. The results are presented as a function of the inverse of the capillary radius $1/R$. A linear relationship is then obtained (Figure 2.18), which corresponds to Eq. (2.63) transformed into

$$\frac{4Q}{\pi R^3} = \dot{\gamma}_{as} + \frac{4V_s}{R} \tag{2.64}$$

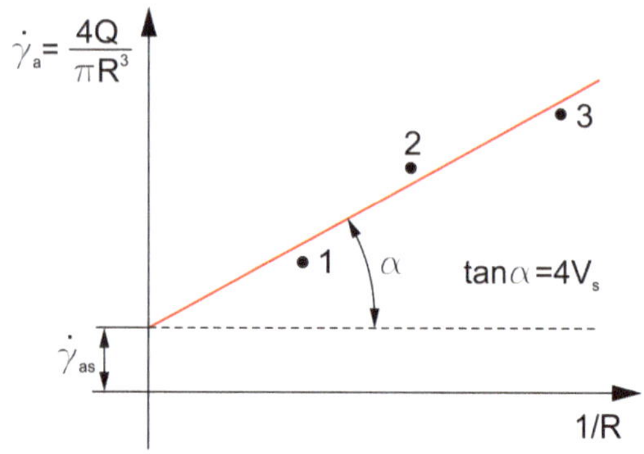

Figure 2.18 Mooney method of determining the slip velocity: 1, 2, 3 - measurement results of non-corrected shear rate ($\dot{\gamma}_a = 4Q/\pi R^3$) for capillaries of different radius, but the same ratio L/R; the dashed line means no slip

The slope of the straight line obtained is equal to $4V_s$, so it determines the value of the slip velocity. Extrapolation of this dependence for $R \to \infty$, i.e. for $1/R = 0$, determines the shear rate $\dot{\gamma}_{as}$, taking into account the occurrence of slip.

2.4.4 Determination of Viscosity

Eqs. (2.14) and (2.43) determine the shear stress and the shear rate of non-Newtonian fluids at the same point, i.e. on the capillary wall, so that they allow determination of the flow curves and the apparent and true viscosity curves. The flow curve expresses the relationship between shear stress and shear rate and can be presented as a function of non-corrected or corrected shear rate, and thus as

$$\tau_w = f(\dot{\gamma}_a) \tag{2.65}$$

or

$$\tau_w = f(\dot{\gamma}_w) \tag{2.66}$$

The apparent viscosity η_a given by Eq. (2.1) is defined by the ratio of the shear stress τ_w and the non-corrected shear rate $\dot{\gamma}_a$ given by Eq. (2.39). Whereas the true viscosity η is defined by the ratio of the shear stress τ_w and the corrected shear rate $\dot{\gamma}_w$ given by Eq. (2.43), i.e.

$$\eta = \frac{\tau_w}{\dot{\gamma}_w} \tag{2.67}$$

Based on Eqs. (2.1), (2.43), and (2.67), the relationship between the true viscosity η and the apparent viscosity η_a can be derived:

$$\eta = \eta_a \left(\frac{4n'}{3n'+1} \right) \tag{2.68}$$

The **true viscosity curve** expresses the dependence of this viscosity on the corrected shear rate, whereas the **apparent viscosity curve** shows the dependence of this viscosity on the non-corrected shear rate. So, these are the functions:

$$\eta = f(\dot{\gamma}_w) \tag{2.69}$$

$$\eta_a = f(\dot{\gamma}_a) \tag{2.70}$$

In order to determine the flow curves and the viscosity curves one should have a relation between the shear stress and the shear rate in the widest possible range of these.

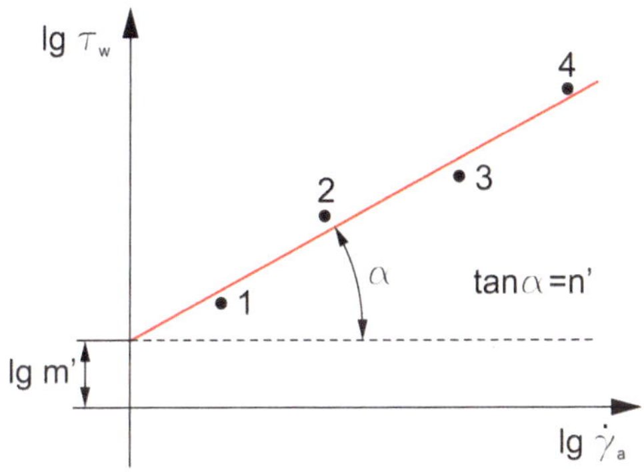

Figure 2.19
Method of determining the flow index n' and the consistency factor m': 1, 2, 3, 4 – measurements results τ_w and $\dot{\gamma}_a$

The procedure for determining these curves is as follows:

- a series of measurements of the pressure drop and the corresponding volume flow rate are made,
- at particular measuring points, the shear stress τ_w is determined according to Eq. (2.14) and the apparent (non-corrected) shear rate $\dot{\gamma}_a$ according to Eq. (2.39),
- the apparent viscosity according to Eq. (2.1) is calculated at each measurement point and the results are presented in the function of apparent shear rate, thus providing the apparent viscosity curve of the non-Newtonian fluid $\eta_a = f(\dot{\gamma}_a)$,
- the dependence of the shear stress on the apparent shear rate is presented in a double logarithmic system (it may be a system of decimal logarithms), i.e. as $\lg \tau_w = f(\lg \dot{\gamma}_a)$; this dependence for polymers is usually linear,

- the characteristic flow index n' is determined according to Eq. (2.42), whereby the natural logarithm can be replaced by the decimal logarithm, and the parameter m' is determined, which is equal to τ_w at the shear rate $\dot{\gamma}_a = 1$ ($\lg \dot{\gamma}_a = 0$), i.e. $m' = \tau_w(\dot{\gamma}_a = 1)$ or $\lg m' = \lg \tau_w(\lg \dot{\gamma}_a = 0)$; in the case of linear dependence of $\lg \tau_w = f(\lg \dot{\gamma}_a)$ the index n' is the slope coefficient of this function,
- the corrected shear rate $\dot{\gamma}_w$ is calculated at each measuring point according to Eq. (2.43),
- the shear stress τ_w is presented as a function of the corrected shear rate $\dot{\gamma}_w$, thus providing the corrected (true) flow curve of the non-Newtonian fluid $\tau_w = f(\dot{\gamma}_w)$,
- the true viscosity is calculated at each measuring point according to Eq. (2.67) and the obtained results are presented in the function of true (corrected) shear rate, resulting in the viscosity curve of the non-Newtonian fluid $\eta = f(\dot{\gamma}_w)$.

The mutual relations of the non-corrected and corrected flow curves as well as the apparent and true viscosity curves are shown in Figure 2.20 and Figure 2.21. These relations result from the difference between the corrected and non-corrected shear rates, which are positive for polymers. Thus, the corrected flow curve is shifted to the right with respect to the non-corrected curve by a section equal to the change in shear rate, i.e. by $\Delta \lg \dot{\gamma} = \lg \dot{\gamma}_w - \lg \dot{\gamma}_a$, equal to the logarithm of the Rabinowitsch correction (Eq. (2.43)).

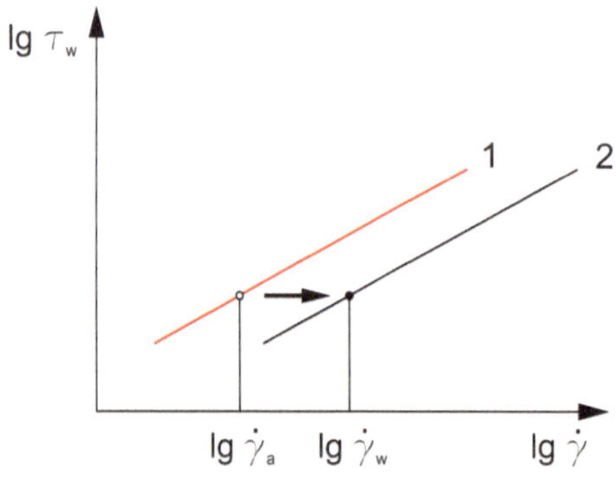

Figure 2.20
Mutual position of the non-corrected flow curve (1) $\lg \tau_w = f(\lg \dot{\gamma}_a)$ and the corrected flow curve (2) $\lg \tau_w = f(\lg \dot{\gamma}_w)$: τ_w - shear stress at the wall, $\dot{\gamma}_a$ - non-corrected shear rate, $\dot{\gamma}_w$ - corrected shear rate, $\dot{\gamma}$ - shear rate

On the other hand, the corrected viscosity curve is shifted from the apparent viscosity curve down by a section equal to the change in viscosity resulting from the change in the shear rate; and to the right, by a section equal to the change in the shear rate. These sections, in the logarithmic system, are equal to each other, so that the offset is along the straight line with the slope coefficient (−1).

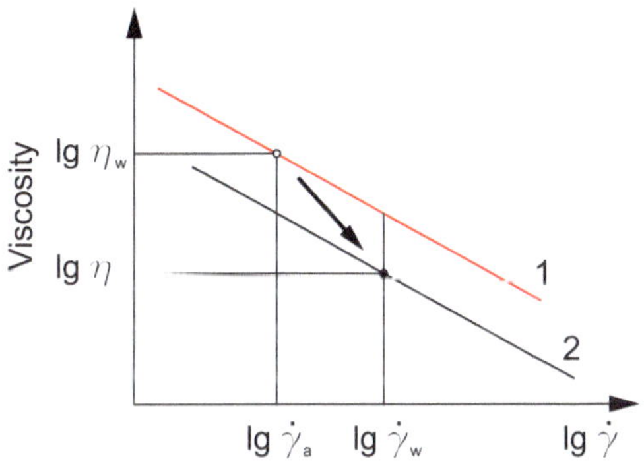

Figure 2.21
Mutual position of the apparent viscosity curve (1) $\lg\eta_a = f(\lg\dot{\gamma}_a)$ and the true viscosity curve (2) $\lg\eta = f(\lg\dot{\gamma}_w)$: τ_w - shear stress at the wall, $\dot{\gamma}_a$ - non-corrected shear rate, $\dot{\gamma}_w$ - corrected shear rate, $\dot{\gamma}$ - shear rate

2.4.5 Determination of Extensional Viscosity

The capillary rheometer enables an evaluation of the extensional viscosity. The basis for this evaluation is the analysis of the inlet pressure loss Δp_e in the rheometer proposed by Cogswell [31]. From this analysis, the following relationships determining the extensional viscosity η_e and the extension rate corresponding to this viscosity $\dot{\varepsilon}$ are derived:

$$\eta_e = \frac{9\left(n'+1\right)^2 \Delta p_e^2}{32\eta_a\dot{\gamma}_a^2} \tag{2.71}$$

$$\dot{\varepsilon} = \frac{4\eta_a\dot{\gamma}_a^2}{3\left(n'+1\right)\Delta p_e} \tag{2.72}$$

where n' is the flow index (Eq. (2.42)); η_a is the apparent viscosity (Eq. (2.1)); $\dot{\gamma}_a$ is the apparent shear rate (Eq. (2.39)); Δp_e is the pressure inlet loss.

The inlet loss can be determined here based on the Bagley correction. From Eqs. (2.14), (2.48), and (2.56), it follows that

$$\tau_w = \frac{\left(\Delta p + \Delta p_e\right)R}{2\left(L + eR\right)} \tag{2.73}$$

and this expression makes it possible to determine the inlet loss as

$$\Delta p_e = 2e\tau_w \tag{2.74}$$

where τ_w is the shear stress at the channel wall (Eq. (2.14)).

The Cogswell method is approximate, but it gives good results and it is used in practical problems of polymer processing.

2.4.6 Determination of Normal Stresses

The capillary rheometer also enables the evaluation of the first difference in normal stresses. It is approximate, and it is made on the basis of measurement of the degree of swelling of the extrudate flowing out of the capillary. The **swelling degree** is characterized by the ratio of the diameter of the extrudate d to the diameter of the capillary D (Figure 2.11) as

$$B = d / D \tag{2.75}$$

The basis of the assessment is the relationship between the swelling degree and the normal stresses (Eqs. (1.276), (1.277), (1.278)), proposed by Tanner [14] (see Section 1.3.4.3). According to this

$$B = 0.13 + \left[1 + \left(\frac{N_{1w}}{2\sqrt{2}\tau_w}\right)^2\right]^{1/6} \tag{2.76}$$

where B is the swelling degree; N_{1w} is the first normal stress difference; τ_w is the shear stress at the capillary wall.

After transforming Eq. (2.76), the formula is obtained that makes it possible to determine the first difference of normal stresses as

$$N_{1w} = 2\sqrt{2}\tau_w\left[(B - 0.13)^6 - 1\right]^{1/2} \tag{2.77}$$

The basic formulas necessary to determine the rheological properties with the capillary rheometer, i.e. the flow curve, the shear viscosity, the extensional viscosity, and the difference in normal stresses, are given in Table 2.1.

Table 2.1 Calculation Formulas for Rheometry Measurements

Rheometer Type	Calculation Formulas		
Capillary rheometer	Shear stress at the capillary wall	$\tau_w = -\frac{\Delta p R}{2L}$	(2.14)
	Shear rate at the capillary wall	$\dot{\gamma}_a = \frac{4Q}{\pi R^3}$	(2.39)
		$\dot{\gamma}_w = \dot{\gamma}_a\left(\frac{3n'+1}{4n'}\right)$	(2.43)
		$n' = \frac{d\ln\tau_w}{d\ln\dot{\gamma}_a}$	(2.42)
	First normal stress difference	$N_{1w} = 2\sqrt{2}\tau_w\left[(B-0.13)^6 - 1\right]^{1/2}$	(2.77)

Rheometer Type	Calculation Formulas		
	Extensional viscosity	$\eta_e = \dfrac{9(n'+1)^2 \Delta p_e^2}{32\eta_a \dot{\gamma}_a^2}$	(2.71)
	Extension rate	$\dot{\varepsilon} = \dfrac{4\eta_a \dot{\gamma}_a^2}{3(n'+1)\Delta p_e}$	(2.72)
	Inlet loss	$\Delta p_e = 2e\tau_w$	(2.74)
Cone-plate rheometer	Shear stress	$\tau_{\theta\varphi} = \dfrac{3M}{2\pi R^3}$	(2.91)
	Shear rate	$\dot{\gamma} = \Omega / \alpha$	(2.84)
	First normal stress difference	$N_1 = \dfrac{2F}{\pi R^2}$	(2.113)
Extensional rheometer	Normal stress	$\sigma_n = \dfrac{F \exp(\dot{\varepsilon}_0 t)}{\pi R_0^2}$	(2.122)
	Extension rate	$\dot{\varepsilon} = U / L$	(2.115)

2.5 Cone-Plate Rheometers

2.5.1 Principle of Operation

The principle of operation of the rotating cone-and-plate rheometer is to shear the material in the gap between the cone and the plate. One of them is rotating, and the other is stationary. The measurement consists in determining the relationship between the angular velocity of the rotating measuring element and the torque that is associated with this rotation. Knowledge of this dependence makes it possible to determine the relationship between the shear rate and the shear stress, and thus to determine the flow curve of the tested material. Cone-and-plate rheometers also allow the determination of the differences in normal stresses by measuring the total normal force occurring in the measurement system or by measuring the distribution of normal forces in this system. The dynamic characteristics of the material can also be determined using these rheometers.

The main advantage of cone-and-plate rheometers is a shear uniformity in the area of the gap, i.e. a constant shear rate and shear stress, which is not found in other rotating rheometers. The limitation is the possibility of using them only in the range of low shear rates (Figure 2.7).

2.5.2 Theoretical Basics

General Assumptions

The schematic of the cone-plate rheometer is shown in the spherical coordinate system r, θ, φ in Figure 2.22. The basic assumptions of the flow are as follows:

- the flow is steady, laminar, and isothermal,
- there is no slip on the cone and plate,
- gravity and inertia forces are neglected,
- the fluid is incompressible and its viscosity does not depend on the pressure,
- $\alpha < 0.05$ rad ($\approx 3°$),
- the velocity field has the form

$$v_r = 0, \quad v_\theta = 0, \quad v_\varphi = v_\varphi(r,\theta) \tag{2.78}$$

- shearing occurs on conical surfaces $\theta\varphi$ of the same apex as the cone of the measuring system, rotating at an angular velocity $\omega = \omega(\theta)$ around the axis of this system, such that

$$v_\varphi = r\omega(\theta)\sin\theta \tag{2.79}$$

where, on the basis of the non-slip condition

$$v_\varphi(\pi/2) = 0 \tag{2.80}$$

$$v_\varphi(\pi/2-\alpha) = \Omega r\sin(\pi/2-\alpha) \tag{2.81}$$

where Ω is the angular velocity (the frequency) of the cone.

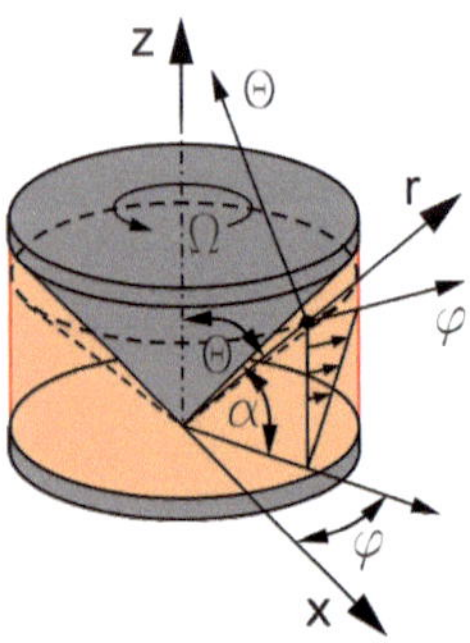

Figure 2.22
Cone-and-plate rheometer: φ - flow direction, θ - direction of velocity change, r - neutral direction

Under assumed flow conditions, the shear rate $\dot{\gamma}$ is expressed as the component $2D_{\theta\varphi}$ of the strain rate tensor $2\,\underset{\sim}{\boldsymbol{D}}$. In the general case, this component is equal to [16]

$$D_{\theta\varphi} = \frac{1}{2}\left[\frac{\sin\theta}{r}\frac{\partial}{\partial\theta}\left(\frac{v_\varphi}{\sin\theta}\right) + \frac{1}{r\sin\theta}\frac{\partial v_\theta}{\partial\varphi}\right] \tag{2.82}$$

Taking into account that $v_\theta = 0$ (Eq. (2.78)) and applying Eq. (2.79), the shear rate is determined by

$$\dot{\gamma} = 2D_{\theta\varphi} = \frac{d\omega}{d\theta}\sin\theta \tag{2.83}$$

Assuming that the angle α is very small, the shear rate may be approximately expressed by

$$\dot{\gamma} \approx \frac{\Omega}{\alpha}\sin(\pi/2) \approx \frac{\Omega}{\alpha} \tag{2.84}$$

It is therefore independent of the coordinate θ.

The constant shear rate in the gap means that the shear stress should also be constant. Determination of this stress will make it possible to determine the viscosity for this shear rate.

Determination of Viscosity

In order to determine the shear stress, the equation of motion in the direction of coordinate φ can be considered [20]. Under assumed flow conditions, this equation is expressed by

$$0 = \frac{1}{r}\frac{\partial \tau_{\theta\varphi}}{\partial\theta} + \frac{2}{r}\tau_{\theta\varphi}\cot\theta \tag{2.85}$$

and, after replacing partial derivatives with ordinary derivatives and simplification, it takes the form

$$\frac{d\tau_{\theta\varphi}}{d\theta} = -2\tau_{\theta\varphi}\cot\theta \tag{2.86}$$

Integrating this expression leads to

$$\tau_{\theta\varphi} = \frac{C}{\sin^2\theta} \tag{2.87}$$

where the constant C is equal to the stress $\tau_{\theta\varphi}|_{\pi/2}$ on the plate surface, since $\tau_{\theta\varphi}(\theta = \pi/2) = \tau_{\theta\varphi}|_{\pi/2}$.

The stress $\tau_{\theta\varphi}|_{\pi/2}$ can be determined from the torque balance on the surface of the plate of radius R. This balance is made by integrating relative to this surface the product of stress $\tau_{\theta\varphi}|_{\pi/2}$ and infinitesimal surface of the plate $r dr d\varphi$ and radius r, so that this torque expresses as

$$M = \int_0^{2\pi}\int_0^R \tau_{\theta\varphi}\,|_{\pi/2}\, r^2 \mathrm{d}r\mathrm{d}\varphi \tag{2.88}$$

and, after integration, it amounts to

$$M = \frac{2\pi R^3}{3}\tau_{\theta\varphi}\,|_{\pi/2} \tag{2.89}$$

On the basis of Eqs. (2.87) and (2.89), the stress $\tau_{\theta\varphi}$ is determined by

$$\tau_{\theta\varphi}(\theta) = \frac{3M}{2\pi R^3 \sin^2\theta} \tag{2.90}$$

which, after taking that $\theta \approx \pi/2$ ($\alpha < 0.05$ rad), expresses the constant shear stress in the measuring gap, according to the formula

$$\tau_{\theta\varphi} = \frac{3M}{2\pi R^3} \tag{2.91}$$

It can be seen that in the cone-and-plate rheometer the shear stress and the shear rate are independent of the position in the measuring gap. They can be easily calculated from Eqs. (2.91) and (2.84) based on the dimensions of measurement system, i.e. the gap angle α and the plate radius R, and on the basis of the results of measurements of the cone angular velocity Ω and the torque M on the plate. Thus, the true flow curve can be obtained directly from the measurement data without the need to differentiate these data, such as in the case of a capillary rheometer or other rotating rheometers.

Determination of Normal Stress Differences

In order to determine the differences in normal stresses in the cone-and-plate rheometer, the equation of motion in the direction of the radius r can be considered [20, 26]. At assumed flow conditions, it is expressed by

$$0 = -\frac{\partial p}{\partial r} + \frac{1}{r^2}\frac{\partial}{\partial r}\left(r^2\tau_{rr}\right) - \frac{\tau_{\theta\theta} + \tau_{\varphi\varphi}}{r} \tag{2.92}$$

This can be presented as a function of the components of the total strain tensor $\underset{\sim}{\sigma}$ given by Eq. (1.7), which are defined by

$$\sigma_{rr} = -p + \tau_{rr}, \quad \sigma_{\theta\theta} = -p + \tau_{\theta\theta}, \quad \sigma_{\varphi\varphi} = -p + \tau_{\varphi\varphi} \tag{2.93}$$

Then, it takes the form

$$0 = \frac{1}{r^2}\frac{\partial}{\partial r}\left(r^2\sigma_{rr}\right) - \frac{\sigma_{\theta\theta} + \sigma_{\varphi\varphi}}{r} \tag{2.94}$$

and, after transformation

$$r\frac{\partial \sigma_{rr}}{\partial r} = \sigma_{\varphi\varphi} + \sigma_{\theta\theta} - 2\sigma_{rr} \tag{2.95}$$

In this flow, the normal stresses (see Section 1.1) are defined by

$$N_1 = \sigma_{\varphi\varphi} - \sigma_{\theta\theta} = \tau_{\varphi\varphi} - \tau_{\theta\theta} \tag{2.96}$$

$$N_2 = \sigma_{\theta\theta} - \sigma_{rr} = \tau_{\theta\theta} - \tau_{rr} \tag{2.97}$$

and Eq. (2.95) can be represented as

$$r\frac{\partial \sigma_{rr}}{\partial r} = N_1 + 2N_2 \tag{2.98}$$

Since the second difference in normal stresses N_2 is a function of the shear rate $\dot{\gamma}$, which according to Eq. (2.84) does not depend on the radius r, after differentiation of Eq. (2.97) we have

$$\frac{\partial N_2}{\partial r} = \frac{\partial \sigma_{\theta\theta}}{\partial r} - \frac{\partial \sigma_{rr}}{\partial r} = 0 \tag{2.99}$$

that is

$$\frac{\partial \sigma_{\theta\theta}}{\partial r} = \frac{\partial \sigma_{rr}}{\partial r} \tag{2.100}$$

Eq. (2.98) can now be represented as

$$\frac{\partial \sigma_{\theta\theta}}{\partial r} = \frac{1}{r}\left(N_1 + 2N_2\right) \tag{2.101}$$

or

$$\frac{\partial \sigma_{\theta\theta}}{\partial \ln r} = N_1 + 2N_2 \tag{2.102}$$

After integrating Eq. (2.102) in the limits from the plate edge R to the radius r, the expression is obtained that determines the distribution of normal stresses $\sigma_{\theta\theta}(r)$:

$$\sigma_{\theta\theta}(r) = \sigma_{\theta\theta}(R) + \left(N_1 + 2N_2\right)\ln\left(\frac{r}{R}\right) \tag{2.103}$$

According to Eq. (2.97), the stress at the edge of plate $\sigma_{\theta\theta}(R)$ can be evaluated by the formula

$$N_2 = \sigma_{\theta\theta}(R) - \sigma_{rr}(R) \tag{2.104}$$

assuming that the radial stress on the edge, i.e. $\sigma_{rr}(R)$, is balanced by the atmospheric pressure p_a, that is

$$\sigma_{rr}(R) = -p_a \tag{2.105}$$

Then we have

$$\sigma_{\theta\theta}(R) = N_2 - p_a \tag{2.106}$$

and Eq. (2.103) takes the form

$$\sigma_{\theta\theta}(r) + p_a = N_2 + (N_1 + 2N_2)\ln\left(\frac{r}{R}\right) \tag{2.107}$$

The stress $\sigma_{\theta\theta}(r)$ can be measured on the surface of the plate or the cone using pressure sensors. The distribution of this stress in the function $[-\ln(r/R)]$ is linear and is shown in Figure 2.23. It can be seen that the stress increases from the edge of the measuring gap to its axis, going in the axis to infinity.

The slope coefficient of the resulting linear equation (2.107) is a combination of the first and the second difference of normal stresses, and the free term is the second difference of normal stresses.

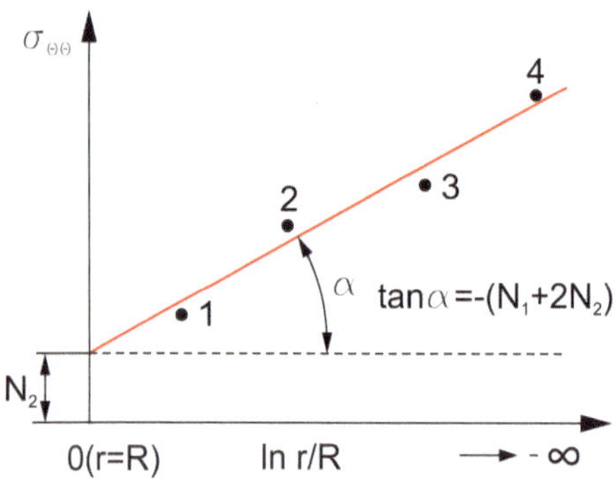

Figure 2.23
Method of determining the normal stress differences N_1, N_2 in the cone-and-plate rheometer on the basis of stress distribution $\sigma_{\theta\theta}(r)$ in the function $[-\ln(r/R)]$: 1, 2, 3, 4 – measurements results $\sigma_{\theta\theta}(r)$

It is possible to determine both the first and the second difference of normal stresses using Eq. (2.107). However, the measurement of stress distribution $\sigma_{\theta\theta}(r)$ is technically difficult and therefore rheometers are usually constructed so that the total normal force F can be measured. The force F can be determined from the balance of forces acting on the plate, according to which

$$F = -\int_0^R 2\pi r \sigma_{\theta\theta}(r)\,\mathrm{d}r - p_a \pi R^2 \tag{2.108}$$

After integrating by parts the first term of this expression we obtain

$$F = -\pi r^2 \sigma_{\theta\theta}\Big|_0^R + \int_0^R \pi r^2 \left(\frac{\partial \sigma_{\theta\theta}}{\partial r}\right)\mathrm{d}r - p_a \pi R^2 \tag{2.109}$$

Then, using Eqs. (2.95) and (2.100), and the fact that the expression ($\sigma_{\varphi\varphi}$ + $\sigma_{\theta\theta}$ − $2\sigma_{rr}$) is a function of the shear rate $\dot{\gamma}$ only, that does not depend on the radius r, we have

$$F = -\pi R^2 \left[\sigma_{\theta\theta}(R) - \frac{\sigma_{\varphi\varphi} + \sigma_{\theta\theta} - 2\sigma_{rr}}{2} + p_a\right] \tag{2.110}$$

and, after transformation

$$F = \pi R^2 \left(\frac{\sigma_{\varphi\varphi} - \sigma_{\theta\theta}}{2} - \sigma_{rr} - p_a\right) \tag{2.111}$$

Finally, after applying Eq. (2.105), the desired formula is obtained as

$$F = \frac{\pi R^2}{2}\left(\sigma_{\varphi\varphi} - \sigma_{\theta\theta}\right) = \frac{\pi R^2}{2} N_1 \tag{2.112}$$

or

$$N_1 = \frac{2F}{\pi R^2} \tag{2.113}$$

Thus, by measuring the total normal force F, the first difference in normal stresses can be determined in the cone-and-plate rheometer.

The basic formulas necessary to determine the rheological properties with the cone-and-plate rheometer, i.e. the flow curve and the first difference in normal stresses, are given in Table 2.1.

■ 2.6 Extensional Rheometers

2.6.1 Principle of Operation

Extensional rheometers allow the measurement of rheological properties under stretching conditions. These measurements are difficult due to the technical problems of purely extensional flows, especially in the case of low viscosity fluids. There are many methods for implementing extensional flows [20], among which the most important are

- simple extension,
- compression,
- multi-axis extension,
- fiber spinning,
- bubble collapse,

- stagnation flows,
- entrance flows.

The first three methods (Figure 2.8) are derived from the methods applied for solids. They ensure the implementation of pure elongation, but since there is a need to capture the sample during the measurement, these can be used in the case of highly viscous materials only. The other methods are an attempt to make measurements with respect to fluids with low viscosity.

The basic type of flow using in the extensional rheometry is simple extension. This can be implemented in clamps moving relative to each other or in rotating clamps. In both cases one of the clamps can be fixed. The samples usually have a circular cross-section, but they can also have a square or rectangular cross-section.

The condition for ensuring a steady uniaxial extension is the constant extension rate of the sample. In order to obtain it, the velocity of the sample ends must be variable in time and appropriately programmed. The deformation rate is therefore set here, and the tensile force that determines the tensile stress is measured.

2.6.2 Theoretical Basics

In order to determine the extensional viscosity η_e, the steady uniaxial extension of the cylindrical sample with initial length L_0 and initial radius R_0 is considered (Figure 2.24). One end of the sample is fixed and the other is moving at the speed U. As a result, the length and the radius of the sample will change, and they are a function of the stretching time, i. e. $L = L(t)$ and $R = R(t)$.

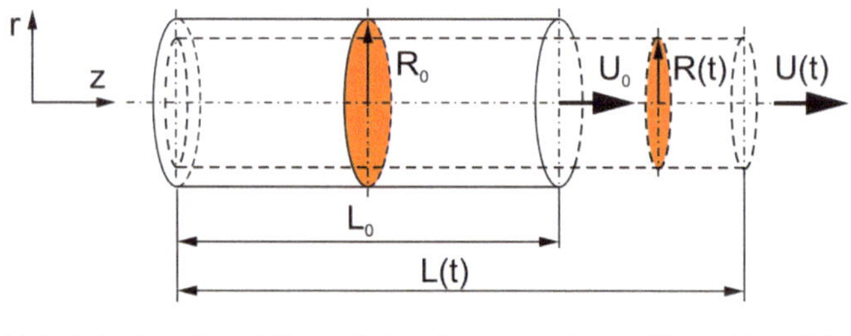

Figure 2.24 Uniaxial extension of the cylindrical sample: R_0 - initial radius of the sample, $R(t)$ - radius after time t, L - initial length of the sample, $L(t)$ - length after time t, U_0 - initial velocity of the sample end, $U(t)$ - velocity of the sample end after time t

The extensional viscosity is defined by Eq. (1.144), and here it takes the form

$$\eta_e = \frac{\sigma_n}{\dot{\varepsilon}} = \frac{\sigma_{zz}}{\mathrm{d}v_z / \mathrm{d}z} \tag{2.114}$$

where η_e is the extensional viscosity.

The extension rate $\dot{\varepsilon}$ is expressed by Eq. (1.143), which here has the form

$$\dot{\varepsilon} = \frac{d\varepsilon}{dt} = \frac{dv_z}{dz} = \frac{dL}{L}\frac{1}{dt} = \frac{U}{L} \qquad (2.115)$$

where $L = L(t)$.

The condition for implementing the steady uniaxial extension flow is the constant extension rate $\dot{\varepsilon}$, i.e.

$$\dot{\varepsilon}(t) = \dot{\varepsilon}_0 = const \qquad (2.116)$$

and from this, on the basis of Eq. (2.115), it follows that

$$\frac{dL}{dt} = \dot{\varepsilon}_0 L \qquad (2.117)$$

Integration of this expression in limits from L_0 to L, where L_0 is the initial length of the sample, i.e. $L_0 = L(t = 0)$, leads to

$$L(t) = L_0 \exp(\dot{\varepsilon}_0 t) \qquad (2.118)$$

In order to ensure extension at the constant rate $\dot{\varepsilon}_0$, the end of the sample must be moved with variable velocity $U(t)$, which using Eq. (2.115) expresses as

$$U(t) = \dot{\varepsilon}_0 L_0 \exp(\dot{\varepsilon}_0 t) \qquad (2.119)$$

The tensile stress σ_n in the sample will be equal to

$$\sigma_n = \frac{F}{\pi R^2} \qquad (2.120)$$

where F is the tensile force.

During extension, both the force F and the sample radius R will change, and with the continuity equation (for incompressible fluids) we have

$$\pi R(t)^2 L(t) = \pi R_0^2 L_0 \qquad (2.121)$$

Hence, on the basis of Eqs. (2.118) and (2.120), an expression is obtained which determines the tensile stress σ_n:

$$\sigma_n = \frac{F \exp(\dot{\varepsilon}_0 t)}{\pi R_0^2} \qquad (2.122)$$

Finally, the extensional viscosity η_e (Eq. (2.114)) is expressed by

$$\eta_e = \frac{F \exp(\dot{\varepsilon}_0 t)}{\dot{\varepsilon}_0 \pi R_0^2} \qquad (2.123)$$

where F is the tensile force that is measured during the test when moving the sample end with the velocity $U(t)$ according to Eq. (2.119), which ensures the constant extension rate $\dot{\varepsilon}_0$.

Basic calculation formulas resulting from this analysis are collected in Table 2.1.

References

1. Reiner, M., *Twelve Lectures on Theoretical Rheology* (1949) North Holland Publishing Co., Amsterdam
2. Eirich, F. R. (Ed.), *Rheology. Theory and Applications* v.1 (1956), v.2 (1958), v.3 (1960) Academic Press Inc., New York
3. Bird, R. B., Stewart, W. E., Lightfoot, E. N., *Transport Phenomena* (1960) John Wiley & Sons, New York
4. Wilkinson, W. L., *Non-Newtonian Fluids* (1960) Pergamon Press Ltd., London
5. Plajer, O., *Praktische Rheologie für Kunststoffschmelzen* (1970) Zechner and Hüthig, Zürich
6. Houwink, R., de Decker, H. K., *Elasticity, Plasticity and Structure of Matter* (1971) Cambridge University Press, Cambridge
7. Lenk, R. S., *Rheologie der Kunststoffe* (1971) Carl Hanser Verlag, Munich
8. Kembłowski, Z., *Rheometry of non-Newtonian Fluids* (1973) WNT, Warsaw (in Polish)
9. Astarita, G., Marucci, G., *Principles of Non-Newtonian Fluid Mechanics* (1974) McGraw Hill, London
10. Han, Ch. D., *Rheology in Polymer Processing* (1976) Academic Press, New York
11. Ferry, J., *Viscoelastic Properties of Polymers* (1980) John Wiley & Sons, New York
12. Zahorski, S., *Mechanics of Viscoelastic Fluids* (1981) Martinus Nijhoff, The Hague, (1978) PWN, Warsaw (in Polish).
13. Pahl, M. H., *Praktische Rheologie der Kunststoffschmelzen und Lösungen* (1983) VDI Verlag GmbH, Düsseldorf
14. Tanner, R. I., *Engineering Rheology* (1985) Clarendon Press, Oxford
15. Bird, R. B., Armstrong, R. C., Hassager, O., *Dynamic of Polymeric Liquids* (1987) John Wiley & Sons, New York
16. White, J. L, *Principles of Polymer Engineering Rheology* (1990) John Wiley & Sons, New York
17. Collyer, A. A., Utracki, L. A., *Polymer Rheology and Processing* (1990) Elsevier, London
18. Ferguson, J., Kembłowski, Z., *Applied Fluid Rheology* (1991) Elsevier, London
19. Barnes, H. A., Hutton, J. F., Walters, K., *An Introduction to Rheology* (1993) Elsevier, Amsterdam
20. Macosko, C. W., *Rheology: Principles, Measurements and Applications* (1994) VCH Publishers, New York
21. Malkin, A. Y., *Rheology Fundamentals* (1994) ChemTec Publishing, Toronto
22. Gordon, G. V., Shaw, M., *Computer Programs for Rheologists* (1994) Hanser Publishers, Munich
23. Schramm, G., *A Practical Approach to Rheology and Rheometry* (1994) Haake GmbH, Karlsruhe
24. Dealy, J. M., Wissbrun, K. F., *Melt Rheology and its Role in Plastics Processing* (1995) Chapman & Hall, London
25. Rohn, Ch. L., *Analytical Polymer Rheology* (1995) Carl Hanser Verlag, Munich
26. Carreau, P. J., De Kee, D. C. R., Chhabra, R. P., *Rheology of Polymeric Systems* (1997) Hanser Publishers, Munich
27. Barnes, H. A., *A Handbook of Elementary Rheology* (2000), University of Wales, Cardiff
28. Morrison, F. A., *Understanding Rheology* (2001), Oxford University Press, New York

29. Han, Ch. D., *Rheology and Processing of Polymeric Materials,* v.1. *Polymer Rheology*, v.2. *Polymer Processing* (2007), Oxford University Press, Oxford
30. Chhabra, R.P., Richardson, J.F., *Non-Newtonian Flow and Applied Rheology* (2008) Elsevier, Amsterdam
31. Vlachopoulos, J., *Polymer Rheology and Extrusion* (2011) McMaster University, Hamilton, ON
32. Osswald, T.A., Menges, G., *Materials Science of Polymers for Engineers* (2012) Carl Hanser Verlag, Munich
33. Malkin, A.Y., Isayev, A.I., *Rheology: Concepts, Methods, and Applications* (2012) ChemTec Publishing, Toronto
34. Shaw, M.T., *Introduction to Polymer Rheology* (2012) John Wiley & Sons, Hoboken
35. Osswald, T., Rudolph, N., *Polymer Rheology. Fundamentals and Applications* (2015) Carl Hanser Verlag, Munich
36. Vlachopoulos, J., *Fundamentals of Fluid Mechanics* (2016) Polydynamics, Hamilton
37. Münstedt, H., *Rheological and Morphological Properties of Dispersed Polymeric Materials* (2016) Carl Hanser Verlag, Munich
38. Agassant, J.F., Avenas, P., Carreau, P.J., Vergnes, B., Vincent, M., *Polymer Processing. Principles and Modeling* (2017) Carl Hanser Verlag, Munich
39. Hopmann, Ch., Michaeli, W., *Extrusion Dies for Plastics and Rubber* (2016) Carl Hanser Verlag, Munich
40. Vlachopoulos, J., Polychronopoulos, N.D., *Understanding Rheology and Technology of Polymer Extrusion* (2019) Polydynamics Inc., Dundas

3 Polymer Processing

In the polymer processing industry, the following steps of processing can be distinguished:

- pre-processing or compounding in which chemically pure polymers are prepared for processing, i.e. such as mixing, pelletizing, drying, etc.,
- polymer processing in which products are formed, e.g. extrusion, injection molding, thermoforming, compression molding, calendering,
- post-processing or complementary processing, i.e. finishing treatment of manufactured products, e.g. dyeing, metallizing, printing, etc.

An immense number of different types of polymeric materials and the variety of their applications require the use of various processing methods among which extrusion and injection molding play a special role. Extrusion, which is used for profiles production, is the most used technology by volume in the polymer processing industry since more than half of all plastics are extruded. Extrusion is also fundamental for compounding, i.e. mixing, pelletizing, filling, reinforcing, etc. Injection molding is also very important. About one-third of all thermoplastics are injection molded and more than half of all polymer processing equipment is for injection molding. It is ideally suited for mass production of parts of complex shapes and precise dimensions.

3.1 Extrusion

3.1.1 Introduction

Extrusion is the process of continuous forming of polymer products, so-called profiles. Profiles are long products of a constant cross-section, e.g. pipes, rods, plates, films, etc. Extrusion involves continuous melting of a polymer in the plasticating unit of the extruder and then pushing (extruding) this melted material, under the pressure developed in the plasticating unit, through the openings of the forming tool that is called the die. Melting occurs as a result of the polymer heating up by the extruder heating system and as a result of energy dissipation in the extruded material. Extrusion is performed in an extrusion line, which includes an extruder and a die, calibrating and cooling units, as well as take-off and cut- off (or winding) units. Extrusion can be used in the processing of thermoplastics and elastomers.

A schematic diagram of the conventional single screw extruder is shown in Figure 3.1.

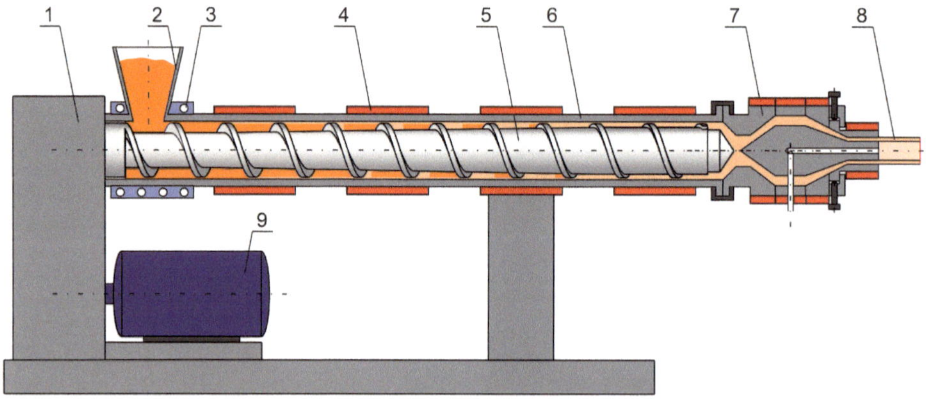

Figure 3.1 Single screw extruder: 1 - gearbox, 2 - hopper, 3 - cooling the feeding section, 4 - heaters, 5 - screw, 6 - barrel, 7 - die, 8 - extrudate (pipe), 9 - motor

An extruder normally consists of:

- a drive system (motor, gearbox),
- a plasticating unit (hopper, barrel, screw, heaters),
- a control cabinet (power supply, temperature control system, etc.).

The plasticating unit is the basic component of the extruder, and it performs the following functions:

- conveying the polymer from the hopper to the die, which is performed by the rotating screw,
- heating up and melting the polymer,
- compression of the polymer to develop the pressure to extrude the polymer through the die,
- mixing to homogenize the material (composition, thermal and mechanical properties, material structure).

Heating up the polymer in the extrusion process takes place due to the heat supplied to the material being processed by the heaters and by the heat produced in the material as a result of energy dissipation.

In summary, five basic functions of extruder can be distinguished:

- transport of the polymer from the hopper to the die (it is usually a screw-type transport),
- melting the polymer so that it can be formed,
- pressure generation to extrude the polymer through the die,
- mixing the material, for thermal and mechanical homogenization,
- forming the product in the extrusion die.

The extrusion process is performed in the extrusion line, which includes an extruder and a die, calibration and cooling units, take-off and cut-off units, as well as a storage unit (Figure 3.2). The design of the line is dependent on the shape and dimensions of the product, the properties of the material, the desired quality, and the extrusion throughput. The most important are the extrusion lines for extruding pipes and other profiles, plates and films, coating wires and cables, and extrusion blow molding.

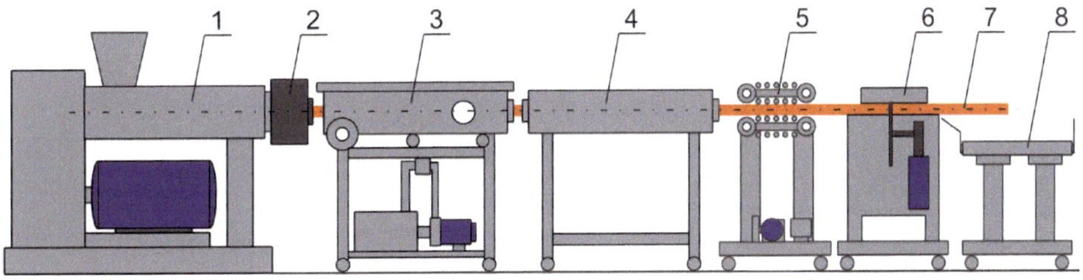

Figure 3.2 Extrusion line for profiles: 1 - extruder, 2 - die, 3 - calibration unit, 4 - cooling unit, 5 - take-off unit, 6 - cut-off unit, 7 - profile, 8 - storage unit

The calibration and cooling units are used to finally shape the product, i.e. to fix the shape and dimension.

Calibration can be divided into external and internal methods; however, external calibration is essential. It involves the extruded profile (e.g. pipe) being passed through the cooled calibration rings of specific cross-sections and dimensions. Due to the difference in temperature between the polymer and the rings, the material is cooled and the shape and outside dimensions of the profile are fixed. The extruded profile is pressed against the calibrating rings, either by pressure or by vacuum. When calibrating by pressure, the pressurized air is introduced into the profile (for hollow profiles). When calibrating by vacuum, a vacuum is produced.

On calibration, the extruded profile is cooled, either directly or indirectly. Usually, the cooling units are long tubs with a cooling water inlet and outlet, often with rollers on which the extruded profile, immersed in water or sprayed with water, is moved.

The take-off devices are used to draw out the extruded profile from the die, carry it through calibrating and cooling devices, and deliver it to the next components of the extrusion line, i.e. cutting and storage devices. The profile is moved using the friction forces between the profile and the drive elements. These elements can be rollers, belts, or tracks.

It is worth paying attention to extrusion with granulation, which is an important stage in preparing the polymer for processing. Granulation involves extruding a bundle of polymer strands and then cutting these into granules (pellets).

Extruders can be divided into single screw, twin screw, and multi-screw extruders. Special extruders can also be distinguished. Single screw extruders may have smooth or grooved barrels and may be equipped with conventional screws (with a continuous channel along the entire screw length) or non-conventional screws (e.g. with elements intensifying melting or mixing action). Twin screw extruders can be co-rotating or counter-rotating, with varying degree of screw meshing. The group of special extruders consists of piston, disc, planetary, cascade, gear extruders, etc. It is worth adding that, historically, the first extruders were the piston extruders.

Extruders are usually fed with solid polymers (plasticating extruders), although they can also be fed with molten polymers (melt extruders).

Extrusion can be performed with gravitational feeding or feeding with metering (Figure 3.3). Single screw extruders are usually fed by gravity (flood fed extruders), while twin screw extruders are fed by metering (starve fed extruders).

When gravitational feeding (Figure 3.3(a)), the screw takes up the polymer from the hopper without limitation, so that it is completely filled with the polymer. Pressure is developed in the feed zone, which causes the polymer to compact, and a uniform, continuous layer of solid material is formed (solid bed). The polymer melts slowly, and it is poorly mixed. In this case, the extrusion throughput mainly depends on the rotational speed of the screw.

When an extruder is metered fed (Figure 3.3(b)), the screw is not completely filled with the polymer in the initial section of the extruder, and pressure is not developed here. A continuous solid polymer layer is not formed. The polymer particles maintain their individuality during melting and form a suspension in the molten material. This promotes faster melting and improved mixing action. In this case, the extrusion throughput does not depend on the rotational speed of the screw, but it is determined by the feeding device.

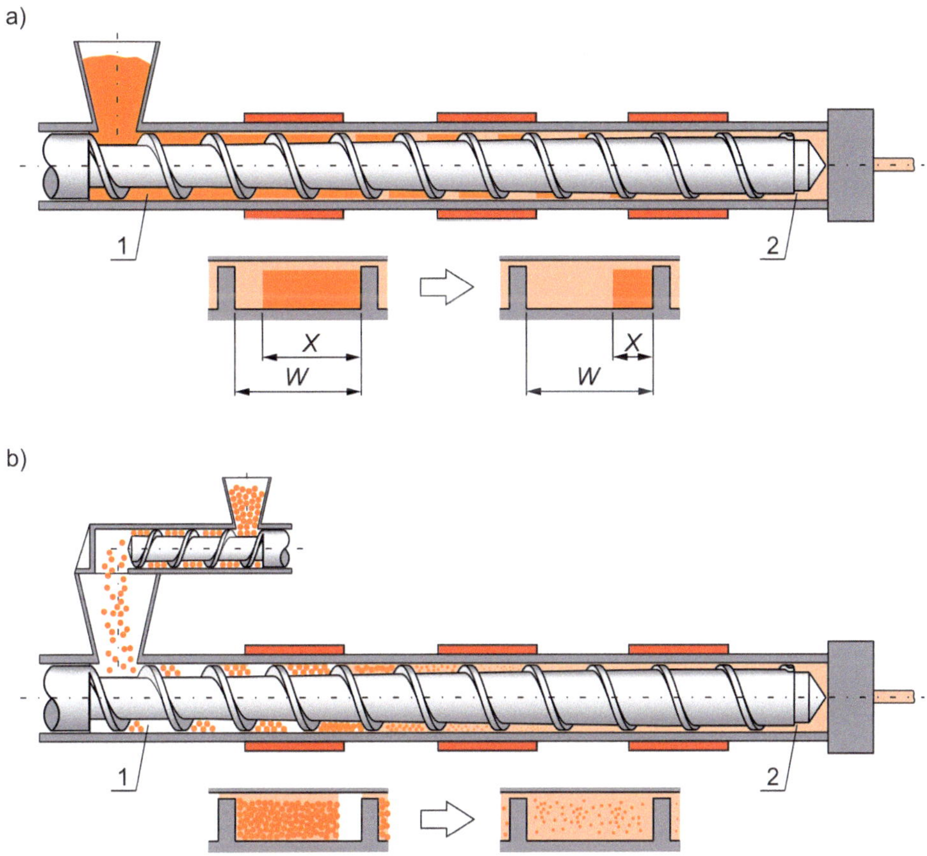

Figure 3.3 Extruder feeding systems: a) gravitational feeding (flood fed), b) metered feeding (starve fed); 1 – solid polymer, 2 – molten polymer, X – solid bed width, W – screw channel width

3.1.2 Single Screw Extrusion

Single screw extrusion is the basic and the oldest extrusion technology. A schematic of a typical single screw extruder is shown in Figure 3.1. The polymer is supplied to the extruder through a hopper (2), usually in the form of a granulate (pellets). It is then transported by means of a rotating screw (5) towards the die (7) in which the product (8) is formed. The screw is placed in the barrel (6), which is equipped with the heating and cooling system (4). This is composed of heaters and fans which facilitate the control of thermal processes in the extruder. The screw is driven by the motor (9) via the transmission system (1). The screw is sometimes drilled and cooled with water or air. Also, the feeding area is sometimes cooled by water (3), so that the polymer granules in the hopper do not melt and do not stick together, which would make it difficult for them to be picked up by the screw. This cooling also isolates the drive system from the plasticating unit, which prevents the heat transfer from the plasticating unit to the drive system.

The fundamental condition for the polymer transport in the single screw extruder is the greater friction on the barrel than on the screw, i. e. the condition that:

$$\frac{f_b}{f_s} > 1 \qquad (3.1)$$

where f_b is the coefficient of friction between the polymer and the barrel, and f_s is the coefficient of friction between the polymer and the screw.

This condition can be explained on the basis of an analysis of the screw/nut action, where the nut represents the polymer. When rotating the screw with the nut, regardless of the direction of the screw rotation, the nut does not move relative to the screw. In other words, it corresponds to the case in which the polymer sticks to the screw and rotates with it, but does not move relative to the screw. When the friction appears on the nut surface and the screw rotates, the nut will start to move relative to the screw, in other words the polymer starts to move relative to the screw. The direction of movement of the nut (polymer) depends on the screw winding direction and on the direction of the screw rotation.

In the case of a right-handed screw, the nut (polymer) will move to the right when the screw rotates to the left and vice versa: the nut will move to the left when the screw rotates to the right. In the case of a left-handed screw, the nut (polymer) will move to the right when the screw rotates to the right and vice versa: the nut will move to the left when the screw rotates counter-clockwise.

The basic component of an extruder is the screw that transports the polymer and enables the pressure development. In single screw extruders, the screws are generally divided into conventional and non-conventional screws.

The conventional three-sectional screw is depicted in Figure 3.4. It is characterized by the following geometrical parameters:

- the outer screw diameter D,
- the screw lead (pitch) t,
- the screw flight angle (helix angle) φ,
- the screw length L,
- the feeding zone length L_f,
- the compression zone length L_c,
- the metering zone length L_m,
- the screw channel depth in the feeding zone H_f,
- the screw channel depth in the metering zone H_m,
- the screw channel width W,

- the screw flight width e,
- the clearance between the tips of the screw flight and the barrel h_f.

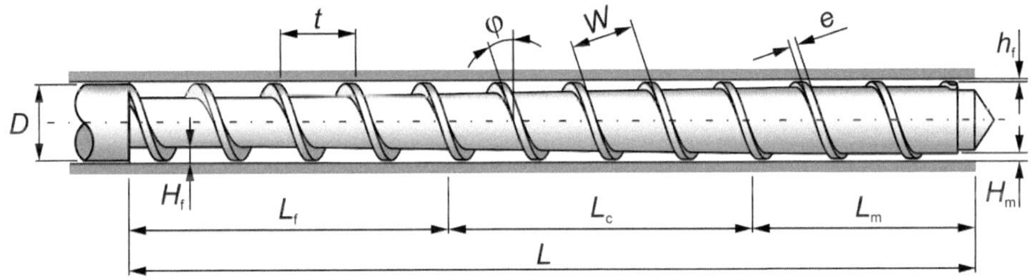

Figure 3.4 Geometry of the conventional three-sectional screw: D - screw diameter, L - screw length, L_f - feeding zone length, L_c - compression zone length, L_m - metering zone length, H_f - channel depth in the feeding zone, H_m - channel depth in the metering zone, t - screw lead (pitch), W - channel width, e - flight width, φ - screw flight angle (helix angle), h_f - clearance (gap)

The basic sizes characterizing the screw are

- the compression ratio $CR = H_f/H_m$, i.e. the ratio of the channel depth in the feeding zone to the channel depth in the metering zone,
- the ratio of the screw length to the screw diameter L/D.

In general, the compression ratio should be at least equal to the ratio of the polymer melt density to the polymer bulk density in order to ensure the continuity of the material flow.

Screws may have various geometries, depending on the intended use and extrusion throughput, for example:

- the compression ratio of screws for extrusion of thermoplastics may be $CR = 1.3{-}5$, while for extrusion of thermosets $CR = 1{-}1.5$,
- the ratio of the screw length to the screw diameter $L/D = 6{-}30$, although most often $L/D = 20{-}30$,
- the screw diameter D may range from a dozen or so mm to several hundred mm, sometimes even up to 800 mm.

In general, the geometrical zones of the conventional three-sectional screw, i.e, feeding zone, compression zone, and metering zone, perform the following functions:

- the feeding zone fulfills the function of taking up the polymer from the hopper, and then compacting and conveying the polymer,
- the compression zone fulfills the functions of polymer melting, compression, and transporting,
- the metering zone fulfills the function of homogenizing the molten polymer and metering it into the die at a given pressure and temperature.

It can be roughly assumed that the screw geometrical zones correspond to the process physical zones, which are characterized by the different physical states of the polymer, i.e.

- the feeding zone corresponds to the solid state zone in which the polymer is in the solid state, usually in the form of a granulate (pellets), more or less compacted,
- the compression zone corresponds to the polymer melting zone in which the melting takes place and the polymer is in the solid and molten state,
- the metering zone corresponds to the polymer melt zone in which the fully molten polymer is transported towards the die.

Conventional screws are usually designed so that the compression zone corresponds to the melting zone. Typically, however, melting takes place over a larger area than the compression zone. In addition, melting is strongly dependent on the operating conditions of the process, which is why it is difficult to fulfill this condition.

The polymer flow in an extruder results from the relative movement of the screw and the barrel as well as from the pressure gradient developed there (Figure 3.5).

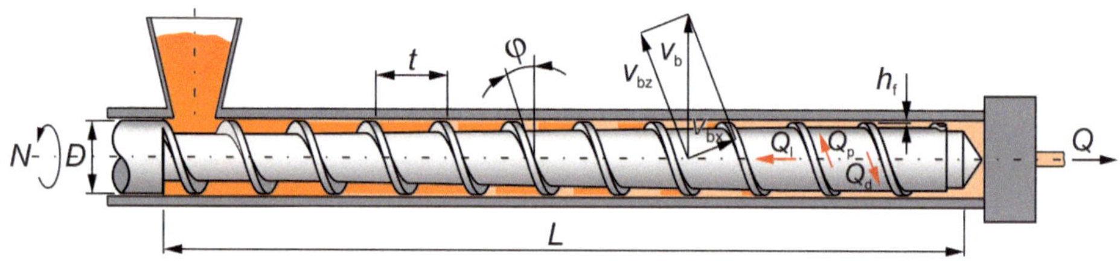

Figure 3.5 The polymer flow in the single screw extruder: Q - polymer flow rate, D - screw diameter, L - screw length, t - screw lead (pitch), φ - screw flight angle (helix angle), V_b - circumferential velocity of the screw, V_{bz} - longitudinal (down-channel) velocity component (parallel to the screw flight), V_{bx} - transverse (cross-channel) velocity component (perpendicular to the screw flight), Q_d - drag flow, Q_p - pressure flow, Q_l - leakage flow (in the gap), h_f - clearance (gap)

According to Figure 3.5, the circumferential velocity of the screw V_b is composed of two components, V_{bz} - in the direction of the channel length (down-channel) - and V_{bx} - in the direction of the channel width (cross-channel). The velocity V_{bz} causes the drag flow that determines the polymer flow rate, i.e. the extrusion output, while the velocity V_{bx} causes the transverse flow that does not affect the flow rate, but fulfills the important function of polymer mixing. The drag flow is superimposed on the pressure flow Q_p, which is the result of pressure gradient in the screw channel caused by the flow resistance through the die. This gradient can be positive or negative, depending on the extrusion conditions. The third component determining the polymer flow rate is the leakage flow Q_l, which is the result of pressure gradient in the extruder, but occurs in the gap between the screw flight and the barrel.

In order to improve the efficiency of mixing and melting in single screw extruders, many different screw designs, called non-conventional screws, have been developed (Figure 3.6). These are equipped with shearing or mixing sections that intensify shearing and mixing actions of the screw. For example, the following screws can be distinguished:

- barrier screws with additional flights (transversal or diagonal) which improve polymer melting, e. g. the screws of Maillefer, Barr, Dray-Lawrence, or Housz,
- mixing screws
  - with shearing elements which force dispersive mixing of the polymer, e. g. the screws of Maddock or Egan (equipped with grooved sections),
  - with mixing elements which force distributive mixing of the polymer, e. g. club-type screws.

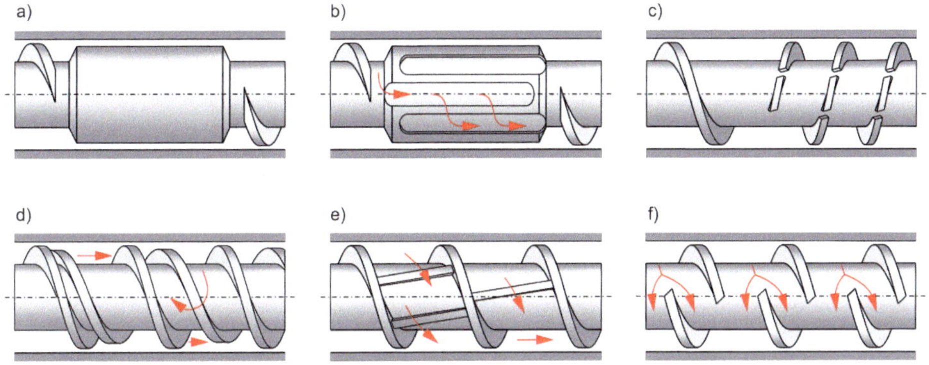

Figure 3.6 Mixing and shearing sections of non-conventional screws: a) torpedo section, b) Maddock section, c) mixing section, d) Maillefer section, e) Barr section, f) Rheotoc section

Extruders may be equipped with classical smooth barrels or grooved barrels. Grooved barrels have in the initial part of the feeding zone (on the length of approximately $L = 3D$) longitudinal or helical grooves, made for technological reasons in the sleeve that is pushed into the barrel. The grooves have a variable depth, the largest at the beginning of the barrel, and gradually disappearing along the feeding zone.

The grooves intensify the transport of the polymer, because they increase the friction on the barrel surface. They prevent the polymer from rotating with the screw and generate more pressure in the feed zone, and as a result increase the polymer flow rate. Grooved sections should be intensively cooled in order to prevent premature polymer melting.

The design of the barrel essentially determines the operation of an extruder. In extruders with smooth barrels, the pressure increases from the feeding zone to the metering zone, so that the latter zone determines the output of the extrusion process. Extruders of this type are characterized in that the extrusion output depends on the pressure drop (flow resistance) in the die. This is not favorable because it means a low stability of the process.

In extruders with grooved barrels, the pressure is developed in the grooved zone and this zone determines the extrusion output. These extruders are characterized by higher extrusion output that is practically independent of the pressure drop in the die, which means greater process stability. In addition, they enable the extrusion of difficult-to-process polymers, e. g. powder polymers. However, these extruders have significant drawbacks. They require the use of a stronger drive and consume more energy. They are characterized by a high wear of grooves and there is the possibility of polymer remaining in the grooves. It is also difficult to ensure adequate polymer melting and mixing, due to the higher process output and the different polymer flow characteristics. Finally, there is a need to use different screws than in the case of smooth barrels, which means those having a shallower feed zone and a deeper metering zone.

3.1.3 Twin Screw Extrusion

Twin screw extrusion is divided according to the direction of the screws rotation into:

- co-rotating extrusion, where the screws are rotating in the same direction (Figure 3.7),
- counter-rotating extrusion, where the screws are rotating in opposite directions (Figure 3.8).

Twin screw extrusion can be performed with varying degrees of meshing. In particular, the following categorizations can be made:

- extruders with non-intermeshing screws, tangential or separated,
- extruders with intermeshing screws, partially or closely intermeshed (with special types of self-wiping screws).

Three basic, practically used, twin screw extrusion systems can be distinguished:

- intermeshing co-rotating extruders (including self-wiping),
- non-intermeshing counter-rotating extruders (tangential),
- intermeshing counter-rotating extruders.

Twin screw extruders are used mainly for extrusion of thermally sensitive polymers, e.g. polyvinyl chloride (PVC), and for specially operations such as compounding, mixing, devolatilization, polymer filling and reinforcing, or reactive extrusion.

The polymer flow mechanism in twin screw extruders is different than that in single screw extruders and to a large extent depends on the meshing degree. In single screw extruders, the polymer flow is a drag flow, while in twin screw extruders there is also a positive displacement mechanism. The degree of positive displacement depends on the degree of screw meshing. It occurs most fully in the case of intermeshing counter-rotating extruders. In the case of non-intermeshing screws, this mechanism does not occur. Generally, the closer the meshing is, the more different is the transport mechanism. Non-intermeshing twin screw extruders resemble basically single screw extruders and therefore do not have such good transport ability as the closely intermeshing screws.

Another important difference is the polymer velocity distribution in the extruder. In twin screw extruders, this distribution is very complex and difficult to describe mathematically. However, this complicated nature of the flow means that twin screw extruders have good mixing and devolatilizing abilities, and are characterized by good heat exchange and fast polymer melting. The difficulty of describing such a flow, however, means that twin screw extrusion theory is less developed than single screw extrusion theory.

Twin screw extruders have significant disadvantages alongside their advantages. They are expensive, structurally complex, and difficult to manufacture.

3.1.3.1 Co-Rotating Twin Screw Extrusion

Co-rotating twin screw extrusion is mainly used in compounding, e.g. polymer mixing, devolatilization, filling, reinforcing, or reactive extrusion. A schematic of co-rotating extrusion is depicted in Figure 3.7.

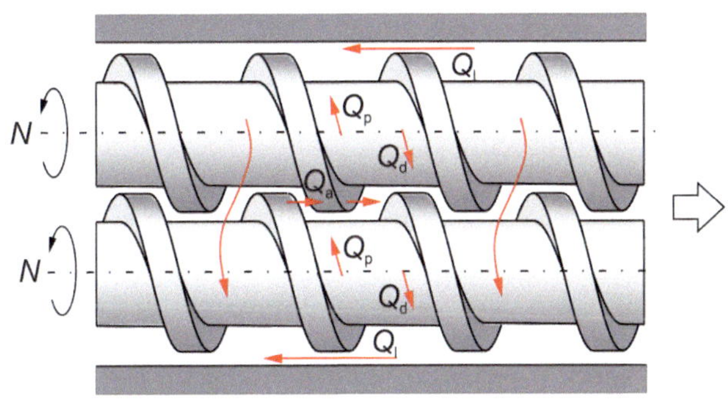

Figure 3.7 The polymer flow in the co-rotating twin screw extruder: Q_d - drag flow, Q_p - pressure flow, Q_a - axial flow, Q_l - leakage flow

In this process, the polymer flows from one screw to the other and moves in an open figure-eight pattern, along the so-called twisted eight. There is a counter-rotating movement in the inter-screw gap, as a result of which high shear stresses are produced in this area. Therefore, this method is used in compounding.

Co-rotating twin screw extrusion is usually performed with metered feeding (starve feeding), which makes the pressure profile in the extruder not continuous and it contains sections where the pressure is reduced to zero. In these areas, i. e. in areas where pressure is not produced, the feeders of various additives may be placed, e. g. for polymer filling, polymer reinforcing, or modifying. In these areas, it is also possible to place the devices for devolatilization or removal of chemical reaction products during reactive extrusion.

In co-rotating extruders, during each screw's rotation, the polymer is transferred from the channel of one screw to the channel of the other one. The flow of polymer is dragged, so that it resembles the flow in a single screw extruder in this respect. But, when moving between the screws, the polymer travels a longer path and undergoes higher shearing.

The flow of polymers in co-rotating extruders is the result of the relative movement of the screws and the barrel and the pressure gradient in the extruder. It is a pressure-drag flow which takes place in open channels between screws (alternately one and the other) and the barrel.

3.1.3.2 Counter-Rotating Twin Screw Extrusion

Counter-rotating twin screw extruders are used mainly for extruding profiles from thermally sensitive materials, e. g. polyvinyl chloride (PVC). A schematic of counter-rotating twin screw extrusion is shown in Figure 3.8.

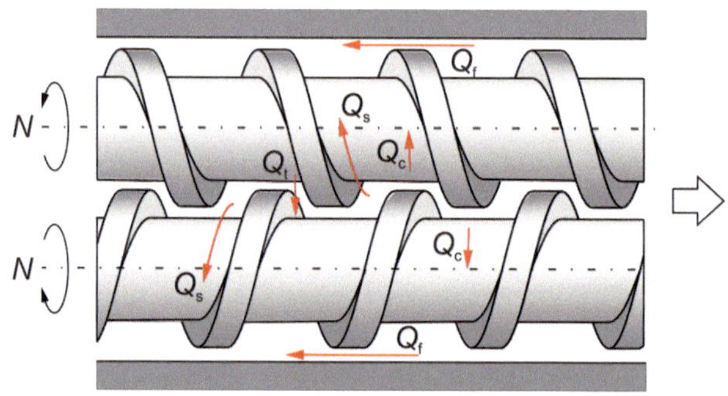

Figure 3.8 The polymer flow in the counter-rotating twin screw extruder: Q_c - calendering leakage flow, Q_f - flight leakage flow, Q_t - pressure leakage flow, Q_s - side leakage flow

Counter-rotating twin screw extruders, in comparison with single screw extruders, provide better feeding of the extruder with polymer (especially in the case of polymers in the form of powder or polymers exhibiting slip properties), and provide a shorter and less varied residence time of the polymer particles in the extruder.

In counter-rotating extruders, the polymer essentially does not flow from one screw to the other, as in the co-rotating one. There is a co-rotating movement in the inter-screw gap, so that high shear stresses are not produced in this area, as in the co-rotating extrusion.

Counter-rotating extrusion, like co-rotating extrusion, is usually performed with metered feeding (starve feeding), which results in the pressure profile in the extruder being non-continuous and having sections where the pressure is reduced to zero. In areas where pressure is not produced, the feeders for filling, reinforcing, or modifying additives may be placed.

The flow of polymer in the counter-rotating twin screw extruder is totally different from that in the single screw extruder, but also different than that in the co-rotating twin screw extruder. The essence of this flow is the so-called positive displacement mechanism, which does not occur in other extrusion variants. The degree of positive displacement depends on the degree of screws meshing. It occurs most fully in the closely intermeshing counter-rotating extruders. There is no drag flow here, so that the energy dissipation is here insignificant.

The flow in the counter-rotating extruder is mainly performed in a closed space, in the C-shaped chamber, the so-called C-chamber, shown in Figure 3.9. There are also various types of leakage flows here, schematically shown in Figure 3.8.

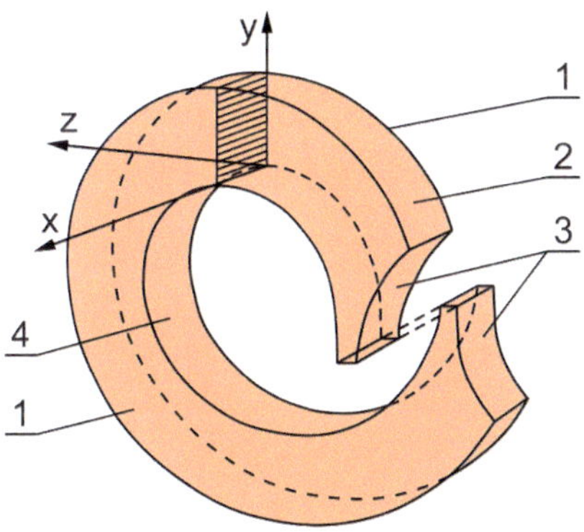

Figure 3.9
C-shaped chamber of the counter-rotating extruder: 1 - side surface of the screw flight, 2 - barrel surface, 3 - front surface of the screw flight, 4 - surface of the screw

The C-chamber is bounded by six surfaces, i.e. the surface of the screw root, the barrel surface, the side surface of the screw flight (twice), and the front surface of the screw flight (twice). There are leakage flows between these surfaces, i.e.

- the calendering leakage flow Q_c between the screw root and the screw flight,
- the flight leakage flow Q_f over the screw flights, through the gap between the screw flight and the barrel,

- the back pressure inter-screw leakage flow Q_t (tetrahedron flow) through the tetrahedral gap between the flight flanks of screws, in the radial direction,
- the side leakage flow Q_s through the side gap between the flight flanks of the screws, in the tangential direction.

3.1.4 Extrusion Dies

Extrusion dies are designed to form the product with desired shape and dimensions, ensuring thermal and mechanical homogeneity in the entire volume, and to produce the appropriate flow resistance (pressure drop) enabling the extrusion process.

Extrusion dies can be classified in various ways. For example, depending on the application they can be classified into the dies for full, empty, or open profiles, or the dies for extruding pipes, films, plates, and fibers, or the dies for granulation and wire coating.

Extrusion dies can also be divided according to the geometry of the product forming section, i.e. the cross-section of the channel at the die exit. Thus, the following die types can distinguished:

- dies with a circular exit cross-section,
- dies with a flat slit exit cross-section,
- dies with an annular exit cross-section,
- profile dies with an irregular exit cross-section.

The basic task of the die, that is, transformation of the circular melt stream of the polymer flowing out of an extruder into the polymer stream of desired shape and dimensions, is performed by gradually changing the cross-section of the flow channel or by using special distribution channels.

Dies with a circular cross-section of the forming part are used for the production of pellets, extrusion of rods, strands, wires, and fibers. Typically, they consist of elements with a circular or annular cross-section of the channel. A typical die for extruding rods is shown in Figure 3.10. Forming the product takes place in the nozzle (3). The die is equipped with a screen (filter) and a breaker plate (6), whose task is to pressurize the polymer flow, stop contaminants and incompletely molten polymer particles, and reduce pressure and flow rate pulsations.

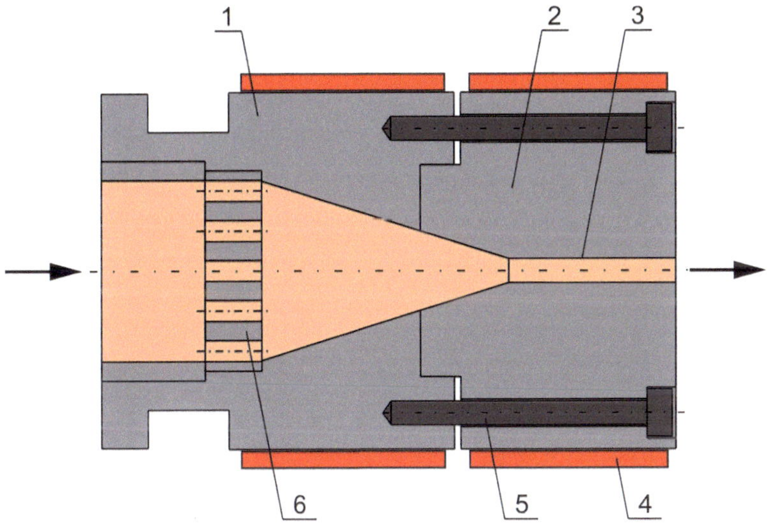

Figure 3.10 Circular die for extruding rods: 1 - body of the die, 2 - die tip, 3 - nozzle, 4 - heaters, 5 - mounting bolts, 6 - screen (breaker plate)

Dies with a flat slit exit cross-section, i. e. flat slit dies, are used for extruding films and sheets, as well as for coating paper or fabrics. The distinction between films and sheets can be set at the thickness of about 0.5–0.7 mm. In the case of slit dies, it is necessary to convert the circular melt stream of polymer flowing out of an extruder into the flat sheet. This is achieved by means of the properly designed distribution channel (the manifold). The task of the manifold is to distribute the polymer evenly along the die width so that the outflow velocity will be independent of the position and the polymer properties will be uniform.

An example of a flat slit die for extruding sheets is shown in Figure 3.11. The polymer is fed into the die by a central channel, coaxial with the extruder barrel, and then distributed symmetrically by two tapered distribution channels (1) to the forming gap (4). The die is equipped with a choker bar (3), sometimes of segmented structure, whose appropriate setting causes various throttling on the lip width, so that it is possible to maintain the same polymer flow rate. The final shaping of the product takes place in the forming gap (the lip), between two bars (4), stationary and movable (adjustable). The position of the adjustable bar is adjusted by screws to obtain the desired thickness of the sheet.

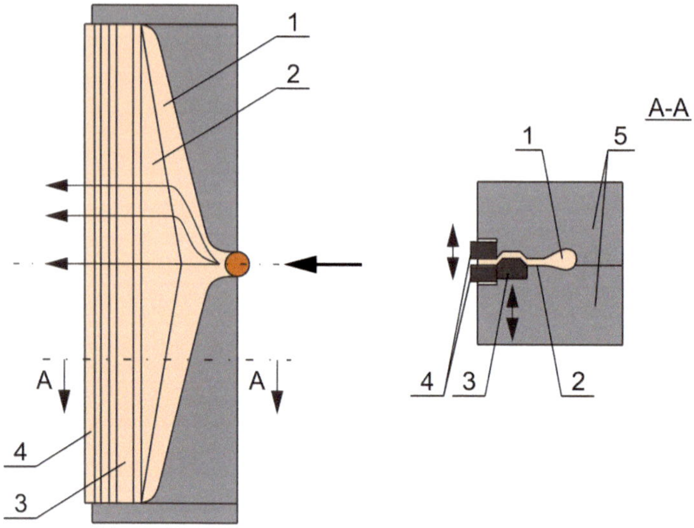

Figure 3.11 Flat slit die for sheet extrusion: 1 - manifold, 2 - island, 3 - choker bar (locally adjustable), 4 - lip, 5 - body of the die

An example of a flat slit die for extruding films is shown in Figure 3.12. In dies of this type, the flow rate control is carried out by locally changing the position of the adjustable bar (4) of the forming gap.

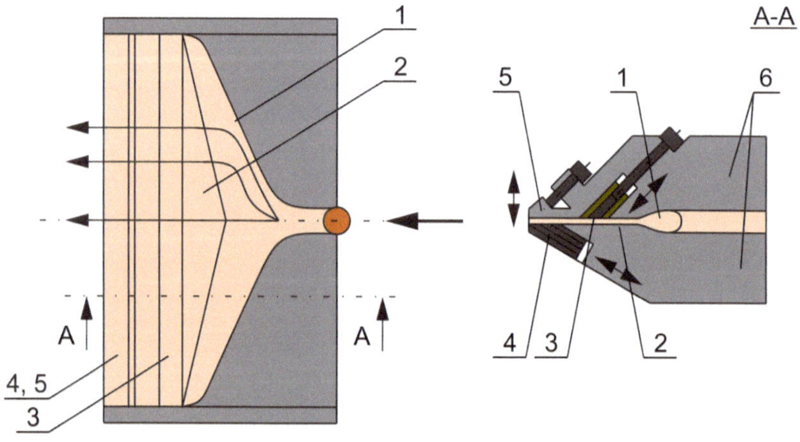

Figure 3.12 Flat slit die for film extrusion: 1 - manifold, 2 - island, 3 - choker bar (locally adjustable), 4 - lip (adjustable), 5 - flex lip, 6 - body of the die

Annular dies, i.e. dies with an annular exit cross-section (forming section), are used for the extrusion of pipes, hoses, tubular film, and tubular parisons, for the extrusion blow molding as well as for cable and pipe jacketing. These dies can be center fed and then the extruder axis coincides with the die axis (e.g. dies for pipe extrusion) or side fed when the extruder axis forms a certain angle with the die (e.g. wire coating dies).

A typical die for extruding pipes is shown in Figure 3.13. The polymer melt stream is fed centrally to the die by a cylindrical channel, and then it is transformed using the mandrel (1) into an annular stream. The mandrel is fixed in the die by means of spider legs (6), which are shaped so that an even flow of the polymer is ensured. The concentricity of the mandrel (1) and the forming ring (3) is set using the centering screws (5). The disadvantage of this type of die is the formation of flow marks, i.e. weld lines. These are caused by the division of the melt in the mandrel into several melt streams that flow around the spider legs and then come together once again. The dies with a spiral mandrel do not have these disadvantages.

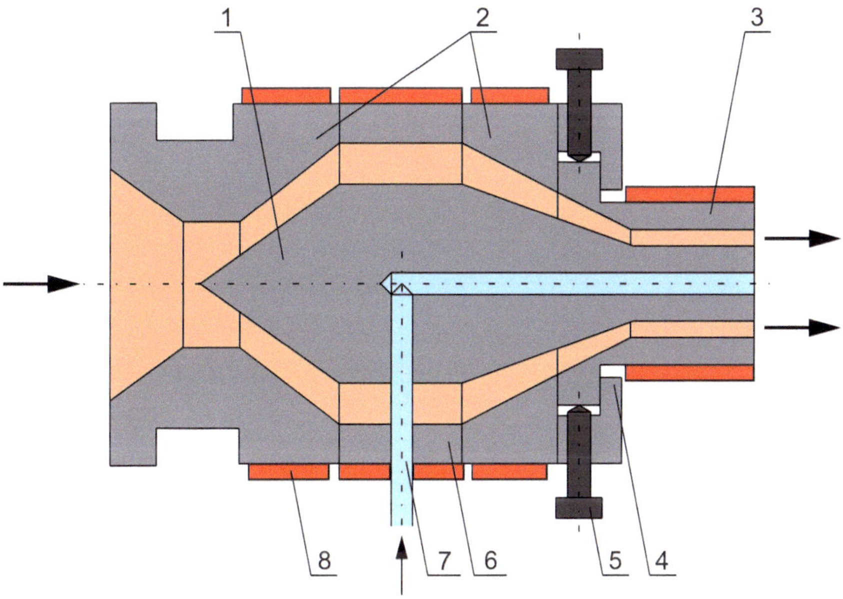

Figure 3.13 Annular die for extruding pipes: 1 - mandrel, 2 - body of the die, 3 - forming ring, 4 - centering ring, 5 - centering screw, 6 - spider legs, 7 - air supply channel, 8 - heaters

Four types of annular dies can be distinguished (Figure 3.14), namely mandrel support dies, screen pack dies, side fed mandrel dies with a distribution channel, and dies with a spiral distributor (mandrel).

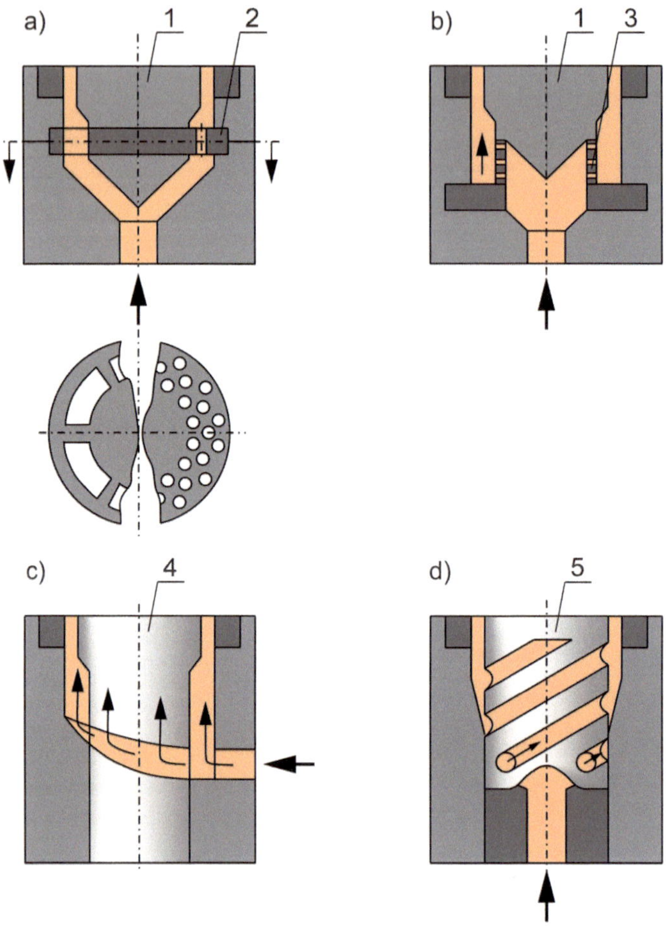

Figure 3.14 Dies with an annular exit cross-section: a) mandrel support die, b) screen pack die, c) side fed mandrel die, d) spiral mandrel die; 1 - mandrel, 2 - mandrel support, 3 - screen pack, 4 - distributing mandrel, 5 - spiral mandrel

Wire-coating dies also belong to the group of extrusion dies with an annular exit cross-section (Figure 3.15). In this type of die, the polymer melt flows around the hollow mandrel (6), which simultaneously fulfills the role of guiding the coated wire.

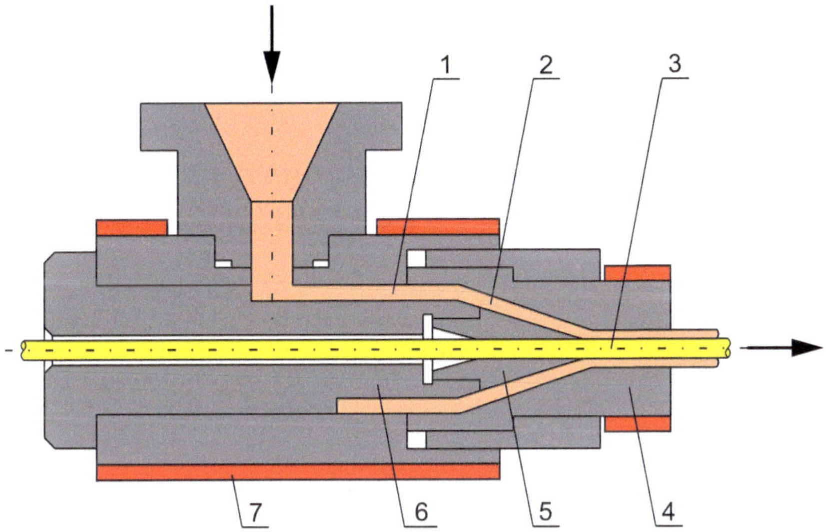

Figure 3.15 Wire-coating die: 1 - manifold, 2 - annular slit, 3 - wire, 4 - die ring, 5 - torpedo tip guiding the wire, 6 - torpedo (hollow mandrel), 7 - heaters

Dies with irregular exit cross-section, i.e. other than circular, annular, and rectangular cross-sections, are commonly called profiles. A rough division can be made between solid and hollow profiles which contain closed hollows. Generally, three basic groups of profile dies are distinguished:

- plate dies (one-step dies),
- stepped dies,
- profile dies with a gradual change in the cross-section.

In one-step dies, profile is formed in a forming plate of a specific cross-section. There are very sharp changes in the flow channel geometry, which can lead to polymer deposition, i.e. dead spots, and consequently to polymer degradation. Stepped dies consist of several forming plates that form a flow channel with a gradually changing cross-section. This reduces polymer deposition, but does not fully eliminate it since it can occur at the contact points of the forming plates. From this point of view, the best solution is the third type of die, i.e. dies with gradual change in the cross-section of the flow channel, in which there are no places of potential polymer deposition.

The extrudate is shaped in the last geometrical section of the die. As a result of the extrudate swelling, i.e. the Barus effect, the cross-section of the extrudate is not equal to the cross section of the forming orifice (Figure 3.16). For this reason, it is necessary to calibrate the product, i.e. to give it the final, desired shape and dimension of the cross-section (Figure 3.2). It is also possible to make a correction of the cross-section of the forming orifice (Figure 3.17). This correction is usually performed experimentally, although computer simulation may also be applied.

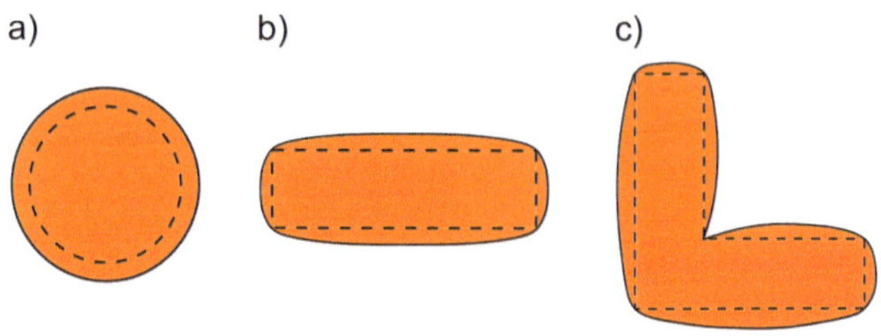

Figure 3.16 Barus effect: a) circular cross-section, b) rectangular cross-section, c) irregular cross-section

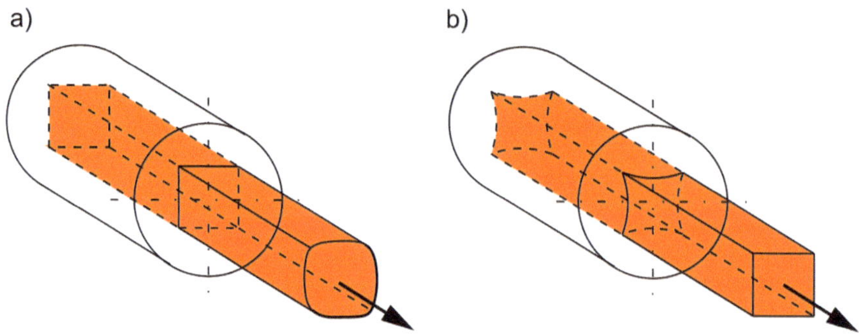

Figure 3.17 Barus effect in the case of a square cross-section: a) non-corrected cross-section, b) corrected cross-section

The Barus effect is usually characterized by the ratio of the dimension of the extrudate to the size of the orifice forming it. In the case of the flow through a cylindrical channel, the degree of swelling is determined by the ratio of the extrudate diameter to the diameter of the orifice forming it.

$$B = \frac{d}{D} \tag{3.2}$$

where B is the extrudate swelling, d is the extrudate diameter, and D is the orifice diameter.

3.2 Injection Molding

3.2.1 Introduction

In addition to extrusion, injection molding is the most important technology in the polymer processing industry, mainly due to the immense variety of manufactured products of various shapes, dimensions, and designs that can be produced, as well as due to the large number of injection techniques used.

Injection molding is a cyclic process of forming polymer products, called moldings. It consists in cyclic melting of a polymer in the plasticating unit of the injection molding machine and then injecting the molten polymer, under the pressure developed in the plasticating unit, into the injection mold, shaping the product. Melting occurs as a result of the polymer heating up by the heating system and as a result of energy dissipation in the processed material.

An injection molding machine normally consists of:

- a drive system,
- a plasticating unit (hopper, barrel, screw, heaters),
- an injection mold,
- a clamping unit
- a control cabinet (power supply, temperature control system, etc.).

A schematic diagram of the injection molding machine is depicted in Figure 3.18, where the basic functional units are distinguished. The fundamental parameters that characterize the injection molding machine are the clamping force, the injection volume, and the injection pressure.

It is worth adding that, historically, the first injection molding machines were the piston injection molding machines.

The plasticating unit is the basic component of the injection molding machine, and it performs functions similar to those in the extruder. However, there are essential differences here. The rotating screw conveys the molten polymer to the screw front, and then the polymer is injected into the mold through axial displacement of the screw. The screws are usually equipped with non-return valves to prevent backflow of the polymer during injection.

The injection mold is attached to the fastening platens of the injection molding machine. One of these platens, the fixed (stationary) platen, is attached to the body of the injection molding machine, whereas the other, the movable platen, is moved along the guiding columns.

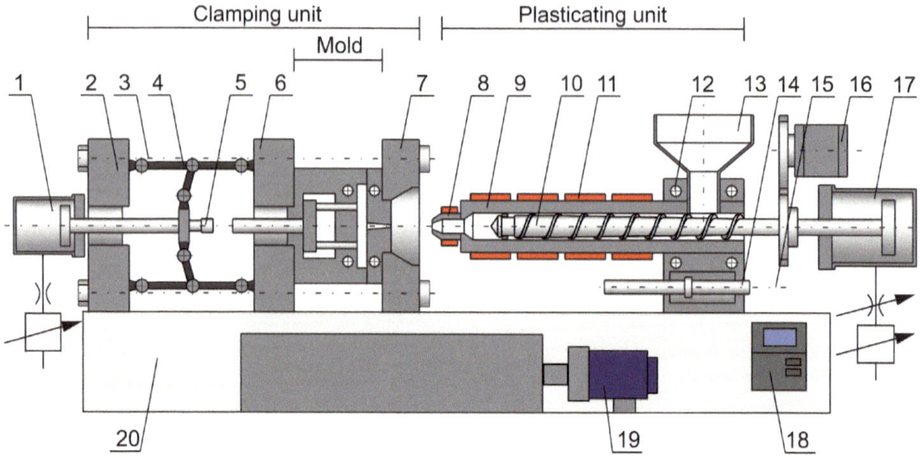

Figure 3.18 Injection molding machine: 1 - clamp cylinder, 2 - rear fixed (stationary) platen (adjustable), 3 - guiding columns (tie bars), 4 - toggle mechanism (column-lever assembly), 5 - fixed stop (bumper), 6 - movable platen, 7 - front fixed (stationary) platen, 8 - nozzle, 9 - barrel, 10 - screw, 11 - heaters, 12 - cooling the feeding section, 13 - hopper, 14 - cylinder for movement of injection unit, 15 - injection unit guides, 16 - motor for screw rotation, 17 - injection cylinder, 18 - control cabinet, 19 - drive system, 20 - body of the machine

3.2.2 Injection Molding Process

Injection molding is a cyclic process. The course of the injection molding process is shown in Figure 3.19 and the phases of the injection molding cycle are highlighted in Figure 3.20.

Just as in an extruder, the polymer is conveyed by the rotating screw from the hopper to the screw tip. The cycle starts when the mold closes, followed by the injection of the molten polymer into the mold cavity. Once the cavity is filled, a holding phase begins with the pressure maintained to compensate for polymer shrinkage. Both these phases are performed by axial movement of the screw, which acts as a piston.

In the next step, the screw rotates, conveying the next shot of molten polymer to the front of the screw. During this plasticating (recovery) phase, the nozzle at the end of the injection unit is closed, usually by a special device called a shutoff nozzle. Since the polymer conveyed to the screw tip requires space, the screw is retracted by the accumulating melt. The space in front of the screw fills with melt. Melt is conveyed until the screw reaches a position specified by the control unit. The distance moved by the screw to reach this point is called the recovery stroke, and the volume of the polymer conveyed during this phase is called the shot volume (injection volume).

The molten polymer injected to the mold is cooled (when thermoplastic), and once the part is sufficiently cooled, the mold opens and the part (the molding) is ejected. Figure 3.20 shows that the cycle time is dominated by the cooling time of the part inside the mold. The removal of the part from the mold requires that the two machine plates, between which the mold is located, be moved apart. This function is performed by the clamping unit, which is responsible for opening and closing the mold. In addition, the auxiliary equipment necessary for part removal (ejection) is also located in the clamping unit.

Only one machine plate is moved when opening/closing the mold. The plate close to the injection unit is firmly attached to the machine frame and is thus immovable. It contains a circular opening in which the mold is positioned by its locating ring. The nozzle of the injection unit advances to the mold through the opening in the plate. This plate is called the fixed (or stationary) plate, while the opposite plate is the movable plate. As a rule, the movable plate is guided on four tie bars during its motion.

When two mold halves bring into contact, the clamping force is applied by the clamping system. This force prevents the mold from opening as a result of the forces developed inside the cavity mold by the pressure acting on the mold halves.

When the part has been removed from the mold, the mold closes, the injection unit moves to the mold, and the process cycle is repeated.

Injection molding can be used in the processing of thermoplastics, elastomers, and thermosets. In case of thermoplastics, the mold is cooled, while in case of thermosets and elastomers, which are crosslinking materials, the mold is heated and thus initiates the crosslinking reactions.

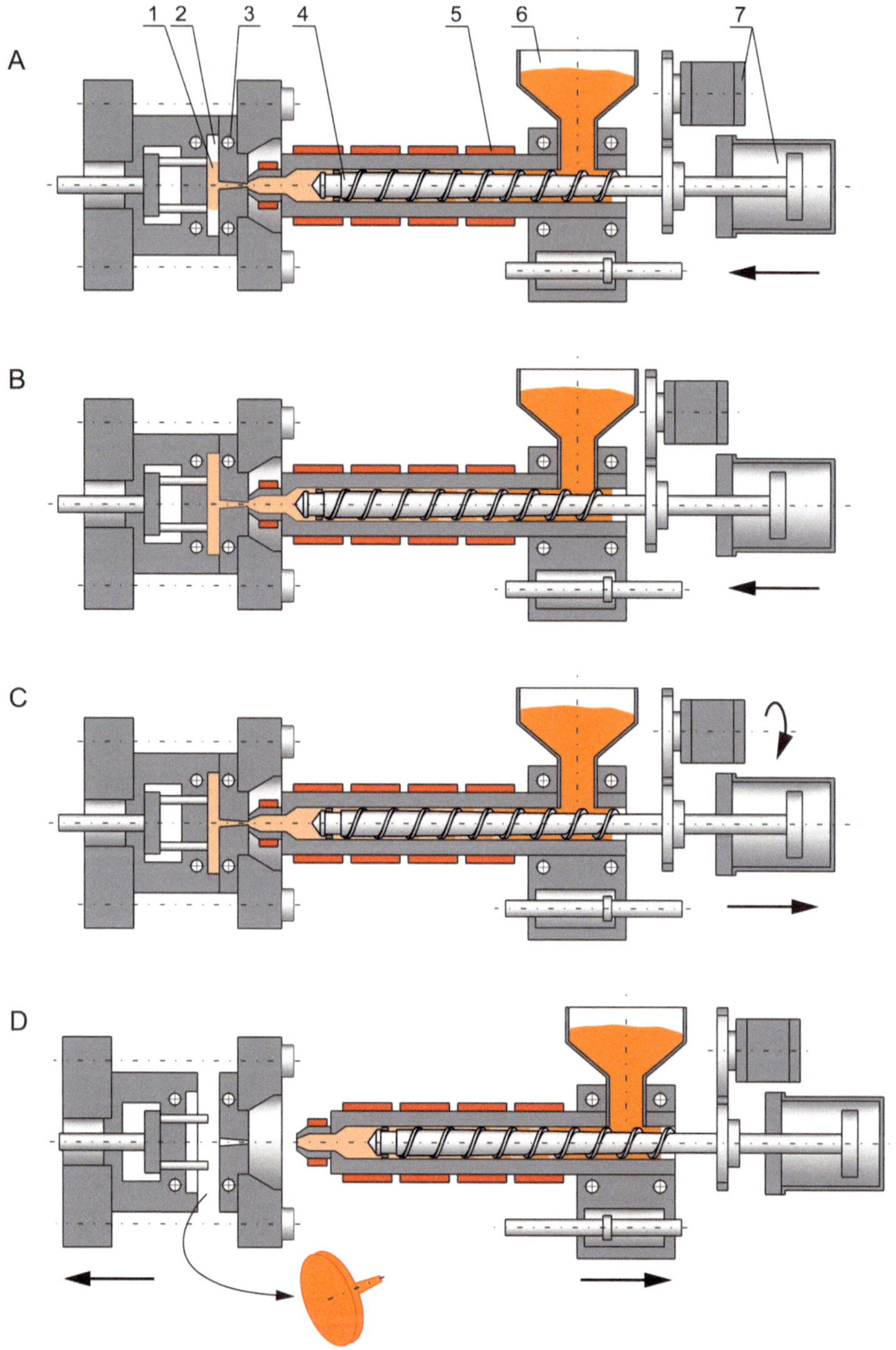

Figure 3.19 Injection molding process sequence: A – injection (mold filling), B – holding, C – plasticating, D – mold opening and molding ejection; 1 – molding (part), 2 – mold cavity, 3 – cooling channels, 4 – screw, 5 – heaters, 6 – hopper, 7 – screw drive system

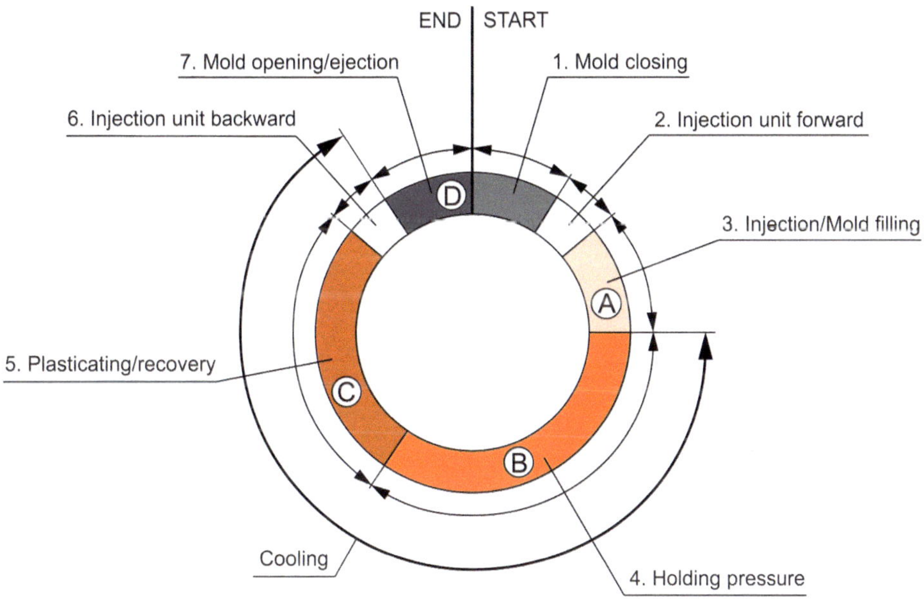

Figure 3.20 Injection molding cycle: A - injection (mold filling), B - holding, C - plasticating, D - mold opening and molding ejection

3.2.3 Injection Molds

Injection molds are designed to form the product with the desired shape and dimensions. The mold distributes a polymer melt into the cavities, where the parts are shaped and cooled. After cooling, they are ejected from the mold. An example of the mold is depicted in Figure 3.21.

The mold normally consists of:

- clamping and mold platens,
- sprue and runner systems,
- a mold cavity,
- a cooling system (thermoplastics),
- an ejection system.

The polymer flows through the sprue and is distributed into the cavities by the runners. It can be injected directly to the cavity or indirectly using an intermediate plate as depicted in Figure 3.22. There are two types of runner systems, cold and hot. Cold runners are ejected with the part and are trimmed after part removal. Hot runners keep the polymer at its melt temperature. The melt stays in the system after ejection, and it is injected into the cavity in the next cycle. This eliminates trimming.

Examples of cavity arrangements are depicted in Figure 3.23. The runner system (layout) can be symmetric or asymmetric. In the symmetric system, all flow paths are the same, and all cavities are filled at the same time causing the polymer to fill all cavities in the same way. This is a geometrically balanced system. In the asymmetric unbalanced system, the flow paths are different, and the cavities are not uniformly filled, which leads to parts of different quality. However, the balanced filling can be achieved by varying runner diameters.

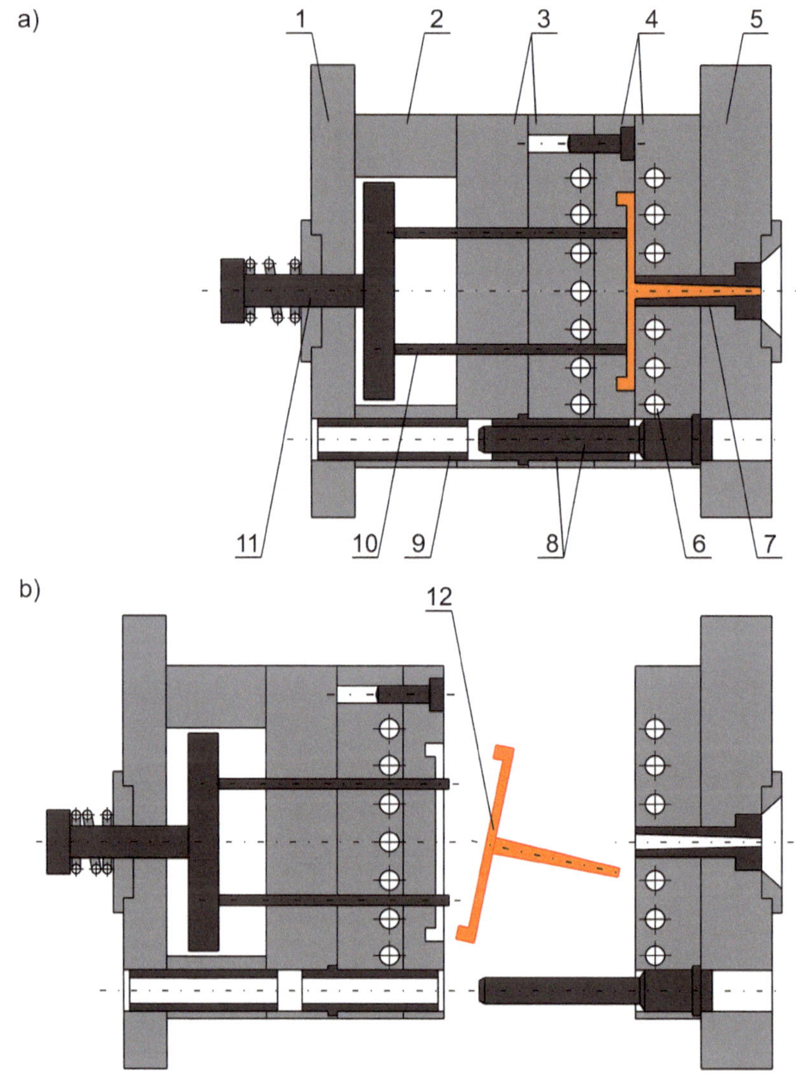

Figure 3.21 Injection mold: a) mold closed (mold filling and cooling), b) mold open (molding ejection); 1 - clamping plate (for movable part of the mold), 2 - distance strip (riser), 3 - backing plates, 4 - mold plates, 5 - clamping plate (for stationary part of the mold), 6 - cooling channels, 7 - sprue bushing, 8 - locating guide pillar (guide pin with guide bushing), 9 - guide bushing, 10 - ejector, 11 - ejector rod, 12 - injected part (molding)

a)

b)

c)

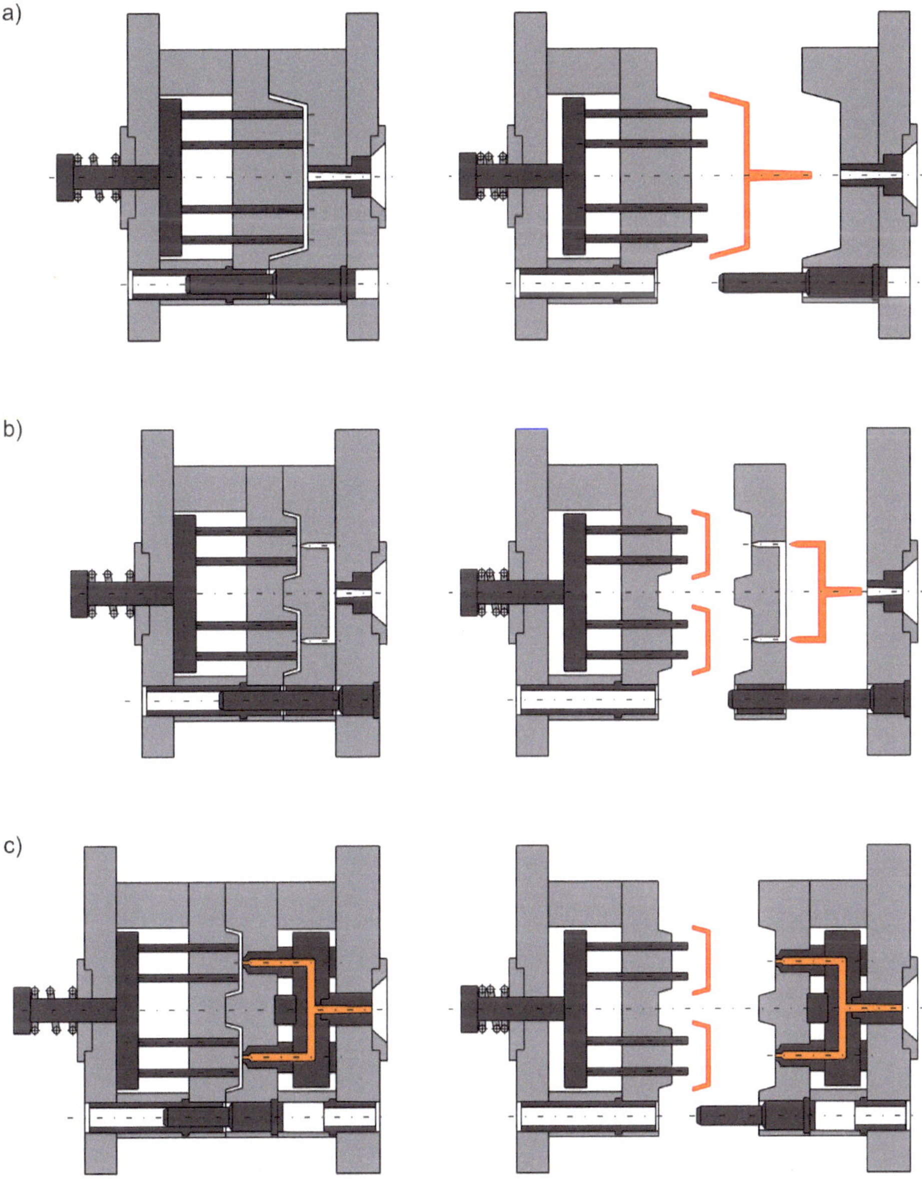

Figure 3.22 Sprue layouts of injection molds: a) direct sprue (one-cavity mold), b) indirect sprue (two-cavity mold), c) hot-runner manifold (two-cavity mold)

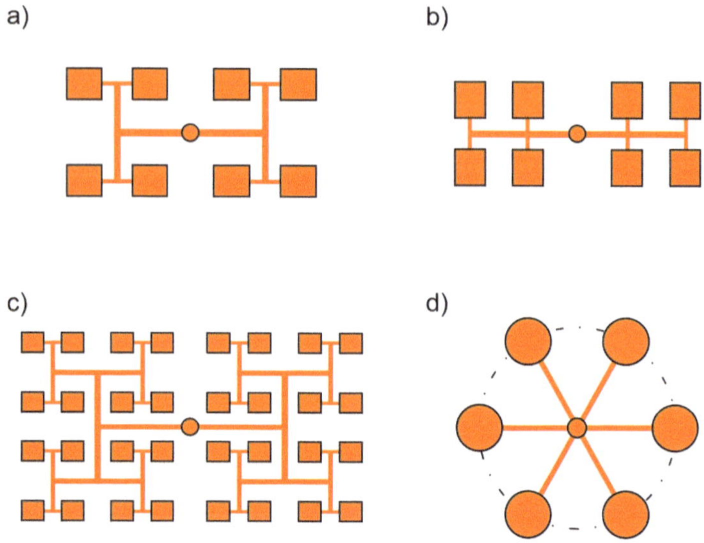

Figure 3.23 Arrangement of mold cavities: a) balanced "H-type" layout, b) unbalanced "H-type" layout c) multi "H-type" layout, d) radial (starry) layout

3.2.4 Special Injection Molding Processes

Although most injection molding processes are covered by the conventional process, there are several important variations including, for example, multi-color or multi-component injection molding, co-injection molding, gas- or water-assisted injection molding, injection-compression molding, microcellular injection molding, reaction injection molding, and many others.

In multi-component or multi-color injection molding, a portion of the molded part is initially formed by the first component, and this is transferred to the other mold cavity that is filled with the next component.

In contrast to multi-component injection molding, co-injection molding yields a sandwich structure in the molded part consisting of different materials. The first component forms the outer skin of the molded part, and the second component forms the interior (core) of the part.

Gas- or water-assisted injection molding processes are comparable to two-component injection molding in which the second component, gas (nitrogen) or water, is injected into the polymer melt, blowing the melt out of the way and depositing it against the mold surfaces. With this method, thick-walled parts can be produced with a simultaneous savings in material and cycle time.

In injection-compression molding, in contrast to conventional injection molding, the polymer is injected into the partially opened mold, and then it is squeezed by closing the mold. This method is used for products that require a high quality surface finish, such as compact discs.

In microcellular injection molding (known as MuCell technology), a polymer melt is blended with a supercritical gas (carbon dioxide or nitrogen) in the machine barrel to create a single phase solution. During the molding the gas forms microscale cells (bubbles), and the pressure arises from the foaming which improves the dimensional stability. This method reduces the amount of plastic used without sacrificing the mechanical properties.

Reaction injection molding (RIM) involves mixing of two reacting liquids in a mixing head before injecting the mixture into the mold cavity. The liquids react in the mold to form a cross-linked part.

3.3 Blow Molding

3.3.1 Introduction

Blow molding involves blowing the pre-formed semi-finished product into the form of a finished product. It can be implemented in a continuous or a cyclic way, i. e. in a continuous extrusion process or in a cyclic injection molding process.

A continuous blow molding can be performed with blowing of the extruded profile in the mold or without any limitation of the blowing process, i. e. without the mold, as a free blow molding. As a result of free blowing, tubular films made of thermoplastics are obtained. On the other hand, as a result of blowing in the mold, mainly hollow articles (containers) made of thermoplastics are obtained.

Injection blow molding involves blowing in the mold a semi-finished product premade by injection molding, called a preform. As a result of injection blow molding, thermoplastic hollow articles (containers) are obtained, as in the case of extrusion blow molding.

The following blow molding methods can be distinguished:

- extrusion with free blowing, called film blowing,
- extrusion with blowing in the mold, called extrusion blow molding (EBM),
- extrusion with blowing and mechanical stretching in the mold, called extrusion stretch blow molding (ESBM),
- injection molding with blowing in the mold, called injection blow molding (IBM),
- injection molding with blowing and mechanical stretching in the mold, called injection stretch blow molding (ISBM).

3.3.2 Film Blowing

Film blowing consists in extruding a thin-walled tube through an annular die and blowing it with a low-pressure air stream, followed by drawing by means of a take-off device.

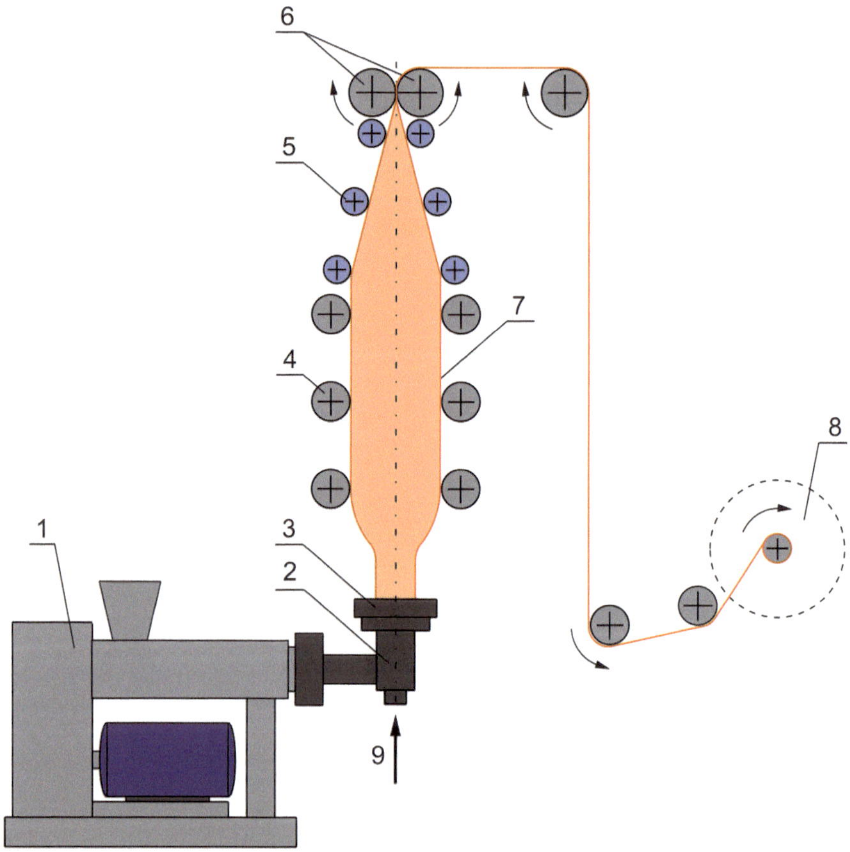

Figure 3.24 Film blowing process: 1 - extruder, 2 - die, 3 - air cooling unit, 4 - guide rolls, 5 - bubble collapsing rolls, 6 - take-off unit (pinch rolls), 7 - blown tube (tubular film, bubble), 8 - wind-up system 9 - blowing air supply

The extrusion film blowing line which is depicted in Figure 3.24 consists of an extruder, a die (normally with a spiral distributor), a cooling inflating device, flattening and take-off units, and a wind-up system. The blowing air is introduced into the extruded tube through a small orifice in the die mandrel. The tube is drawn by the rollers of the take-off device and flattened between the collapsing rolls. The extruded tubular profile passes through an air ring to cool the material, and it is drawn and inflated until the freezing line is reached. The air is retained in a tubular film by collapsing the film well above its freeze-off point and tightly pinching it between the rollers. The size of a tubular film is calibrated between the air ring and the collapsing rolls. During blowing there is primarily stretching in the transverse direction, while when drawing the stretching is in the longitudinal direction. This biaxial stretching renders a strong and low-permeable film.

3.3.3 Extrusion Blow Molding

Extrusion blow molding can be performed in one or two steps. A parison or a tubular profile is extruded and inflated in a mold cavity with the specified geometry. The blown article is held inside the mold until it is sufficiently cool.

In one-step extrusion blow molding, a tubular profile (polymer hose) is extruded directly into the blow mold cavity with the shape and dimensions corresponding to the produced article (container). The polymer hose is then welded and cut off at its ends when closing the mold, and after closing it is blown until it adopts the shape of the forming cavity. Blowing is performed by inserting the blow pin into the hose, which also calibrates the neck of the container. Then, the material solidifies in the mold after which the mold opens and the container is removed from the cavity. When blowing, the material is stretched in the transverse direction and, to a lesser extent, in the longitudinal direction, but the molecular (biaxial) orientation is usually small. The schematic of one-step extrusion molding is depicted in Figure 3.25.

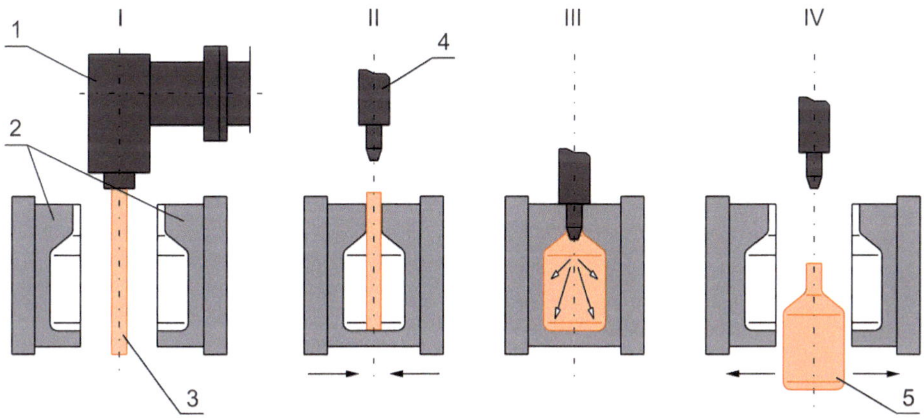

Figure 3.25 One-step extrusion blow molding: I - extrusion of polymer hose (tubular profile), II - mold closing and hose cutting, III - inserting the blow pin and blowing, IV - mold opening and article removal; 1 - die, 2 - blow mold, 3 - hose, 4 - blow pin, 5 - molded article (hollow container)

Two-step extrusion blow molding consists in:

- pre-forming, i.e. extrusion of the polymer hose, as in the one-step process, but in order to obtain not a finished container, but a preform with smaller dimensions and a thicker wall,
- basic molding, i.e. blowing with optional longitudinal mechanical stretching (using a pivot) to obtain a finished container.

In this process, during blowing, the material is significantly stretched in the transverse direction and in the longitudinal direction, such that a large biaxial molecular orientation occurs. In this way, the mechanical properties are improved substantially. The schematic of two-step extrusion molding is depicted in Figure 3.26.

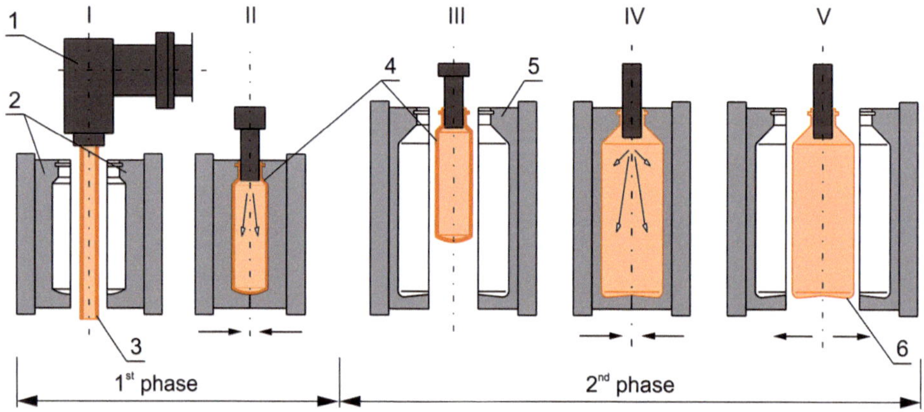

Figure 3.26 Two-step extrusion blow molding: I, II - 1st phase (parison forming), III, IV, V - 2nd phase (finished article forming in the blow mold), I - extrusion of polymer hose (tubular profile), II - parison forming, III - inserting the parison into a blow mold, IV - article forming (blowing), V - mold opening and article removal; 1 - die, 2 - mold, 3 - polymer hose, 4 - parison, 5 - blow mold, 6 - molded article (hollow container)

3.3.4 Injection Blow Molding

Injection blow molding is somewhat similar to extrusion blow molding. It consists in injection molding of the preform, and then blowing it in the blow mold. This is a two-step process with optional longitudinal mechanical stretching. The schematic of injection blow molding is depicted in Figure 3.27.

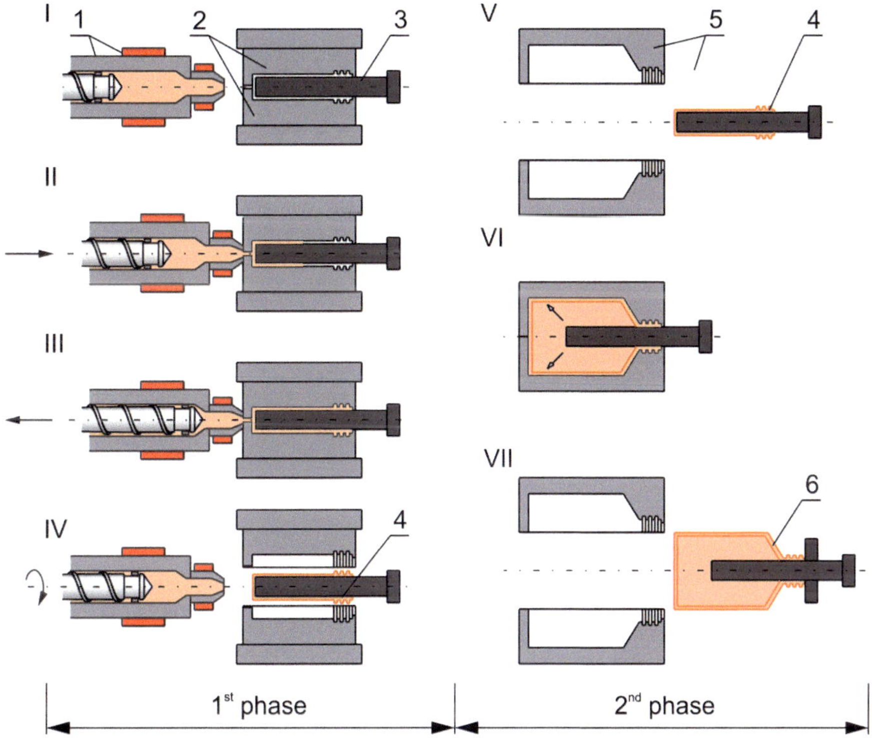

Figure 3.27 Injection blow molding: I, II, III, IV - 1st phase (pre-forming), V, VI, VII - 2nd phase (finished article forming in the blow mold), I - inserting the core into an injection mold, II - polymer injection, III - mold cavity filled, IV - mold opening and preform removal, V - inserting the preform into a blow mold, VI - article forming (blowing), VII - mold opening and article removal; 1 - injection unit, 2 - injection mold, 3 - core, 4 - preform, 5 - blow mold, 6 - molded article (hollow container)

3.4 Thermoforming

3.4.1 Introduction

Thermoforming consists in softening of thermoplastic sheets or films by heat, followed by forming with application vacuum, pressure, or a moving plug. The sheet may be stretched over a male mold (positive forming) or into a female mold (negative forming). On contact with the mold, heat is lost and the material regains stiffness as it cools. Geometries of thermoformed products are usually simple, e. g. boxes, containers, etc.

The main advantages of this process compared with injection molding and blow molding are the relatively low cost of thermoforming machines and the very low cost of molds, as well as the ease of forming the thin-walled large area parts. Disadvantages are the limited product shapes and difficulties in obtaining the required thickness distribution.

When thermoforming, a basic distinction is made between positive (male mold) and negative (female mold) forming. Using a positive or a negative method depends on which side of the part must exhibit exact replication. Only the side in contact with the mold is dimensionally accurate.

3.4.2 Negative Thermoforming

In negative thermoforming, the heated material is pressed into the concave mold cavity (female mold) by the application of vacuum, pressure, or a moving plug. The wall thickness distribution is generally not improved by the application of pressure, but the corners and edges are better formed. When thermoforming deeper articles, mechanical assistance is applied to have uniformity in the wall thickness. Compared with simple negative forming, the wall thickness distribution is significantly improved. Examples of negative thermoforming are depicted in Figure 3.28 and Figure 3.29.

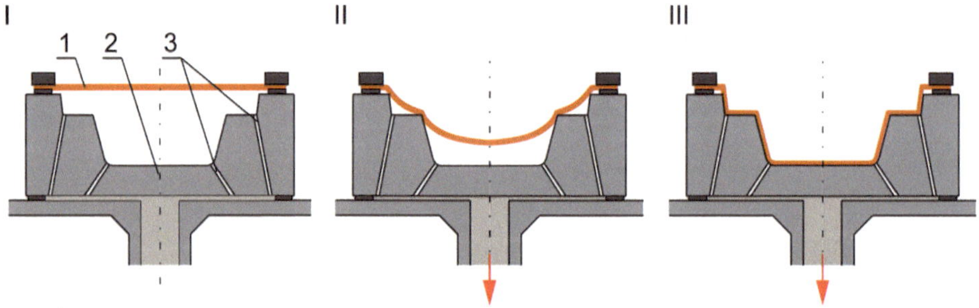

Figure 3.28 Negative thermoforming: I - sheet heating, II - sheet drawing by vacuum, III - forming finished; 1 - polymer sheet, 2 - negative mold (female mold), 3 - vacuum channels

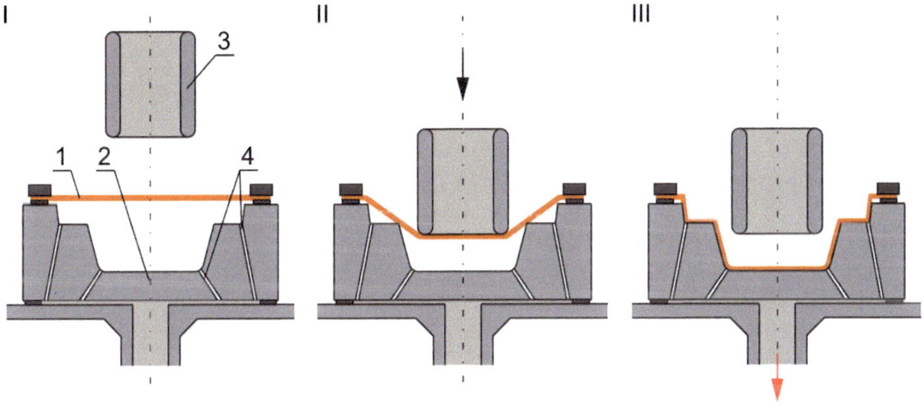

Figure 3.29 Negative thermoforming with mechanical stretching (assisting plug): I - sheet heating, II - mechanical stretching, III - forming by vacuum; 1 - polymer sheet, 2 - negative mold (female mold), 3 - plug, 4 - vacuum channels

3.4.3 Positive Thermoforming

In positive thermoforming, the heated material is forced against the convex mold (male mold) and adheres completely and tightly to the male mold. Vacuum and mechanical stretching can be applied, as depicted in Figure 3.30 and Figure 3.31.

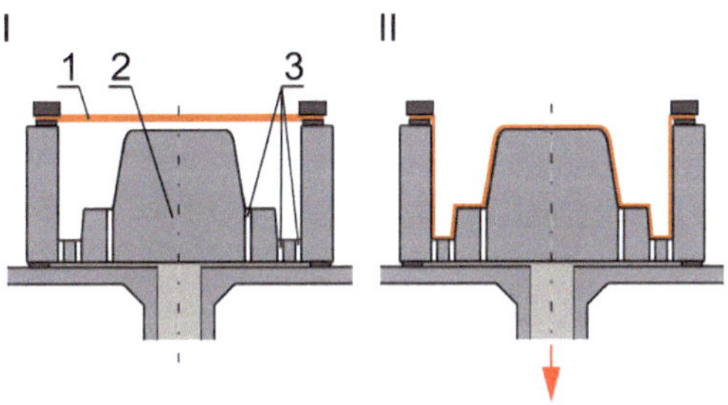

Figure 3.30 Positive thermoforming: I - sheet heating, II - forming by vacuum; 1 - polymer sheet, 2 - positive mold (male mold), 3 - vacuum channels

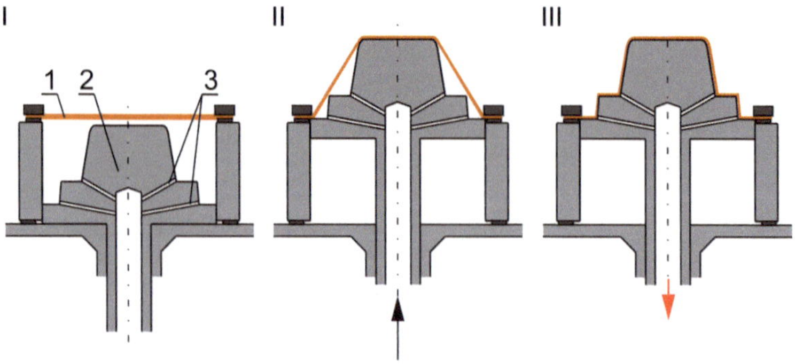

Figure 3.31 Positive thermoforming with mechanical stretching: I - sheet heating, II - mechanical stretching, III - forming by vacuum; 1 - polymer sheet, 2 - positive mold (male mold), 3 - vacuum channels

To reduce thickness variations in the product, the sheet can be pre-stretched by forming a bubble at the beginning of the process, which is depicted in Figure 3.32. The mold is raised into the bubble, and a vacuum finishes the forming process.

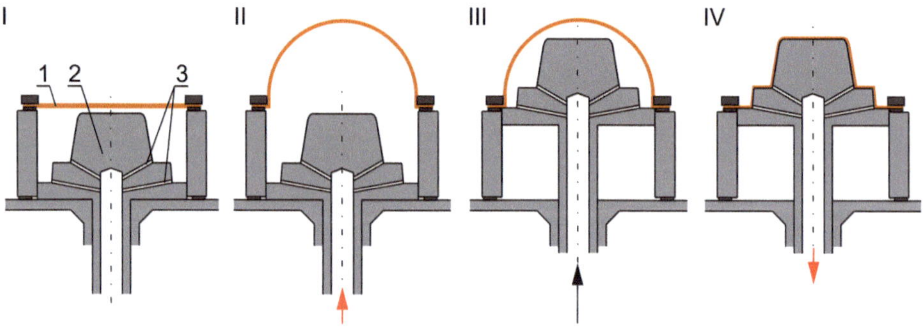

Figure 3.32 Positive forming with pressure stretching: I - sheet heating, II - pressure stretching (bubble forming), III - mold raising, IV - forming by vacuum; 1 - polymer sheet, 2 - positive mold (male mold), 3 - vacuum channels

■ 3.5 Calendering

Calendering is a continuous process used for the production of films and sheets as well as for coating. In a calender line, the polymer melt is squeezed between two or more counter-rotating heated calenders (rolls) and transformed into final products. The thermoplastic melt is fed behind the minimum gap between the two calenders as depicted in Figure 3.33. The rotational movement forces the material to flow in the machine direction (downstream) and in the lateral direction, with subsequent detachment from the calender surface at a specific thickness. Two-roll or four-roll calendering lines are usually used in the polymer processing industry, which can be arranged in different ways, as depicted in Figure 3.34, according to specific applications.

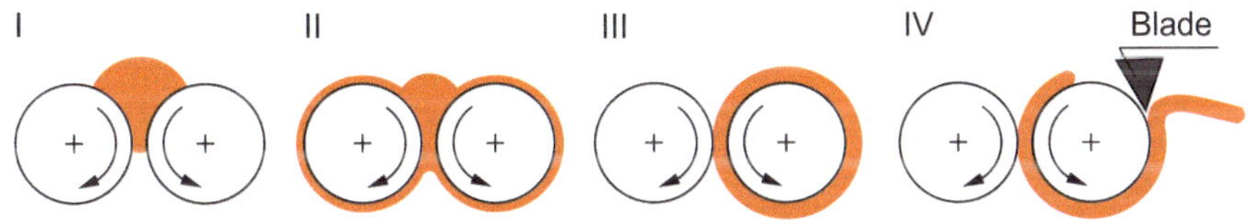

Figure 3.33 Calendering scheme: I - polymer supply to the calendering gap, II - calendering, III - calendering finished, IV - material cutting-off

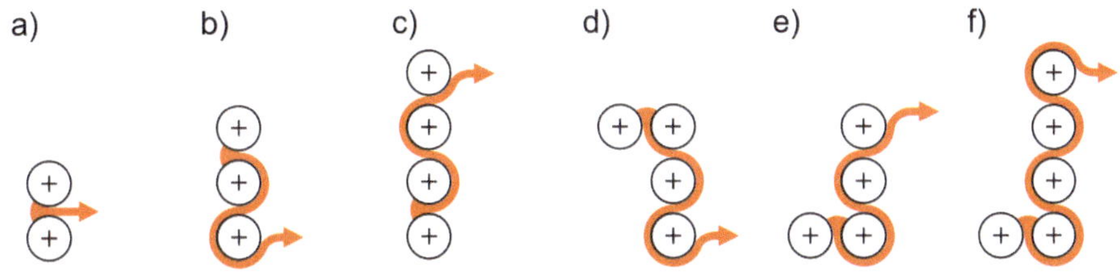

Figure 3.34 Calender arrangements: a), b), c) I-type, d) F-type, e), f) L-type

An example of a calendering line is depicted in Figure 3.35. In the plasticating unit, the material is melted and mixed by an internal batch mixer or a roller-mill before it is fed into the nip of the first two rollers. The first pair of rollers controls the feeding rate, while subsequent ones calibrate the sheet thickness. After the main calender, the sheet can be passed through a secondary operation for embossing. The sheet is then passed through a series of cooling rollers. After cooling, the sheet is wound.

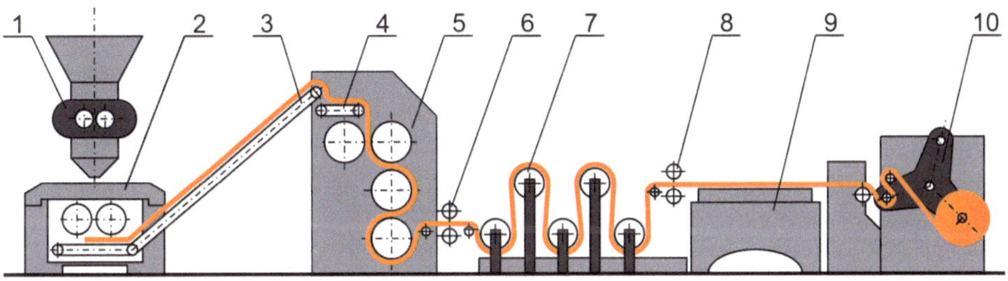

Figure 3.35 Calendering line: 1 - internal batch mixer, 2 - roll-mill, 3 - belt conveyer, 4 - strip feeder, 5 - calender, 6 - embosser, 7 - cooling rolls, 8 - take-off rolls, 9 - control area, 10 - wind-up station

3.6 Compression Molding

Compression molding is the oldest technique for the production of polymer products and is mainly used for thermosets. In this process the compound is pressed in the mold by the heated platens of a hydraulic press. Injection molding of polymers has replaced compression molding for some polymers, because of the advantages in material handling and automation. However, compression molding has an advantage in the processing of reinforced polymers. Owing to the modest level of deformation and stress involved in compression molding, the reinforcing fibers are not damaged. Very high fiber concentrations and longer fibers can be included in compression-molded products.

When compression molding, a basic distinction is made between compression pressing (compression molding) and transfer pressing (transfer compression molding), which differ in the way the compound is loaded into the mold cavity.

In compression molding, the compound charge is loaded directly into the mold cavity, as depicted in Figure 3.36.

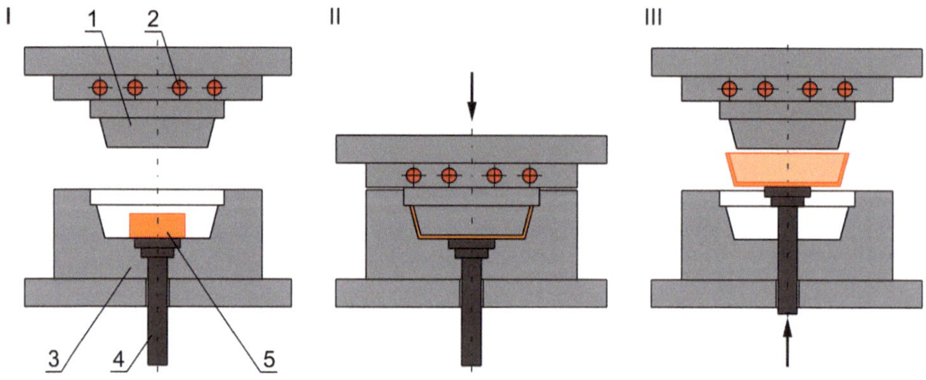

Figure 3.36 Compression molding: I - compound loading, II - forming, III - part removal; 1 - punch, 2 - heating elements, 3 - forming mold half, 4 - ejector, 5 - compound charge

In transfer compression molding, the compound charge is not directly loaded into the mold cavity but into the chamber where it is softened. Then pressure is applied to force a soft material to flow into the heated mold, where it cures.

There are two variants of transfer compression molding:

- pot transfer molding (Figure 3.37), when a charge is injected from the chamber (pot) through the vertical sprue channel into the mold cavity,
- plunger transfer molding (Figure 3.38), when the plunger injects a charge from the heated chamber through channels into the mold cavity.

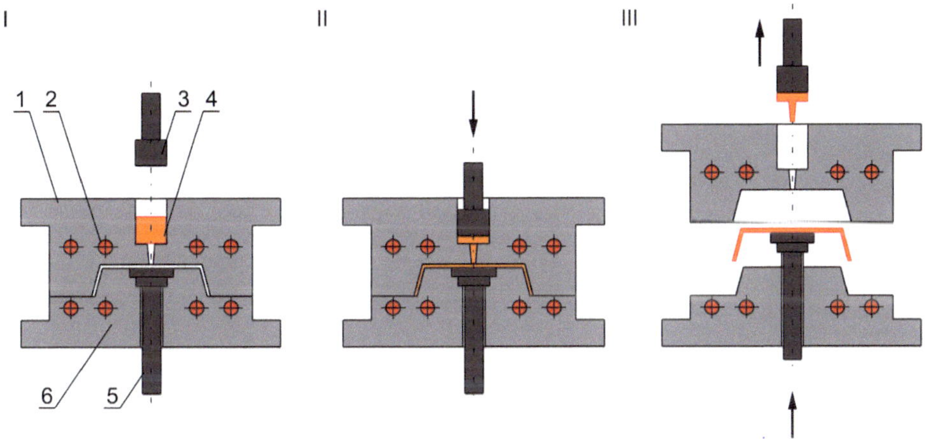

Figure 3.37 Pot transfer compression molding: I - compound loading (transferring), II - forming, III - part removal; 1 - upper mold half (matrix platen, female platen), 2 - heating elements, 3 - transfer ram (piston), 4 - compound charge, 5 - ejector, 6 - lower mold half (punch platen, male platen)

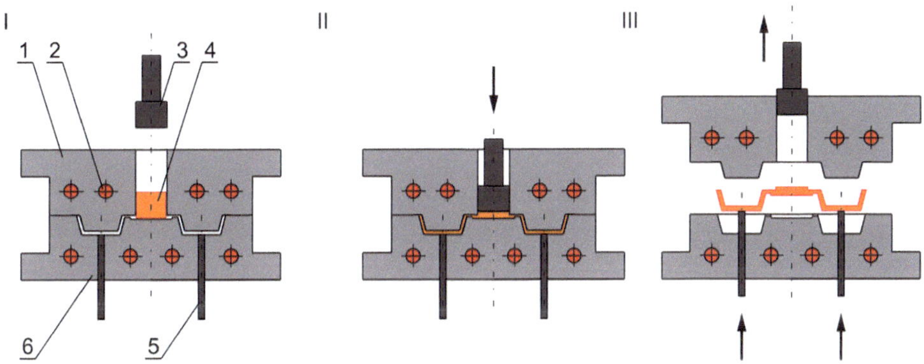

Figure 3.38 Plunger transfer compression molding: I - compound loading (transferring), II - forming, III - part removal; 1 - upper mold half (punch platen, male platen), 2 - heating elements, 3 - transfer ram (piston), 4 - compound charge, 5 - ejector, 6 - lower mold half (matrix platen, female platen)

References

1. Bernhardt, E.C. (Ed.), *Processing of Thermoplastic Materials* (1959) Van Nostrand Reinhold, New York
2. Schenkel, G., *Schneckenpressen für Kunststoffe* (1959) Carl Hanser Verlag, Munich
3. McKelvey, J.M., *Polymer Processing* (1962) John Wiley & Sons, New York
4. Schenkel, G., *Kunststoff Extruder Technik* (1963), Carl Hanser Verlag, Munich
5. Mink, W., *Grundzüge der Extrudertechnik* (1963) Rudolf Zechner Verlag, Speyer am Rhein
6. Schenkel, G., *Plastics Extrusion. Technology and Theory* (1966) Illife Books Ltd., London
7. Fenner, R.T., *Extruder Screw Design* (1966) Illife-Butterworth, London

8. Mink, W., *Grundzüge der Spritzgiesstechnik* (1979) Rudolf Zechner Verlag, Speyer am Rhein
9. Stevens, M.J., *Extruder Principles and Operation* (1985) Elsevier, London
10. Hensen, F., Knappe, W., Potente, H., *Handbuch der Kunststoff-Extrusionstechnik. Grundlagen* (1989) Carl Hanser Verlag, Munich
11. Hensen, F., Knappe, W., Potente, H., *Handbuch der Kunststoff-Extrusionstechnik. Extrusionsanlagen* (1989) Carl Hanser Verlag, Munich
12. Charrier, J.M., *Polymeric Materials and Processing* (1990) Carl Hanser Verlag, Munich
13. White, J.L., *Twin Screw Extrusion* (1990) Carl Hanser Verlag, Munich
14. Knappe, W., Lampl, A., Heuel, O., *Kunststoff-Verarbeitung und Werkzeugbau* (1992) Carl Hanser Verlag, Munich
15. Rosato, D.V., *Rosato's Plastics Encyclopedia and Dictionary* (1993) Carl Hanser Verlag, Munich
16. Belofsky, H., *Plastics: Product Design and Process Engineering* (1995) Carl Hanser Verlag, Munich
17. White, J.L., *Rubber Processing* (1995) Carl Hanser Verlag, Munich
18. Griskey, R.G., *Polymer Process Engineering* (1995) Chapman and Hall, New York
19. Michaeli, W., *Plastics Processing* (1995) Carl Hanser Verlag, New York
20. Osswald, T.A., *Polymer Processing Fundamentals* (1998) Carl Hanser Verlag, Munich
21. Beall, G., *Rotational Molding* (1998) Carl Hanser Verlag, Munich
22. Menges, G., Michaeli, W., Mohren, P., *How to Make Injection Molds* (2001) Carl Hanser Verlag, Munich
23. Rees, H., *Understanding Injection Mold Design* (2001) Carl Hanser Verlag, Munich
24. White, J.L., Potente, H., *Screw Extrusion* (2002) Carl Hanser Verlag, Munich
25. Dave, R., Loos, A., *Compression Molding* (2003) Carl Hanser Verlag, Munich
26. Rosato, D., Rosato, A., DiMattia D., *Blow Molding Handbook* (2004) Carl Hanser Verlag, Munich
27. Tadmor, Z., Gogos, C.G., *Principles of Polymer Processing* (2006) John Wiley & Sons, New Jersey
28. Unger, P., *Gastrow Injection Molds* (2006) Carl Hanser Verlag, Munich
29. Pötsch, G., Michaeli, W., *Injection Molding. An Introduction* (2007) Carl Hanser Verlag, Munich
30. Johannaber, F., *Injection Molding Machines* (2007) Carl Hanser Verlag, Munich
31. Lee, N., *Understanding Blow Molding* (2007) Carl Hanser Verlag, Munich
32. Lee, N., *Blow Molding Design Guide* (2008) Carl Hanser Verlag, Munich
33. Throne, J., *Understanding Thermoforming* (2008) Carl Hanser Verlag, Munich
34. Kamal, M.R., Isayev, A.I., Liu, S.-J. (Eds.), *Injection Molding. Technology and Fundamentals* (2009) Carl Hanser Verlag, Munich
35. Manas-Zloczower, I. (Ed.), *Mixing and Compounding of Polymers. Theory and Practice* (2009) Carl Hanser Verlag, Munich
36. Osswald, T., Turng, L.-S., Gramann, P., *Injection Molding Handbook* (2009) Carl Hanser Verlag, Munich
37. White, J.L., Kim, E.K., *Twin Screw Extrusion* (2010) Carl Hanser Verlag, Munich
38. Rauwendaal, C., *Polymer Extrusion* (2013) Carl Hanser Verlag, New York
39. Mennig, G., Stoeckhert, K., *Mold-Making Handbook* (2013) Carl Hanser Verlag, Munich
40. Campbell, G.A., Spalding, M.A., *Analyzing and Troubleshooting Single-Screw Extruders* (2014) Carl Hanser Verlag, Munich
41. Kazmer, D.O., *Injection Mold Design Engineering* (2016) Carl Hanser Verlag, Munich
42. Hopmann, Ch., Michaeli, W., *Extrusion Dies for Plastics and Rubber* (2016) Carl Hanser Verlag, Munich

43. Agassant, J.-F., Avenas, P., Carreau, P.J., Vergnes, B., Vincent, M., *Polymer Processing. Principles and Modeling* (2017) Carl Hanser Verlag, Munich
44. Osswald, T.A., *Understanding Polymer Processing* (2017) Carl Hanser Verlag, Munich
45. Baur, E., Osswald, T.A., Rudolph, N., *Plastics Handbook* (2019) Carl Hanser Verlag, Munich
46. Vlachopoulos, J., Polychronopoulos, N.D., *Understanding Rheology and Technology of Polymer Extrusion* (2019) Polydynamics Inc., Dundas
47. Cantor, K., *Blown Film Extrusion* (2019) Carl Hanser Verlag, Munich
48. Turng, L., *Advanced Injection Molding Technologies* (2019) Carl Hanser Verlag, Munich
49. Rauwendaal, C., *Understanding Extrusion* (2019) Carl Hanser Verlag, Munich
50. Illig (Ed.), Schwarzmann, P., *Thermoforming* (2019) Carl Hanser Verlag, Munich
51. Bonten, C., *Plastics Technology. Introduction and Fundamentals* (2020) Carl Hanser Verlag, Munich
52. Chung, Ch. I., *Extrusion of Polymers. Theory and Practice* (2020) Carl Hanser Verlag, Munich
53. Pruner, H., Nesch, W., *Understanding Injection Molds* (2020) Carl Hanser Verlag, Munich
54. Kohlgrüber, K., *Co-Rotating Twin Screw Extruders: Fundamentals* (2020) Carl Hanser Verlag, Munich
55. Kohlgrüber, K., *Co-Rotating Twin Screw Extruders: Applications* (2020) Carl Hanser Verlag, Munich

4 Process Modeling

4.1 Introduction

A model of the process is its simplified physical or mathematical representation. A physical model is a model where a simplified representation of the process is constructed. Physical models are used to understand the physics of the process, as well as to validate mathematical models of this. A mathematical model describes the process with equations. Mathematical models are developed using material and balance equations, along with assumptions that simplify the process. Usually, the more complex the mathematical model, the more accurately it describes the process.

We model polymer processing to better understand the manufacturing processes, to solve the existing problems, and to improve the processes. Furthermore, mathematical models can be used to optimize polymer processing.

Process modeling is based on the fundamentals of fluid dynamics. The basic problem of fluid mechanics (Section 1.5) is to determine the motion of fluid in a given flow geometry. It consists in solving the set of differential equations of conservation of continuity (Eq. (1.36)), momentum (Eq. (1.65)), and energy (Eq. (1.90), with the corresponding constitutive equation of the material (and possibly the equation of state) and the corresponding boundary and initial conditions.

In order to solve the problem (i.e. to describe the flow) some appropriate assumptions should be made regarding the field of physical quantities that describe it, e.g. velocity, temperature, as well as assumptions that determine the properties of the fluid.

The basic classification of the flows results from the following:

- dependencies of physical quantities describing the flow from time (steady or unsteady flow),
- the number of space coordinates defining the flow (one-dimensional flow, two-dimensional flow, or three-dimensional flow),

- the variability of the flow temperature (isothermal or non-isothermal flow),
- the compressibility of the fluid (incompressible or compressible flow),
- the type of the constitutive equation of the fluid (Newtonian or non-Newtonian flow).

Solving the specific problems of fluid mechanics consists in the following:

- formulating assumptions describing the problem concerning the flow geometry, the physical flow fields (including boundary and initial conditions), and the fluid properties (including the constitutive equation of the fluid),
- reducing the equations of conservation and the constitutive equation of the fluid according to the assumptions made,
- solving these simplified equations (the set of equations) with assumed boundary and initial conditions, analytically or numerically.

There are only a few exact or analytical solutions of the equations of conservation, and most of them are for unidirectional (viscometric) flows, that is, the flows that have only one velocity component dependent on the one coordinate that does not coincide with the direction of flow (see Section 1.3.4.2).

Most of the polymer processing models are represented by sets of non-linear partial differential equations and/or integral equations that do not have analytical solutions. Numerical analysis will provide methods for obtaining solutions to these problems.

4.2 Simple Flows

The basic classes of flows in polymer processing are the pressure flows and the drag flows. The selected flows presented in the further part of the chapter belong to the group of classical flows, i.e. historically the first that were solved in an exact manner.

4.2.1 Pressure Flows

A **pressure flow** is defined as the flow that takes place in a closed channel due to the pressure difference occurring along this channel (i.e. pressure driven flow). Pressure flows are an important class of flows in rheology and polymer processing. These flows occur in capillary rheometers, extrusion dies, and injection molds, as well as in plasticating units of extruders and injection molding machines. Some examples of pressure flows are depicted in Figure 2.2.

In the following part, some of the most important pressure flows are considered. The basic assumptions of these flows are as follows:

- the flow is steady, laminar, and isothermal,
- the flow is fully developed,
- there is no slip on the channel wall,
- gravity and inertia forces are neglected,
- the fluid is incompressible and its viscosity does not depend on the pressure.

4.2.1.1 Flow between Parallel Plates

Isothermal Newtonian Flow

The geometry of flow between two parallel plates is shown in the Cartesian system *xyz* in Figure 4.1. It is conventionally assumed that this flow takes place in a channel with a very large length L, a practically infinite width W, and a small height H.

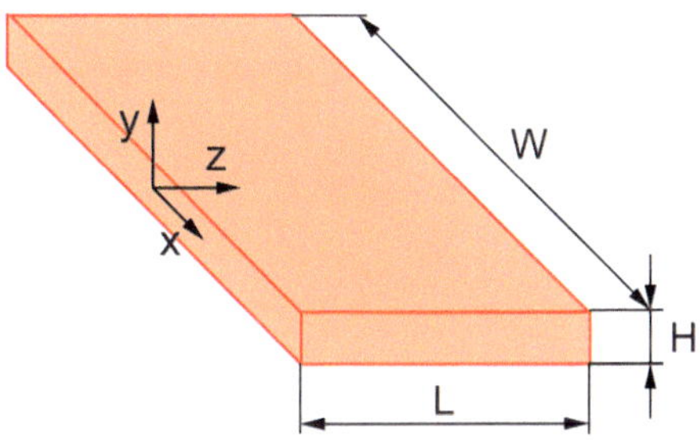

Figure 4.1 Flow geometry in a flat channel (between two parallel plates)

It is assumed that the velocity field has the form

$$v_x = 0, \quad v_y = 0, \quad v_z = v_z(y) \tag{4.1}$$

Under such conditions, the equation of motion in the flow direction, i.e. in the z-axis direction (Table 1.3), is expressed as

$$0 = -\frac{\partial p}{\partial z} + \frac{\partial \tau_{yz}}{\partial y} \tag{4.2}$$

Since the pressure in this case is a function of the z-coordinate only, and the stress depends on the y-coordinate only, Eq. (4.2) can be presented as

$$0 = -\frac{\mathrm{d}p}{\mathrm{d}z} + \frac{\mathrm{d}\tau_{yz}}{\mathrm{d}y} \tag{4.3}$$

The pressure gradient in this flow is constant and equals to

$$-\frac{\mathrm{d}p}{\mathrm{d}z} = \frac{\Delta p}{L} \tag{4.4}$$

where $\Delta p = p_0 - p_L$ is the pressure drop along the channel (along z-axis); p_0 is the pressure at the beginning of the channel; p_L is the pressure at the end of the channel; L is the channel length.

Thus, Eq. (4.3) can be transformed into

$$\frac{d\tau_{yz}}{dy} = -\frac{\Delta p}{L} \tag{4.5}$$

and after integrating Eq. (4.5) we have

$$\tau_{yz} = -\frac{\Delta p}{L} y + C_1 \tag{4.6}$$

In the case of the Newtonian fluid, under assumed flow conditions, the component τ_{yz} (Table 1.7) is equal to

$$\tau_{yz} = \mu \frac{dv_z}{dy} \tag{4.7}$$

where the viscosity μ is the Newtonian viscosity.

After substituting the Eq. (4.7) into Eq. (4.6), we obtain

$$\mu \frac{dv_z}{dy} = -\frac{\Delta p}{L} y + C_1 \tag{4.8}$$

The constant C_1 can be determined on the basis of the flow symmetry, from which it follows that for $y = 0$ the derivative of velocity $dv_z/dy = 0$, i.e. $dv_z/dy(y = 0) = 0$. Thus, the constant $C_1 = 0$ and Eq. (4.8) can be written as

$$\frac{dv_z}{dy} = -\frac{1}{\mu} \frac{\Delta p}{L} y \tag{4.9}$$

This equation expresses the **shear rate.**

After integrating Eq. (4.9) we obtain

$$v_z = -\frac{1}{\mu} \frac{\Delta p}{L} \frac{y^2}{2} + C_2 \tag{4.10}$$

The constant C_2 can be determined from the condition of no slip on the channel wall, i.e. for $v_z(y = H/2) = 0$, and is equal to

$$C_2 = \frac{1}{\mu} \frac{\Delta p}{L} \frac{H^2}{8} \tag{4.11}$$

The result is a parabolic **velocity distribution** in the form

$$v_z = \frac{H^2}{8\mu} \frac{\Delta p}{L} \left[1 - \left(\frac{2y}{H} \right)^2 \right] \tag{4.12}$$

where the maximum velocity v_{max} occurs for $y = 0$, i.e. in the channel axis, and it is expressed by

$$v_{max} = \frac{H^2}{8\mu}\frac{\Delta p}{L} \tag{4.13}$$

The flow rate can be determined by considering an elementary flow dQ through a flat channel of a width W and a thickness dy, occurring at the local velocity $v_z(y)$, that is expressed as

$$\mathrm{d}Q = v_z(y)W\mathrm{d}y \tag{4.14}$$

Then, the flow rate in relation to the entire cross-section will be equal to

$$Q = 2W\int_0^{H/2} v_z(y)\mathrm{d}y \tag{4.15}$$

After substituting for v_z from Eq. (4.12) we have

$$Q = 2W\int_0^{H/2} \frac{H^2}{4\mu}\frac{\Delta p}{L}\left[1-\left(\frac{2y}{H}\right)^2\right]\mathrm{d}y \tag{4.16}$$

and after its integration, we obtain a formula defining the **volume flow rate:**

$$Q = \frac{WH^3}{12\mu}\frac{\Delta p}{L} \tag{4.17}$$

The obtained formula can be transformed into a form that allows to calculate the **pressure drop** in the flat channel based on the known flow rate, channel dimensions and viscosity of the material, i.e.

$$\Delta p = \frac{12L}{WH^3}\mu Q \tag{4.18}$$

Such calculations are very useful when designing tools for polymer processing, e.g. for extrusion dies.

Using Eq. (4.18), the shear rate (Eq. (4.9)) can be expressed as a function of the flow rate

$$\frac{\mathrm{d}v_z}{\mathrm{d}y} = -\frac{12Q}{WH^3}y \tag{4.19}$$

and then, the shear rate at the channel wall, i.e. for $y = H/2$, can be expressed as

$$\dot{\gamma}_w = \frac{\mathrm{d}v_z}{\mathrm{d}y}(y = H/2) = -\frac{6Q}{WH^2} \tag{4.20}$$

where the shear rate $\dot{\gamma}_w$ is usually conventionally taken as a positive value (see Section 2.4.2), i.e.

$$\dot{\gamma}_w = \left(-\frac{\mathrm{d}v_z}{\mathrm{d}y}\right)_w = \frac{6Q}{WH^2} \tag{4.21}$$

Distribution of stress in this flow can be determined on the basis of Eq. (4.6) using the condition $C_1 = 0$ as

$$\tau_{yz} = -\frac{\Delta p}{L} y \tag{4.22}$$

According to this formula, the shear stress τ_w at the channel wall is

$$\tau_w = \tau_{yz}\left(y = H/2\right) = -\frac{\Delta pH}{2L} \tag{4.23}$$

so that the **stress distribution** can be represented as

$$\frac{\tau_{yz}}{\tau_w} = \frac{2y}{H} \tag{4.24}$$

Assuming the appropriate stress sign is contractual and usually the stress τ_w is positive (see Section 2.4.2).

Equation (4.24) shows that the stress distribution in the flow is linear and independent on the constitutive equation of the material.

Distributions of shear stress (Eq. (4.24)), velocity (Eq. (4.12)), and shear rate (Eq. (4.19)) are shown in Figure 4.2.

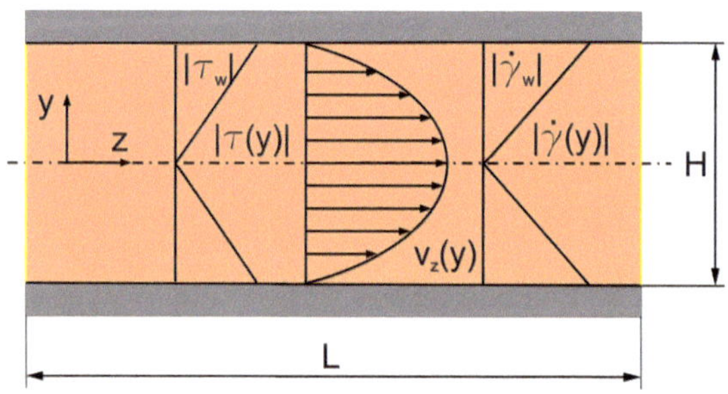

Figure 4.2 Distribution of the shear stress $\tau(y)$, velocity $v_z(y)$ and shear rate $\dot{\gamma}(y)$ in the Newtonian pressure flow in a flat channel (between two parallel plates): τ_w - shear stress at the channel wall, $\dot{\gamma}_w$ - shear rate at the wall

Isothermal Non-Newtonian Flow

A non-Newtonian flow will be considered on the example of the power-law model of Ostwald-de Waele that in the general case, based on Eq. (1.159), has the form

$$\underset{\sim}{\tau} = m\dot{\gamma}_D^{n-1} 2\underset{\sim}{\mathbf{D}} \tag{4.25}$$

where: $\underset{\sim}{\tau}$ is the extra-stress tensor; m is the consistency factor; n is the flow exponent; $\underset{\sim}{D}$ is the shear rate tensor; $\dot{\gamma}_D$ is the generalized shear rate defined by Eq. (1.157).

The component τ_{yz} of the stress tensor expresses, on the basis of Eq. (4.25), as

$$\tau_{yz} = m\dot{\gamma}_D^{n-1} 2D_{yz} \tag{4.26}$$

Under assumed flow conditions, based on Eq. (1.157) and Table 1.10, the generalized shear reduces to

$$\dot{\gamma}_D = \frac{\mathrm{d}v_z}{\mathrm{d}y} \tag{4.27}$$

and the component of the strain rate tensor D_{yz} is equal (Table 1.1) to

$$D_{yz} = \frac{1}{2}\frac{\mathrm{d}v_z}{\mathrm{d}y} \tag{4.28}$$

Thus, we obtain the component τ_{yz} of the stress tensor as

$$\tau_{yz} = m\left|\frac{\mathrm{d}v_z}{\mathrm{d}y}\right|^{n-1}\frac{\mathrm{d}v_z}{\mathrm{d}y} \tag{4.29}$$

The absolute value sign is necessary in this case because $\mathrm{d}v_z/\mathrm{d}y$ can be negative, and for polymers n is usually less than one, i.e. $n - 1 < 0$.

After substituting Eq. (4.29) into Eq. (4.6) we have

$$m\left|\frac{\mathrm{d}v_z}{\mathrm{d}y}\right|^{n-1}\frac{\mathrm{d}v_z}{\mathrm{d}y} = -\frac{\Delta p}{L}y + C_1 \tag{4.30}$$

where the constant $C_1 = 0$ because the flow is symmetric, i.e. $\mathrm{d}v_z/\mathrm{d}y(y = 0) = 0$.

Thus, Eq. (4.30) can be written as

$$m\left|\frac{\mathrm{d}v_z}{\mathrm{d}y}\right|^{n-1}\frac{\mathrm{d}v_z}{\mathrm{d}y} = -\frac{\Delta p}{L}y \tag{4.31}$$

Since the right side of this equation is negative, the derivative $\mathrm{d}v_z/\mathrm{d}y$ must also be negative, so that after transformation the expression determining the **shear rate** takes the form

$$\frac{\mathrm{d}v_z}{\mathrm{d}y} = -\left(\frac{1}{m}\frac{\Delta p}{L}\right)^{1/n} y^{1/n} \tag{4.32}$$

and after integration

$$v_z = -\frac{1}{(1+1/n)}\left(\frac{\Delta p}{mL}\right)^{1/n} y^{1+(1/n)} + C_2 \tag{4.33}$$

The constant C_2 can be determined from the condition of no slip on the channel wall, i.e. for $v_z(y = H/2) = 0$. Then, it is equal to

$$C_2 = \frac{1}{(1+1/n)} \left(\frac{\Delta p}{mL}\right)^{1/n} \left(\frac{H}{2}\right)^{1+(1/n)} \tag{4.34}$$

As a result, we have **velocity distribution** in the form

$$v_z = \frac{H}{2(1+1/n)} \left(\frac{H\Delta p}{2mL}\right)^{1/n} \left[1 - \left(\frac{2y}{H}\right)^{1+(1/n)}\right] \tag{4.35}$$

Due to the flow symmetry, the absolute value of the y-coordinate must be entered into this equation, that is

$$v_z = \frac{H}{2(1+1/n)} \left(\frac{H\Delta p}{2mL}\right)^{1/n} \left[1 - \left|\frac{2y}{H}\right|^{1+(1/n)}\right] \tag{4.36}$$

The maximum velocity $v_{\max}$ is for $y = 0$, i.e. in the channel axis, and expresses as

$$v_{max} = \frac{H}{2(1+1/n)} \left(\frac{H\Delta p}{2mL}\right)^{1/n} \tag{4.37}$$

The flow rate can be determined by substituting Eq. (4.36) into Eq. (4.15). Then, we have

$$Q = W \int_0^{H/2} \frac{H}{(1+1/n)} \left(\frac{H\Delta p}{2mL}\right)^{1/n} \left[1 - \left(\frac{2y}{H}\right)^{1+(1/n)}\right] \mathrm{d}y \tag{4.38}$$

After integration, we obtain the formula that determines the **volume flow rate:**

$$Q = \frac{WH^2}{2(2+1/n)} \left(\frac{H\Delta p}{2mL}\right)^{1/n} \tag{4.39}$$

This formula can be transformed into a form that allows calculation of the **pressure drop** in the channel, based on the known flow rate, channel dimensions, and viscosity of the material, i.e.

$$\Delta p = \frac{2mL}{H} \left[\frac{2(2+1/n)Q}{WH^2}\right]^n \tag{4.40}$$

Using Eq. (4.40), the shear rate (Eq. (4.32)) can be expressed as a function of the flow rate:

$$\frac{\mathrm{d}v_z}{\mathrm{d}y} = -\frac{2(2+1/n)Q}{WH^2} \left(\frac{2y}{H}\right)^{1/n} \tag{4.41}$$

and then, the shear rate at the channel wall, i. e. for $y = H/2$, can be expressed as

$$\dot{\gamma}_w = \left(-\frac{\mathrm{d}v_z}{\mathrm{d}y}\right)_w = \frac{2(2+1/n)Q}{WH^2} \tag{4.42}$$

Distributions of velocity (Eq. (4.36)) and shear rate (Eq. (4.41)) are shown in Figure 4.3. These distributions are given for different values of the flow index, i. e. for $n < 1, n = 1, n > 1$ (see Section 1.2.2), which correspond to shear thinning (pseudoplastic), Newtonian (compare Figure 4.2), and shear thickening (dilatant) behavior.

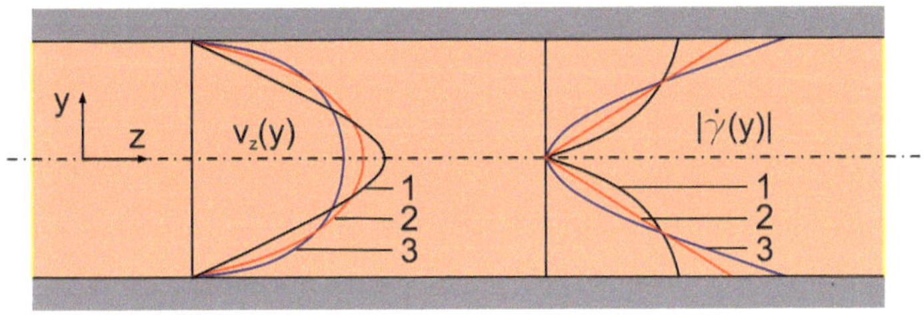

Figure 4.3 Distribution of the velocity $v_z(y)$ and shear rate $\dot{\gamma}(y)$ in non-Newtonian pressure flow in a flat channel (between two parallel plates): 1 - $n > 1$, 2 - $n = 1$, 3 - $n < 1$

It is worth noting that for $n = 1$ and $m = \mu$ the formulas corresponding to Newtonian flow are obtained.

Non-Isothermal Newtonian Flow

In isothermal flows, the temperature is constant throughout the entire flow area. In non-isothermal flows it is a function of the location, and in unsteady flows also a function of the time.

In order to discuss non-isothermal flow between two parallel plates, the case will be considered in which the material with an initial temperature T_0 flows into a channel whose walls have a constant temperature T_0 (Figure 4.4).

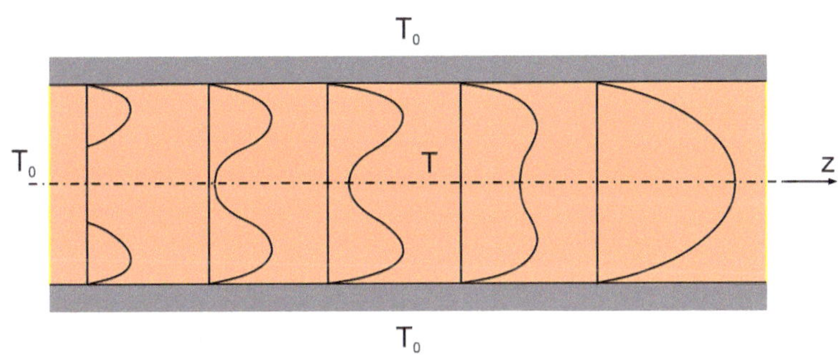

Figure 4.4 Temperature profile T in the pressure flow: T_0 - wall temperature and initial temperature of the material

In the initial section of the channel, the highest temperature prevails near the walls because the highest shear stress is developed in this area (this results from the velocity and shear rate distribution). Then, gradually as a result of heat conduction towards the channel center, the temperature profile changes until an equilibrium is reached in which the maximum of temperature occurs in the channel center.

Generally, three sections can be distinguished in which different thermal conditions occur [35]:

1. In the initial part of the channel, the temperature rise is small, so that the heat transfer by conduction can be neglected. The energy dissipated during the flow is used to heat the material, as if it was isolated from the surrounding environment. These flow conditions are called the adiabatic conditions.
2. In the next section, the temperature rise is such that the heat transfer cannot be neglected. The energy dissipated during the flow is consumed not only for the material heating, but also is transferred to the environment. These are the transition thermal flow conditions.
3. In the further section of the channel, the temperature rise reaches a certain limit and all the dissipated energy is transmitted by conduction to the environment. These flow conditions are called the equilibrium conditions.

To solve the non-isothermal problem it is necessary to simultaneously solve the equations of motion and energy. An exact solution to this problem, even in the case of a simple flow, is complex because a change in the material temperature causes a change in the viscosity, which in turn affects the velocity profile. The equations of motion and energy are inherently coupled and usually should be solved iteratively at each flow point. However, in many practical cases, when separating motion and energy equations and independently solving them, good results (close to the real situation) are also obtained.

Under the flow conditions assumed here, the energy equation will be considered. It is also assumed that the viscosity does not depend on the temperature.

The velocity field is defined by the conditions in Eq. (4.1), and the temperature field is

$$T = T(y,z) \tag{4.43}$$

Then, the energy equation has the form (Table 1.4):

$$\rho c_p v_z \frac{\partial T}{\partial z} = k\left(\frac{\partial^2 T}{\partial y^2} + \frac{\partial^2 T}{\partial z^2}\right) + \tau_{yz}\frac{\partial v_z}{\partial y} \tag{4.44}$$

The first part of this equation $\rho c_p v_z \frac{\partial T}{\partial z}$ represents heat convection, the second

$k\left(\frac{\partial^2 T}{\partial y^2}+\frac{\partial^2 T}{\partial z^2}\right)$ determines heat conduction, and the third one $\tau_{yz}\frac{\partial v_z}{\partial y}$ describes energy dissipation.

In most cases of polymer processing, the heat transfer in the z-axis direction is negligible in relation to the convective part of Eq. (4.44). Thus, this equation can be written as

$$\rho c_p v_z \frac{\partial T}{\partial z} = k\frac{\partial^2 T}{\partial y^2} + \tau_{yz}\frac{\partial v_z}{\partial y} \tag{4.45}$$

In the case of Newtonian fluid, under assumed flow conditions, the component τ_{yz} of the stress tensor is determined by Eq. (4.7), so that, after substituting it into Eq. (4.45), we have

$$\rho c_p v_z \frac{\partial T}{\partial z} = k\frac{\partial^2 T}{\partial y^2} + \mu\left(\frac{\partial v_z}{\partial y}\right)^2 \tag{4.46}$$

The shear rate $\partial v_z/\partial y$ can be determined using Eq. (4.19), and after entering the average flow rate V in the form

$$V = \frac{Q}{WH} \tag{4.47}$$

it can be expressed as

$$\frac{\partial v_z}{\partial y} = -\frac{12V}{H^2}y \tag{4.48}$$

Then, Eq. (4.46) takes the form

$$\rho c_p v_z \frac{\partial T}{\partial z} = k\frac{\partial^2 T}{\partial y^2} + 144\frac{\mu V^2}{H^4}y^2 \tag{4.49}$$

This equation can be considered in two limiting cases, i.e. in the equilibrium conditions and in the adiabatic flow conditions.

In the equilibrium conditions, the convective part of Eq. (4.49) is equal to zero and the equation, after transformation, takes the form

$$\frac{\mathrm{d}^2 T}{\mathrm{d}y^2} = -144\frac{\mu V^2}{kH^4}y^2 \tag{4.50}$$

After integration of Eq. (4.50) we have

$$\frac{\mathrm{d}T}{\mathrm{d}y} = -144\frac{\mu V^2}{kH^4}\frac{y^3}{3} + C_1 \tag{4.51}$$

where the constant C_1 can be determined on the basis of the flow symmetry, from which it follows that for $y = 0$ the derivative of temperature $\mathrm{d}T/\mathrm{d}y = 0$, i.e. $\mathrm{d}T/\mathrm{d}y(y = 0) = 0$. Thus, the constant $C_1 = 0$ and Eq. (4.51) can be written as

$$\frac{\mathrm{d}T}{\mathrm{d}y} = -48\frac{\mu V^2}{kH^4}y^3 \tag{4.52}$$

After further integration we have

$$T = -12\frac{\mu V^2}{kH^4}y^4 + C_2 \tag{4.53}$$

where the constant C_2 can be determined on the basis of the wall temperature condition, i.e. the assumption that $T(y = H/2) = T_0$. Then this constant is equal to

$$C_2 = T_0 + 12\frac{\mu V^2}{kH^4}\left(\frac{H}{2}\right)^4 \tag{4.54}$$

The result is a parabolic **temperature distribution** in the form

$$T = T_0 + \frac{3}{4}\frac{\mu V^2}{k}\left[1 - \left(\frac{2y}{H}\right)^4\right] \tag{4.55}$$

where the maximum temperature $T_{\max}$ is for $y = 0$, i.e. in the channel axis, and expresses as

$$T_{max} = T_0 + \frac{3}{4}\frac{\mu V^2}{k} \tag{4.56}$$

The difference between the maximum temperature $T_{\max}$ and the temperature of the channel wall T_0 is defined as the temperature increase due to the viscous energy dissipation:

$$\Delta T_{max} = T_{max} - T_0 = \frac{3}{4}\frac{\mu V^2}{k} \tag{4.57}$$

It is worth noting that in most polymer processing situations, equilibrium is rarely achieved.

In the adiabatic conditions, the part representing the heat transfer in the energy equation (Eq. (4.49)) is negligibly small, so that the equation takes the form

$$\rho c_p v_z \frac{\mathrm{d}T}{\mathrm{d}z} = 144\frac{\mu V^2}{H^4}y^2 \tag{4.58}$$

The velocity v_z is determined by Eq. (4.12), which after applying Eq. (4.18) and entering the average velocity V (Eq. (4.47)) can be presented as

$$v_z = \frac{3}{2}V\left[1 - \left(\frac{2y}{H}\right)^2\right] \tag{4.59}$$

and, after substitution of Eq. (4.59) into the energy equation (Eq. (4.58)), we have

$$\frac{3}{2}\rho c_p V\left[1-\left(\frac{2y}{H}\right)^2\right]\frac{\mathrm{d}T}{\mathrm{d}z}=144\frac{\mu V^2}{H^4}y^2 \tag{4.60}$$

This equation can be solved by seeking the average temperature T^*. To do this, we need to write this equation in a global form relative to the volume $WH\mathrm{d}z$. Then, after dividing both sides of it by dz, the following relationship is obtained:

$$\frac{3}{2}\rho c_p\frac{\mathrm{d}T^*}{\mathrm{d}z}2W\int_0^{H/2}V\left[1-\left(\frac{2y}{H}\right)^2\right]\mathrm{d}y=144\frac{\mu V^2}{H^4}2W\int_0^{H/2}y^2\mathrm{d}y \tag{4.61}$$

After integration, we have

$$WH\rho c_p\frac{\mathrm{d}T^*}{\mathrm{d}z}=12\frac{W}{H}\mu V^2 \tag{4.62}$$

and, after transformation, we obtain

$$\frac{\mathrm{d}T^*}{\mathrm{d}z}=\frac{12\mu V}{\rho c_p H^2} \tag{4.63}$$

Next, after integrating Eq. (4.63), we have

$$\int_{T_0}^{T^*}\mathrm{d}T^*=\int_0^z\frac{12\mu V}{\rho c_p H^2}\mathrm{d}z \tag{4.64}$$

And finally the formula is obtained that determines the **average temperature:**

$$T^*=T_0+\frac{12\mu V}{\rho c_p H^2}z \tag{4.65}$$

This equation can be transformed using Eqs. (4.18) and (4.47) to the form

$$\Delta T^*=T^*-T_0=\frac{\Delta p}{\rho c_p}\frac{z}{L} \tag{4.66}$$

It follows from Eq. (4.66) that under adiabatic flow conditions in the channel of length L, the total temperature increase is equal to

$$\Delta T^*(z=L)=\frac{\Delta p}{\rho c_p} \tag{4.67}$$

This equation is very useful in solving practical problems of polymer processing, e.g. when calculating the temperature increase in an extrusion die.

4.2.1.2 Flow through a Circular Tube

Isothermal Newtonian Flow

The geometry of flow in a tube of radius R and length L is shown in the cylindrical system $r\varphi z$ in Figure 4.5.

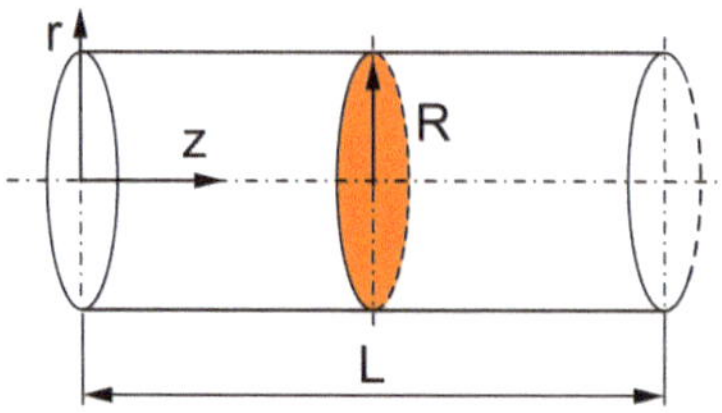

Figure 4.5
Flow geometry in a tube

It is here assumed that the velocity field has the form of Eq. (2.7).

Under these conditions, the equation of motion (Table 1.3) is expressed as

$$0 = -\frac{\partial p}{\partial z} + \frac{1}{r}\frac{\partial}{\partial r}\left(r\tau_{rz}\right) \tag{4.68}$$

and, after replacing the partial derivatives with the ordinary derivatives and appropriate transformations (see Section 2.4.2), as

$$\frac{\mathrm{d}}{\mathrm{d}r}\left(r\tau_{rz}\right) = -\frac{\Delta p}{L} r \tag{4.69}$$

where $\Delta p = p_0 - p_L$ is the pressure drop along the tube; p_0 is the pressure at the beginning of the tube; p_L is the pressure at the end of the tube; L is the tube length.

After integrating Eq. (4.69) we have

$$\tau_{rz} = -\frac{\Delta p}{L}\frac{r}{2} + \frac{C_1}{r} \tag{4.70}$$

In the case of the Newtonian fluid, under assumed flow conditions, the stress component τ_{rz} is equal (Table 1.7) to

$$\tau_{rz} = \mu\frac{\mathrm{d}v_z}{\mathrm{d}r} \tag{4.71}$$

After substituting Eq. (4.71) into Eq. (4.70) we have

$$\mu\frac{\mathrm{d}v_z}{\mathrm{d}r} = -\frac{\Delta p}{L}\frac{r}{2} + \frac{C_1}{r} \tag{4.72}$$

where the constant $C_1 = 0$ since the shear rate in the channel axis $\mathrm{d}v_z/\mathrm{d}r(r = 0)$ cannot be infinite.

Thus, Eq. (4.72) can be written as

$$\frac{dv_z}{dr} = -\frac{1}{\mu}\frac{\Delta p}{L}\frac{r}{2} \tag{4.73}$$

and this expresses the **shear rate.**

After integrating Eq. (4.73) we have

$$v_z = -\frac{1}{\mu}\frac{\Delta p}{L}\frac{r^2}{4} + C_2 \tag{4.74}$$

where the constant C_2 can be determined from the no slip condition at the tube wall, i. e. for $v_z(r = R) = 0$. Then, it is equal to

$$C_2 = \frac{1}{\mu}\frac{\Delta p}{L}\frac{R^2}{4} \tag{4.75}$$

As a result, we have a parabolic **velocity distribution** in the form

$$v_z = \frac{R^2}{4\mu}\frac{\Delta p}{L}\left[1 - \left(\frac{r}{R}\right)^2\right] \tag{4.76}$$

where the maximum velocity $v_{\max}$ occurs for $r = 0$, i. e. in the tube channel axis, and expresses as

$$v_{max} = \frac{R^2}{4\mu}\frac{\Delta p}{L} \tag{4.77}$$

The flow rate can be determined on the basis of Eq. (2.23).

After substituting here Eq. (4.76) and integrating it (see Section 2.4.2), we have the formula determining the **volume flow rate:**

$$Q = \frac{\pi R^4}{8\mu}\frac{\Delta p}{L} \tag{4.78}$$

This formula can be transformed into a form that allows calculation the **pressure drop** in the tube based on the known flow rate, tube dimensions, and viscosity of the material, i. e.

$$\Delta p = \frac{8L}{\pi R^4}\mu Q \tag{4.79}$$

Using Eq. (4.79), the shear rate Eq. (4.73) can be expressed as a function of the flow rate:

$$\frac{dv_z}{dr} = -\frac{4Q}{\pi R^4}r \tag{4.80}$$

and then, the shear rate at the tube wall can be determined, i.e. for $r = R$, according to the formula

$$\dot{\gamma}_w = \left(-\frac{\mathrm{d}v_z}{\mathrm{d}r}\right)_w = \frac{4Q}{\pi R^3} \tag{4.81}$$

The stress distribution in this flow can be determined on the basis of Eq. (4.70), by applying the condition $C_1 = 0$ (see Section 2.4.2), as

$$\tau_{rz} = -\frac{\Delta p}{L}\frac{r}{2} \tag{4.82}$$

According to this equation, the shear stress τ_w at the tube wall is equal to

$$\tau_w = \tau_{rz}(r = R) = -\frac{\Delta pR}{2L} \tag{4.83}$$

so that the **stress distribution** can be expressed as

$$\frac{\tau_{rz}}{\tau_w} = \frac{r}{R} \tag{4.84}$$

Eq. (4.84) shows that the stress distribution in the flow is linear and independent on the constitutive equation of the material.

Distributions of shear stress (Eq. (4.84)), velocity (Eq. (4.76)), as well as shear rate (Eq. (4.80)) are depicted in Figure 2.12.

Isothermal Non-Newtonian Flow

A non-Newtonian flow will be considered on the example of the power-law model of Ostwald-de Waele that in the general case is expressed by Eq. (4.25).

The stress component τ_{rz}, based on Eq. (4.25), has the following form

$$\tau_{rz} = m\dot{\gamma}_D^{n-1} 2D_{rz} \tag{4.85}$$

At assumed flow conditions, based on Eq. (1.157) and Table 1.10, the generalized shear rate reduces to

$$\dot{\gamma}_D = \frac{\mathrm{d}v_z}{\mathrm{d}r} \tag{4.86}$$

and the component D_{rz} of the strain rate tensor is equal (Table 1.1) to

$$D_{rz} = \frac{1}{2}\frac{\mathrm{d}v_z}{\mathrm{d}r} \tag{4.87}$$

Then, an expression is obtained that determines the stress component τ_{rz} as

$$\tau_{rz} = m\left|\frac{\mathrm{d}v_z}{\mathrm{d}r}\right|^{n-1}\frac{\mathrm{d}v_z}{\mathrm{d}r} \tag{4.88}$$

The absolute value sign is necessary in this case since dv_z/dr can be negative, and for polymers n is usually less than one, i. e. $n - 1 < 0$.

After substituting Eq. (4.88) into Eq. (4.70) we have

$$m\left|\frac{dv_z}{dr}\right|^{n-1}\frac{dv_z}{dr}=-\frac{\Delta p}{L}\frac{r}{2}+\frac{C_1}{r} \tag{4.89}$$

where the constant $C_1 = 0$ since the shear rate at the tube axis $dv_z/dr(r = 0)$ cannot be infinite.

Thus, Eq. (4.89) can be written as

$$m\left|\frac{dv_z}{dr}\right|^{n-1}\frac{dv_z}{dr}=-\frac{\Delta p}{L}\frac{r}{2} \tag{4.90}$$

Since the right side of this equation in negative, the derivative dv_z/dr also must be negative, so that after transformation the formula determing the **shear rate** takes the form

$$\frac{dv_z}{dr}=-\left(\frac{\Delta p}{2mL}\right)^{1/n}r^{1/n} \tag{4.91}$$

After integration we have

$$v_z=-\frac{1}{(1+1/n)}\left(\frac{\Delta p}{2mL}\right)^{1/n}r^{1+(1/n)}+C_2 \tag{4.92}$$

where the constant C_2 can be determined from the no slip condition at the tube wall, i. e. for $v_z(r = 0) = 0$. And then it is equal to

$$C_2=\frac{1}{(1+1/n)}\left(\frac{\Delta p}{2mL}\right)^{1/n}R^{1+(1/n)} \tag{4.93}$$

As a result, we have the **velocity distribution** in the form

$$v_z=-\frac{R}{(1+1/n)}\left(\frac{R\Delta p}{2mL}\right)^{1/n}\left[1-\left(\frac{r}{R}\right)^{1+(1/n)}\right] \tag{4.94}$$

The maximum velocity v_{max} is for $r = 0$, i. e. in the tube axis, and expresses as

$$v_{max}=\frac{R}{(1+1/n)}\left(\frac{R\Delta p}{2mL}\right)^{1/n} \tag{4.95}$$

The flow rate can be determined by substituting Eq. (4.94) into Eq. (2.23). Then, we have

$$Q=2\pi\int_0^R\frac{R}{(1+1/n)}\left(\frac{R\Delta p}{2mL}\right)^{1/n}\left[1-\left(\frac{r}{R}\right)^{1+(1/n)}\right]rdr \tag{4.96}$$

and, after integration, the formula determining the **volume flow rate** is obtained:

$$Q = \frac{\pi R^3}{(3+1/n)} \left(\frac{R \Delta p}{2mL} \right)^{1/n} \tag{4.97}$$

Equation (4.97) can be transformed into a form that allows calculation of the **pressure drop** in the tube based on the known flow rate, tube dimensions, and viscosity of the material, i. e.

$$\Delta p = \frac{2mL}{R} \left[\frac{(3+1/n)Q}{\pi R^3} \right]^n \tag{4.98}$$

Using Eq. (4.98), the shear rate (Eq. (4.91)) can be expressed as a function of the flow rate as

$$\frac{dv_z}{dr} = -\frac{(3+1/n)Q}{\pi R^3} \left(\frac{r}{R} \right)^{1/n} \tag{4.99}$$

and then, it can be determined at the tube wall, i. e. for $r = R$, as

$$\dot{\gamma}_w = \left(-\frac{dv_z}{dr} \right)_w = \frac{(3+1/n)Q}{\pi R^3} \tag{4.100}$$

Distributions of velocity (Eq. (4.94)) and shear rate (Eq. (4.99)) are similar to the flow between parallel plates (Figure 4.2).

It is worth noting that for $n = 1$ and $m = \mu$ these formulas transform into the solutions of the Newtonian flow.

Non-Isothermal Newtonian Flow

In order to consider the non-isothermal flow in a tube, the case will be considered in which the material with an initial temperature T_0 flows into a tube whose walls have a constant temperature equal to T_0.

The velocity field is defined by the conditions in Eq. (2.6), and the temperature field is

$$T = T(r,z) \tag{4.101}$$

Then, the energy equation has the form (Table 1.4):

$$\rho c_p v_z \frac{\partial T}{\partial z} = k \left[\frac{1}{r} \frac{\partial}{\partial r} \left(r \frac{\partial T}{\partial r} \right) + \frac{\partial^2 T}{\partial z^2} \right] + \tau_{rz} \frac{\partial v_z}{\partial r} \tag{4.102}$$

The first part of this equation $\rho c_p v_z \frac{\partial T}{\partial z}$ represents heat convection, the second $k \left[\frac{1}{r} \frac{\partial}{\partial r} \left(r \frac{\partial T}{\partial r} \right) + \frac{\partial^2 T}{\partial z^2} \right]$ determines heat conduction, and the third one $\tau_{rz} \frac{\partial v_z}{\partial r}$ describes energy dissipation.

Since the heat transfer in the z-axis direction is negligible in relation to the convective part of Eq. (4.102), this equation can be written as

$$\rho c_p v_z \frac{\partial T}{\partial z} = k\left[\frac{1}{r}\frac{\partial}{\partial r}\left(r\frac{\partial T}{\partial r}\right)\right] + \tau_{rz}\frac{\partial v_z}{\partial r} \tag{4.103}$$

In the case of Newtonian fluid, under assumed flow conditions, the component τ_{rz} of stress tensor is determined by Eq. (4.71), so that, after substituting it into Eq. (4.103), we have

$$\rho c_p v_z \frac{\partial T}{\partial z} = k\left[\frac{1}{r}\frac{\partial}{\partial r}\left(r\frac{\partial T}{\partial r}\right)\right] + \mu\left(\frac{\partial v_z}{\partial r}\right)^2 \tag{4.104}$$

Shear rate $\partial v_z/\partial r$ can be determined using Eq. (4.80) and after entering the average flow rate V in the form

$$V = \frac{Q}{\pi R^2} \tag{4.105}$$

it can be expressed as

$$\frac{\mathrm{d}v_z}{\mathrm{d}r} = -\frac{4V}{R^2}r \tag{4.106}$$

Then, Eq. (4.104) takes the form

$$\rho c_p v_z \frac{\partial T}{\partial z} = k\left[\frac{1}{r}\frac{\partial}{\partial r}\left(r\frac{\partial T}{\partial r}\right)\right] + 16\frac{\mu V^2}{R^4}r^2 \tag{4.107}$$

This equation can be considered in two limiting cases, i. e. in the equilibrium conditions and in the adiabatic flow conditions.

In the equilibrium conditions, the convective part of Eq. (4.107) is equal to zero, and after transformation, this equation takes the form

$$\frac{\mathrm{d}}{\mathrm{d}r}\left(r\frac{\mathrm{d}T}{\mathrm{d}r}\right) = -16\frac{\mu V^2}{kR^4}r^3 \tag{4.108}$$

After integration we have

$$\frac{\mathrm{d}T}{\mathrm{d}r} = -16\frac{\mu V^2}{kR^4}\frac{r^3}{4} + \frac{C_1}{r} \tag{4.109}$$

where the constant $C_1 = 0$ since the temperature gradient in the tube axis $\mathrm{d}T/\mathrm{d}r(r = 0)$ cannot be infinite.

Thus, Eq. (4.109) can be written as

$$\frac{\mathrm{d}T}{\mathrm{d}r} = -4\frac{\mu V^2}{kR^4}r^3 \tag{4.110}$$

After integration we have

$$T=-\frac{\mu V^2}{kR^4}r^4+C_2 \tag{4.111}$$

where the constant C_2 can be determined on the basis of the wall temperature condition, i.e. the assumption that $T(r=R)=T_0$. Then this constant is equal to

$$C_2=T_0+\frac{\mu V^2}{k} \tag{4.112}$$

The result is a parabolic **temperature distribution:**

$$T=T_0+\frac{\mu V^2}{k}\left[1-\left(\frac{r}{R}\right)^4\right] \tag{4.113}$$

where the temperature T_{max} is for $r=0$, i.e. in the tube axis, and expresses as

$$T_{max}=T_0+\frac{\mu V^2}{k} \tag{4.114}$$

The difference between the maximum temperature T_{max} and the temperature of the channel wall T_0 is defined as the temperature increase due to the viscous energy dissipation:

$$\Delta T_{max}=T_{max}-T_0=\frac{\mu V^2}{k} \tag{4.115}$$

Under adiabatic conditions, the part representing the heat transfer in the energy equation (Eq. (4.107)) is negligibly small, so that the equation takes the form

$$\rho c_p v_z\frac{\mathrm{d}T}{\mathrm{d}z}=16\frac{\mu V^2}{R^4}r^2 \tag{4.116}$$

The velocity v_z is determined by Eq. (4.76), which, after applying Eq. (4.79) and entering the average velocity V (Eq. (4.105)), can be presented as

$$v_z=2V\left[1-\left(\frac{r}{R}\right)^2\right] \tag{4.117}$$

and, after substituting Eq. (4.117) into the energy equation (Eq. (4.116)), we have

$$2\rho c_p V\left[1-\left(\frac{r}{R}\right)^2\right]\frac{\mathrm{d}T}{\mathrm{d}z}=16\frac{\mu V^2}{R^4}r^2 \tag{4.118}$$

This equation can be solved by seeking the average temperature T^*. To do this, we need to write this equation in a global form relative to the volume $\pi R^2\mathrm{d}z$. Then, after division the both sides of it by dz, the following relationship is obtained:

$$2\rho c_p \frac{\mathrm{d}T^*}{\mathrm{d}z} 2\pi \int_0^R V\left[1-\left(\frac{r}{R}\right)^2\right] r\mathrm{d}r = 16\frac{\mu V^2}{R^4} 2\pi \int_0^R r^3 \mathrm{d}r \tag{4.119}$$

After integration we have

$$\rho c_p \pi R^2 V \frac{\mathrm{d}T^*}{\mathrm{d}z} = 8\pi\mu V^2 \tag{4.120}$$

and after transformation

$$\frac{\mathrm{d}T^*}{\mathrm{d}z} = \frac{8\mu V}{\rho c_p R^2} \tag{4.121}$$

Next, after integrating Eq. (4.121) we have

$$\int_{T_0}^{T^*} \mathrm{d}T^* = \int_0^z \frac{8\mu V}{\rho c_p R^2} \mathrm{d}z \tag{4.122}$$

and finally the formula is obtained which determines the **average temperature:**

$$T^* = T_0 + \frac{8\mu V}{\rho c_p R^2} z \tag{4.123}$$

This equation can be transformed, using Eqs. (4.79) and (4.105), to the form

$$\Delta T^* = T^* - T_0 = \frac{\Delta p}{\rho c_p} \frac{z}{L} \tag{4.124}$$

It results from Eq. (4.124) that under adiabatic flow conditions in the channel of the length L, the total temperature increase is equal to

$$\Delta T^*(z = L) = \frac{\Delta p}{\rho c_p} \tag{4.125}$$

This equation is very useful in solving practical problems of polymer processing, e. g. when calculating the temperature increase in the extrusion die.

As you can see, the formulas here obtained are the same as Eqs. (4.66) and (4.67) obtained in the case of flow between parallel plates. Thus, in the adiabatic conditions of these flows, no knowledge of the channel geometry and the viscosity of the material is necessary to determine the temperature rise. It is enough to know the pressure drop along the length of the channel and the product of density and specific heat ρc_p.

4.2.1.3 Flow through a Tapered Channel

The schematic of flow in a tapered slit is depicted in the Cartesian system *xyz* in Figure 4.6.

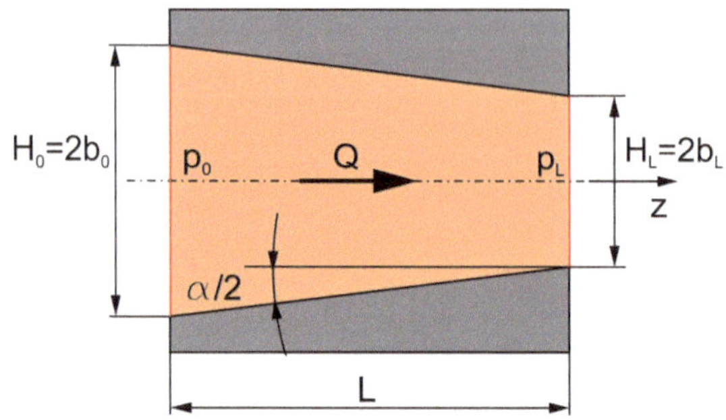

Figure 4.6 Schematic of flow in a tapered slit: Q - volume flow rate, p_0 - pressure at channel inlet, p_L - pressure at channel outlet

The flow will be considered as isothermal and Newtonian. Furthermore, assuming that the angle of convergence $2\alpha < 30°$, it is reasonable to consider this flow as being close to the flow between parallel plates [30].

Based on this assumption, Eq. (4.18) obtained for the flow between parallel plates (see Section 4.2.1.1 - Isothermal Newtonian Flow) is applied. For this case, it will be convenient to write it in the form

$$\Delta p = \frac{3L}{2Wb^3}\mu Q \tag{4.126}$$

where $b = H/2$.

On the basis of Eq. (4.4), Eq. (4.126) can be presented with reference to the infinitesimal length dz as

$$\mathrm{d}p = -\frac{3\mu Q}{2Wb^3}\mathrm{d}z \tag{4.127}$$

where dp is the pressure drop along the length dz.

Dimension b is a function of the location in the z-axis direction and expresses as

$$b = b_0 - \frac{b_0 - b_L}{L} z \tag{4.128}$$

where $b_0 = H_0/2$ and $b_L = H_L/2$ are the dimensions at the inlet and outlet of the channel.

On the basis of Eq. (4.128), the length dz can be determined as

$$\mathrm{d}z = -\frac{L}{b_0 - b_L}\mathrm{d}b \tag{4.129}$$

and Eq. (4.127) can be transformed to

$$\mathrm{d}p = \frac{3\mu Q}{2W}\frac{L}{b_0 - b_L} b^{-3}\mathrm{d}b \tag{4.130}$$

After integrating Eq. (4.130) as follows

$$\int_{p_0}^{p_L} \mathrm{d}p = \frac{3\mu Q}{2W}\frac{L}{b_0 - b_L}\int_{b_0}^{b_L} b^{-3}\mathrm{d}b \tag{4.131}$$

we get

$$p_L - p_0 = -\frac{3\mu Q}{2W}\frac{L}{b_0 - b_L}\frac{1}{2}\left(\frac{1}{b_L^2} - \frac{1}{b_0^2}\right) \tag{4.132}$$

and finally

$$\Delta p = p_0 - p_L = \frac{3\mu QL}{4W}\frac{b_0 + b_L}{b_0^2 b_L^2} \tag{4.133}$$

or

$$\Delta p = \frac{6L}{W}\frac{H_0 + H_L}{H_0^2 H_L^2}\mu Q \tag{4.134}$$

It is easy to check that for $b_0 = b_L = b$ or $H_0 = H_L = H$, i. e. for the parallel channel, this formula reduces to Eq. (4.18) or Eq. (4.126).

4.2.1.4 Flow through a Cone

The schematic of flow in a truncated cone in the cylindrical system $r\varphi z$ is depicted in Figure 4.7.

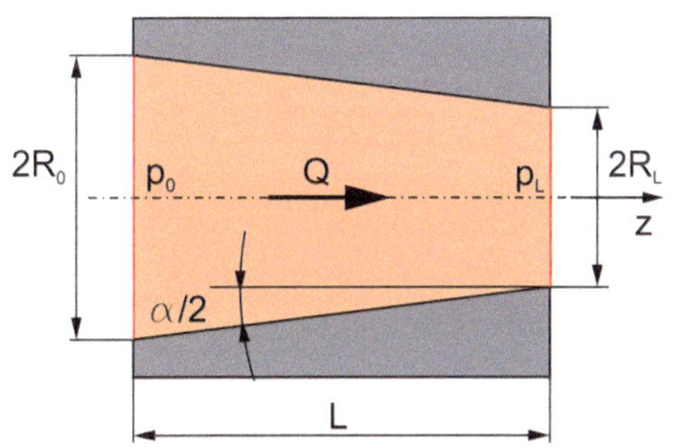

Figure 4.7 Schematic of flow in a truncated cone: Q - volume flow rate, p_0 - pressure at channel inlet, p_L - pressure at channel outlet

The flow will be considered as isothermal and Newtonian. Furthermore, assuming that the angle of convergence $2\alpha < 30°$, it is reasonable to consider this flow as being close to the flow in a tube [30].

Based on this assumption, Eq. (4.79) obtained for the flow in a tube (see Section 4.2.1.2 - Isothermal Newtonian Flow) will be applied.

On the basis of Eq. (4.4), Eq. (4.79) can be presented with reference to the infinitesimal length dz as

$$\mathrm{d}p = -\frac{8\mu Q}{\pi R^4}\mathrm{d}z \tag{4.135}$$

where dp is the pressure drop along the length dz.

Radius R is a function of the location in the z-axis direction and expresses as

$$R = R_0 - \frac{R_0 - R_L}{L}z \tag{4.136}$$

where R_0 and R_L are the dimensions at the inlet and outlet of the channel, respectively.

On the basis of this relationship, the length dz can be determined as

$$\mathrm{d}z = -\frac{L}{R_0 - R_L}\mathrm{d}R \tag{4.137}$$

and, Eq. (4.135) can be transformed to

$$\mathrm{d}p = \frac{8\mu Q}{\pi}\frac{L}{R_0 - R_L}R^{-4}\mathrm{d}R \tag{4.138}$$

After integrating the expression in Eq. (4.138)

$$\int_{p_0}^{p_L}\mathrm{d}p = \frac{8\mu Q}{\pi}\frac{L}{R_0 - R_L}\int_{R_0}^{R_L}R^{-4}\mathrm{d}R \tag{4.139}$$

we get

$$p_L - p_0 = -\frac{8\mu Q}{\pi}\frac{L}{R_0 - R_L}\frac{1}{3}\left(\frac{1}{R_L^3} - \frac{1}{R_0^3}\right) \tag{4.140}$$

and finally

$$\Delta p = p_0 - p_L = \frac{8\mu QL}{3\pi}\frac{R_0^2 + R_0 R_L + R_L^2}{R_0^3 R_L^3} \tag{4.141}$$

It is easy to check that for $R_0 = R_L = R$, i.e. for a tube, this formula reduces to Eq. (4.79).

4.2.1.5 Flow through an Annulus

The geometry of flow in an annulus of length L, outer radius R_z, and inner radius R_w, with $k = R_z/R_w$, is depicted in the cylindrical system $r\varphi z$ in Figure 4.8.

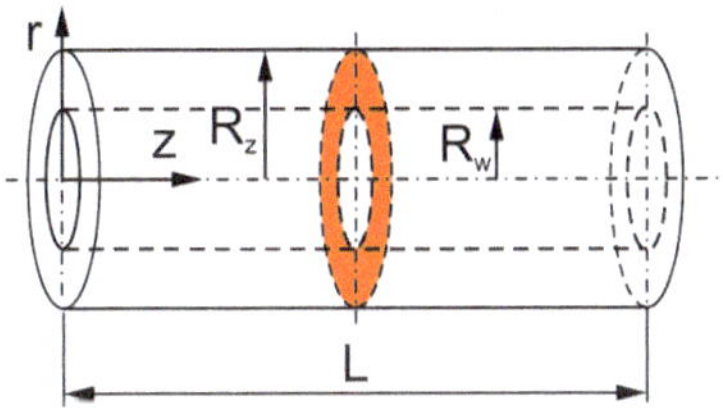

Figure 4.8
Flow geometry in an annulus

The flow will be considered as isothermal and Newtonian [1]. It is assumed that the velocity field has the form of Eq. (2.7).

Under such conditions, the equation of motion in the flow direction (Table 1.3), with other assumptions as in Section 4.2.1.2, is determined by Eq. (4.68). After replacement of the partial derivatives with ordinary derivatives and appropriate transformations (see Section 4.2.1.2 - Isothermal Newtonian Flow) it is expressed by Eq. (4.69).

After integrating Eq. (4.69) we have the relationship given in Eq. (4.70).

The constant C_1 cannot be determined directly in this equation because the conditions on the boundaries of this flow, i.e. for $r = kR$ and $r = R$, are unknown. It is only known that there will be a maximum in the velocity profile, which corresponds to a zero shear rate, and therefore also a zero stress value τ_{rz}. Since the location of this maximum is unknown, it can be assumed that it occurs on the radius $r = \lambda R$, where λ will be unknown here.

Thus, a boundary condition is obtained, according to which $\tau_{rz}(r = \lambda R) = 0$. Based on this condition, we can determine C_1 as

$$C_1 = \frac{\Delta p}{2L}(\lambda R)^2 \tag{4.142}$$

and next, we can write Eq. (4.70) in the form

$$\tau_{rz} = -\frac{\Delta p R}{2L}\left[\left(\frac{r}{R}\right) - \lambda^2\left(\frac{R}{r}\right)\right] \tag{4.143}$$

In the case of the Newtonian fluid, under assumed conditions, the stress component τ_{rz} (Table 1.7) is expressed by Eq. (4.71).

After substituting Eq. (4.71) into Eq. (4.143), we have the equation that expresses the shear rate as

$$\frac{\mathrm{d}v_z}{\mathrm{d}r} = -\frac{\Delta p R}{2\mu L}\left[\left(\frac{r}{R}\right) - \lambda^2\left(\frac{R}{r}\right)\right] \tag{4.144}$$

After integrating Eq. (4.144) we get

$$v_z = -\frac{\Delta p R^2}{2\mu L}\left[\left(\frac{r}{R}\right)^2 - 2\lambda^2 \ln\left(\frac{R}{r}\right) + C_2\right] \tag{4.145}$$

where the constant C_2 can be determined from the no slip condition at the channel walls, i.e. for $v_z(r = kR) = 0$ and $v_z(r = R) = 0$.

Using these boundary conditions yields two equations with two unknows λ and C_2:

$$0 = -\frac{\Delta p R^2}{4\mu L}\left(k^2 - 2\lambda^2 \ln k + C_2\right) \tag{4.146}$$

$$0 = -\frac{\Delta p R^2}{4\mu L}\left(1 + C_2\right) \tag{4.147}$$

The solution of this set of equations is the constant $C_2 = -1$, and the parameter λ that determines the location of the maximum velocity profile (also the zero shear stress) and is expressed by

$$\lambda^2 = \frac{1-k^2}{2\ln(1/k)} \tag{4.148}$$

Introduction the constants C_2 and λ^2 to Eqs. (4.143), (4.144), and (4.145) yields the following relationships:

$$\tau_{rz} = -\frac{\Delta p R}{2L}\left[\left(\frac{r}{R}\right) - \frac{1-k^2}{2\ln(1/k)}\left(\frac{R}{r}\right)\right] \tag{4.149}$$

$$\frac{dv_z}{dr} = -\frac{\Delta p R}{2\mu L}\left[\left(\frac{r}{R}\right) - \frac{1-k^2}{2\ln(1/k)}\left(\frac{R}{r}\right)\right] \tag{4.150}$$

$$v_z = \frac{\Delta p R^2}{4\mu L}\left[1 - \left(\frac{r}{R}\right)^2 + \frac{1-k^2}{\ln(1/k)}\ln\left(\frac{r}{R}\right)\right] \tag{4.151}$$

The maximum velocity $v_{\max}$ is for $r = \lambda R$, and is equal to

$$v_{max} = \frac{\Delta p R^2}{4\mu L}\left[1 - \lambda^2 + \frac{1-k^2}{\ln(1/k)}\ln\lambda\right] \tag{4.152}$$

and, after substituting Eq. (4.148)

$$v_{max} = \frac{\Delta p R^2}{4\mu L}\left\{1 - \left(\frac{1-k^2}{2\ln(1/k)}\right)\left[1 - \ln\left(\frac{1-k^2}{2\ln(1/k)}\right)\right]\right\} \tag{4.153}$$

Shear stress on the inner wall of the channel, i.e. $\tau_{w1} = \tau_{rz}(r = kR)$, is not equal to the shear stress on its outer wall, i.e. $\tau_{w2} = \tau_{rz}(r = R)$. These stresses can be determined on the basis of Eq. (4.149), according to which the **shear stress on the inner wall** is expressed by

$$\tau_{w1} = \tau_{rz}\left(r = kR\right) = -\frac{\Delta p R}{2L}\left[k - \left(\frac{1-k^2}{2\ln(1/k)}\right)\left(\frac{1}{k}\right)\right] \tag{4.154}$$

and the **shear stress on the outer wall** is determined by

$$\tau_{w2} = \tau_{rz}\left(r = R\right) = -\frac{\Delta pR}{2L}\left[1 - \left(\frac{1-k^2}{2\ln\left(1/k\right)}\right)\right] \tag{4.155}$$

Shear rate on the inner wall of the channel, i.e. $\dot{\gamma}_{w1} = \dot{\gamma}_w\,(r = kR)$, also differs from the shear rate on its outer wall, i.e. $\dot{\gamma}_{w2} = \dot{\gamma}_w\,(r = R)$. These shear rates can be determined on the basis of Eq. (4.150), according to which the **shear rate on the inner wall** is equal to

$$\dot{\gamma}_{w1} = \dot{\gamma}_w\left(r = kR\right) = \left(-\frac{dv_z}{dr}\right)_{w1} = \frac{\Delta pR}{2\mu L}\left[k - \left(\frac{1-k^2}{2\ln\left(1/k\right)}\right)\left(\frac{1}{k}\right)\right] \tag{4.156}$$

while the **shear rate on the outer wall** expresses as

$$\dot{\gamma}_{w2} = \dot{\gamma}_w\left(r = R\right) = \left(-\frac{dv_z}{dr}\right)_{w2} = \frac{\Delta pR}{2\mu L}\left[1 - \left(\frac{1-k^2}{2\ln\left(1/k\right)}\right)\right] \tag{4.157}$$

Distributions of shear stress (Eq. (4.149)), velocity (Eq. (4.151)), and shear rate (Eq. (4.150)) are depicted in Figure 4.9.

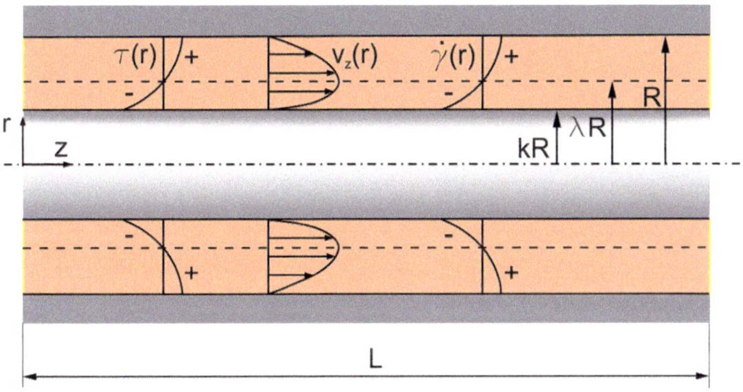

Figure 4.9 Distribution of the shear stress $\tau(r)$, velocity $v_z(r)$, and shear rate $\dot{\gamma}\,(r)$ in the Newtonian flow in an annulus

The flow rate can be determined on the basis of Eq. (2.23) where the lower limit of integration will be $r = kR$, i.e.

$$Q = 2\pi \int_{kR}^{R} v_z\left(r\right) r\,dr \tag{4.158}$$

After substituting here the expression in Eq. (4.151) and integrating it, we have the formula determining the **volume flow rate:**

$$Q = \frac{\pi R^4}{8\mu}\frac{\Delta p}{L}\left[1 - k^4 - \frac{\left(1-k^2\right)^2}{\ln\left(1/k\right)}\right] \tag{4.159}$$

It is worth noting that when $k \to 0$, i.e. when $R_w \to 0$, all the obtained formulas defining shear stress, shear rate, velocity, and volume flow rate are approaching the solutions for the flow through the cylindrical channel (tube). It can also be shown that when $k \to 1$, i.e. when $R_w \to R_z$, these formulas are approaching the solutions for the flow between parallel plates.

The fact that in the case of annular channels of small slit thicknesses the exact solutions are close to the solutions for the flow between parallel plates (and thus practically for the flows in flat channels of small gap thickness) enables the use of simpler solutions for practical calculations. In order to apply these solutions, the following substitutions must be made

$$W = \pi\left(R_w + R_z\right) \quad \text{and} \quad H = R_z - R_w \tag{4.160}$$

where W is the width of the flat channel; H is the height of the flat channel; R_w is the inner radius of the annulus; R_z is the outer radius of the annulus.

For example, the formulas for calculating the pressure drop and the shear rate at the channel wall, i.e. Eqs. (4.18) and (4.21) regarding the Newtonian flow as well as Eqs. (4.40) and (4.42) that relate to the non-Newtonian flow, are very useful.

After substituting Eq. (4.160) into these formulas, we get the following equations that allow the calculation of the flows in annular channels:

$$\Delta p = \frac{12L\mu Q}{\pi\left(R_w + R_z\right)\left(R_z - R_w\right)^3} \tag{4.161}$$

$$\dot{\gamma}_w = \frac{6Q}{\pi\left(R_w + R_z\right)\left(R_z - R_w\right)^2} \tag{4.162}$$

$$\Delta p = \frac{2mL}{R_z - R_w}\left[\frac{2\left(2 + 1/n\right)Q}{\pi\left(R_w + R_z\right)\left(R_z - R_w\right)^2}\right]^n \tag{4.163}$$

$$\dot{\gamma}_w = \frac{2\left(2 + 1/n\right)Q}{\pi\left(R_w + R_z\right)\left(R_z - R_w\right)^2} \tag{4.164}$$

4.2.1.6 Disk Flow

The schematic of disk flow in the cylindrical system $r\varphi z$ is shown in Figure 4.10. It occurs in a gap of small thickness $2H$ between two parallel disks of the inner radius R_0 and the outer radius R.

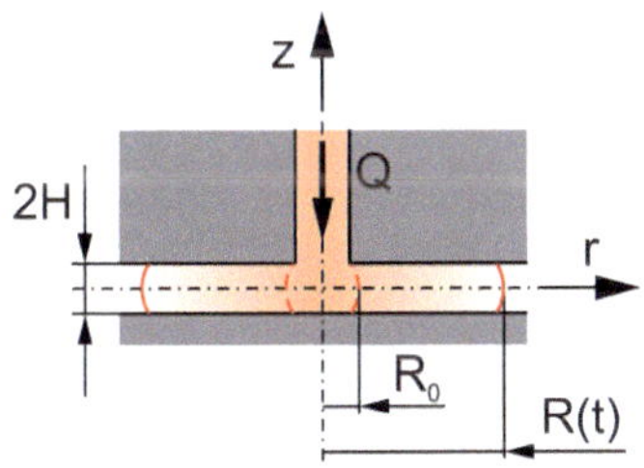

Figure 4.10 Schematic of flow in a disk: Q - volume flow rate, t - time, $2H$ - gap thickness, R_0 - radius of flow front at time $t = 0$, $R(t)$ - radius of flow front at time t

This flow will be considered as isothermal and Newtonian [35]. It is assumed that at time $t = 0$ the polymer begins to flow into the space between the disks at the constant flow rate Q.

It is also assumed that at certain distance from the inflow, the flow of the fluid is symmetrical and radial, so that the velocity field expresses as

$$v_r = v_r(r,z), \quad v_\varphi = 0, \quad v_z = 0 \tag{4.165}$$

Such a flow may represent the phase of filling the injection mold.

Under assumed conditions, the equation of motion in the flow direction (Table 1.3), with other assumptions as in Section 4.2.1.2, expresses as

$$0 = -\frac{\partial p}{\partial r} + \frac{1}{r}\frac{\partial}{\partial r}(r\tau_{rr}) + \frac{\partial \tau_{rz}}{\partial z} \tag{4.166}$$

and the equation of continuity (Table 1.2) takes the form

$$\frac{1}{r}\frac{\partial}{\partial r}(rv_r) = 0 \tag{4.167}$$

It results from the equation of continuity that the product (rv_r) does not depend on the radius r. This means that it is only a function of the coordinate z since, according to the flow field represented by Eq. (4.165), $v_r = v_r(r, z)$. Thus, the velocity component v_r can be represented as

$$v_r(r,z) = \frac{A(z)}{r} \tag{4.168}$$

where $A(z)$ is the unknown function of the variable z.

The strain rate tensor $\underset{\sim}{\boldsymbol{D}}$ is represented by the following components (Table 1.1):

$$\underset{\sim}{\boldsymbol{D}}:\begin{bmatrix} \frac{\partial v_r}{\partial r} & 0 & \frac{1}{2}\frac{\partial v_r}{\partial z} \\ 0 & \frac{v_r}{r} & 0 \\ \frac{1}{2}\frac{\partial v_r}{\partial z} & 0 & 0 \end{bmatrix} \tag{4.169}$$

and, after using Eq. (4.168), it expresses as

$$\underset{\sim}{\boldsymbol{D}}:\begin{bmatrix} -\frac{1}{r^2}A(z) & 0 & \frac{1}{2r}\frac{\mathrm{d}A(z)}{\mathrm{d}z} \\ 0 & \frac{1}{r^2}A(z) & 0 \\ \frac{1}{2r}\frac{\mathrm{d}A(z)}{\mathrm{d}z} & 0 & 0 \end{bmatrix} \tag{4.170}$$

In the case of the Newtonian model assumed here, the extra-stress tensor $\underset{\sim}{\boldsymbol{\tau}}$, determined by Eq. (1.133), is represented by the following components:

$$\underset{\sim}{\boldsymbol{\tau}}:\begin{bmatrix} -\frac{2\mu}{r^2}A(z) & 0 & \frac{\mu}{r}\frac{\mathrm{d}A(z)}{\mathrm{d}z} \\ 0 & \frac{2\mu}{r^2}A(z) & 0 \\ \frac{\mu}{r}\frac{\mathrm{d}A(z)}{\mathrm{d}z} & 0 & 0 \end{bmatrix} \tag{4.171}$$

Thus, the components τ_{rr} and τ_{rz} in Eq. (4.166) are expressed by

$$\tau_{rr} = -\frac{2\mu}{r^2}A(z) \tag{4.172}$$

$$\tau_{rz} = \frac{\mu}{r}\frac{\mathrm{d}A(z)}{\mathrm{d}z} \tag{4.173}$$

For viscous flows, the normal components of the stress tensor are negligibly small compared to the shear components, so that the component τ_{rr} (Eq. (4.172)) in the equation of motion (Eq. (4.166)) can be omitted and then it will take the form

$$0 = -\frac{\partial p}{\partial r} + \frac{\partial \tau_{rz}}{\partial z} \tag{4.174}$$

and after using Eq. (4.173)

$$0=-\frac{\partial p}{\partial r}+\frac{\mu}{r}\frac{\mathrm{d}^2 A(z)}{\mathrm{d}z^2} \tag{4.175}$$

Since the hydrostatic pressure p is only a function of r, i. e. $p = p(r)$, and A depends on z, that is, $A = A(z)$, Eq. (4.175) can be written as

$$\frac{1}{\mu}r\frac{\mathrm{d}p}{\mathrm{d}r}=\frac{\mathrm{d}^2 A(z)}{\mathrm{d}z^2}=C=\mathrm{const} \tag{4.176}$$

After integration the expression

$$\frac{\mathrm{d}^2 A(z)}{\mathrm{d}z^2}=C \tag{4.177}$$

we have

$$\frac{\mathrm{d}A(z)}{\mathrm{d}z}=Cz+C_1 \tag{4.178}$$

where the constant C_1 can be determined on the basis of flow symmetry and using Eq. (4.168), from which it results that

$$A(z)=rv_r(r,z) \tag{4.179}$$

Then, the condition of symmetry leads to the conclusion that for $z = 0$ the derivative of velocity $\mathrm{d}v_r/\mathrm{d}z = 0$, and also $\mathrm{d}A(z)/\mathrm{d}z = 0$. Thus, the constant $C_1 = 0$ and Eq. (4.178) takes the form

$$\frac{\mathrm{d}A(z)}{\mathrm{d}z}=Cz \tag{4.180}$$

After next integration we have

$$A(z)=C\frac{z^2}{2}+C_2 \tag{4.181}$$

where the constant C_2 can be determined from the no slip condition on the channel wall, i. e. for $v_r(z = H) = 0$, and thus also for $A(z = H) = 0$. The constant C_2 is equal to

$$C_2=-C\frac{H^2}{2} \tag{4.182}$$

such that when substituting it to Eq. (4.181), we have

$$A(z)=C\frac{z^2-H^2}{2} \tag{4.183}$$

and finally

$$v_r(r,z) = C\frac{z^2 - H^2}{2r} \tag{4.184}$$

The constant C can be determined based on the known and constant value of the flow rate Q that can be calculated by considering the elementary radial flow dQ through a flat channel with a circumference of $2\pi r$ and a thickness of dz occurring with the local velocity $v_r(r, z)$.

This flow rate expresses as

$$Q = 4\pi r \int_0^H v_r(r,z)\,\mathrm{d}z \tag{4.185}$$

so that, after using here Eq. (4.184) and integration, we get

$$Q = -\frac{4\pi CH^3}{3} \tag{4.186}$$

and, next, the constant C as

$$C = -\frac{3Q}{4\pi H^3} \tag{4.187}$$

As a result, the **velocity distribution** has the following form:

$$v_r(r,z) = \frac{1}{r}\frac{3Q}{8\pi H}\left[1 - \left(\frac{z}{H}\right)^2\right] \tag{4.188}$$

The pressure distribution can be determined using Eq. (4.176), according to which

$$\frac{1}{\mu} r \frac{\mathrm{d}p}{\mathrm{d}r} = C \tag{4.189}$$

After integrating Eq. (4.189) we get

$$p = \mu C \ln(r) + C_1 \tag{4.190}$$

where the constant C_1 can be determined from the zero pressure condition on the front flow, i. e. $p(r = R_f) = 0$, where R_f is the location of the front flow at the current time t. The constant C_1 is equal to

$$C_1 = -\mu C \ln R_f \tag{4.191}$$

As a result, using Eq. (4.187), we get the **pressure distribution** in the form

$$p(r) = \frac{3\mu Q}{4\pi H^3}\ln\left(\frac{R_f}{r}\right) \tag{4.192}$$

where, for $r = R_0$, it is possible to calculate the pressure at the place of polymer feed (e. g. injection pressure) as

$$p(R_0) = \frac{3\mu Q}{4\pi H^3} \ln\left(\frac{R_f}{R_0}\right) \tag{4.193}$$

At a constant flow rate, the position of the flow front R_f varies with time t according to

$$t = \frac{2\pi H\left(R_f^2 - R_0^2\right)}{Q} \tag{4.194}$$

The time to fill the disk-shaped mold with an outer radius R can be expressed as

$$t_w = \frac{2\pi H\left(R^2 - R_0^2\right)}{Q} \tag{4.195}$$

The basic formulas obtained in this chapter that may be useful in simple flow calculations for polymer processing, e. g. in extrusion dies and injection molds, are presented in Table 4.1.

Table 4.1 Calculation Formulas for Pressure Flows

Model	Formula	Equation
Flow between Parallel Plates, $H << W$		
Newtonian model	$\Delta p = \frac{12L}{WH^3}\mu Q$	(4.18)
	$\dot{\gamma}_w = \frac{6Q}{WH^2}$	(4.21)
Non-Newtonian model	$\Delta p = \frac{2mL}{H}\left[\frac{2(2+1/n)Q}{WH^2}\right]^n$	(4.40)
	$\dot{\gamma}_w = \frac{2(2+1/n)Q}{WH^2}$	(4.42)
Flow in a Cylindrical Tube		
Newtonian model	$\Delta p = \frac{8L}{\pi R^4}\mu Q$	(4.79)
	$\dot{\gamma} = \frac{4Q}{\pi R^3}$	(4.81)
non-Newtonian model	$\Delta p = \frac{2mL}{R}\left[\frac{(3+1/n)Q}{\pi R^3}\right]^n$	(4.98)
	$\dot{\gamma}_w = \frac{(3+1/n)Q}{\pi R^3}$	(4.100)

Table 4.1 Calculation Formulas for Pressure Flows *(continued)*

Model	Formula	Equation
Flow in an Annulus, $H << W$		
Newtonian model	Δp - Eq. (4.18) $\dot{\gamma}_w$ - Eq. (4.21)	$H = R_z - R_w$
Non-Newtonian model	Δp - Eq. (4.40) $\dot{\gamma}_w$ - Eq. (4.42)	$W = \pi(R_z + R_w)$
Flow in a Tapered Slit		
Newtonian model	$\Delta p = \frac{6L}{W}\frac{H_0 + H_L}{H_0^2 H_L^2}\mu Q$	(4.134)
Flow in a Truncated Cone		
Newtonian model	$\Delta p = \frac{8L}{3\pi}\frac{R_0^2 + R_0 R_L + R_L^2}{R_0^3 R_L^3}\mu Q$	(4.141)

4.2.1.7 Hele-Shaw Flow

Hele-Shaw flow is defined as the flow in a narrow gap between two nearly parallel plates as shown in Figure 4.11.

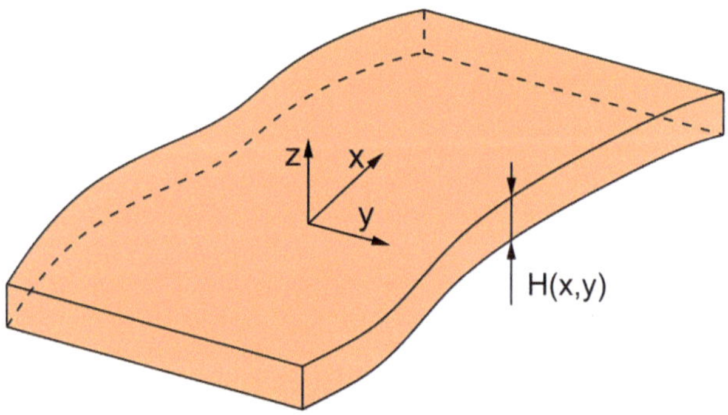

Figure 4.11 Hele-Shaw flow geometry: $H(x, y)$ - gap thickness

Various problems in fluid mechanics can be approximated to Hele-Shaw flows and thus the research of these flows is of importance. The essence of this approximation is that the flow is always locally like pressure driven flow between two parallel plates. It applies to geometries in which the gap thickness varies slowly with position.

It is assumed that the velocity field has the form

$$v_x = v_x(z), \quad v_y = v_y(z), \quad v_z = 0 \tag{4.196}$$

Under such conditions, the equations of motion in the flow directions, i.e. in the x-axis and the y-axis directions (Table 1.3), are expressed as

$$0 = -\frac{\partial p}{\partial x} + \frac{\partial \tau_{zx}}{\partial z} \tag{4.197}$$

$$0 = -\frac{\partial p}{\partial y} + \frac{\partial \tau_{zy}}{\partial z} \tag{4.198}$$

For the generalized Newtonian fluid we have

$$0 = -\frac{\partial p}{\partial x} + \frac{\partial}{\partial z}(\eta \frac{\partial v_x}{\partial z}) \tag{4.199}$$

$$0 = -\frac{\partial p}{\partial y} + \frac{\partial}{\partial z}(\eta \frac{\partial v_y}{\partial z}) \tag{4.200}$$

where v_x and v_y are velocity components in the x- and y-directions, respectively, $p(x, y)$ is the pressure, $\eta(\dot{\gamma}, T)$ is the shear viscosity, and $\dot{\gamma}$ is the shear rate. Under the present assumptions, the shear rate $\dot{\gamma}$ is given by

$$\dot{\gamma} = \left[(\frac{\partial v_x}{\partial z})^2 + (\frac{\partial v_y}{\partial z})^2 \right]^{\frac{1}{2}} \tag{4.201}$$

It is assumed that the velocity is symmetric in the z-direction and the boundary conditions for Eqs. (4.199) and (4.200) are given by

$$v_x = v_y = 0 \quad \text{at} \quad z = \frac{H}{2} \text{ (no slip)} \tag{4.202}$$

$$\frac{\partial v_x}{\partial z} = \frac{\partial v_y}{\partial z} = 0 \quad \text{at} \quad z = 0 \text{ (symmetry)} \tag{4.203}$$

Based on the lubrication approximation, the thickness-average continuity equation results in

$$\frac{\partial \left(H\overline{v}_x \right)}{\partial x} + \frac{\partial \left(H\overline{v}_y \right)}{\partial y} = 0 \tag{4.204}$$

where $\overline{v}_x$ and $\overline{v}_y$ are averaged velocities over z, defined by

$$\overline{v}_x = \frac{1}{\frac{H}{2}} \int_0^{\frac{H}{2}} v_x dz \tag{4.205}$$

$$\overline{v}_y = \frac{1}{\frac{H}{2}} \int_0^{\frac{H}{2}} v_y dz \tag{4.206}$$

Since the pressure is not a function of z, $\partial p/\partial x$ and $\partial p/\partial y$ in Eqs. (4.199) and (4.200) may be treated as parameters, allowing each of these equations to be integrated independently. By integrating Eqs. (4.199) and (4.200) twice with respect to z and using the boundary conditions in Eqs. (4.202) and (4.203), the velocities and shear rate can be obtained as

$$v_x = -\frac{\partial p}{\partial x}\int_z^{\frac{H}{2}} \frac{z'}{\eta} dz' \tag{4.207}$$

$$v_y = -\frac{\partial p}{\partial y}\int_z^{\frac{H}{2}} \frac{z'}{\eta} dz' \tag{4.208}$$

$$\dot{\gamma} = \frac{\left[(-\frac{\partial p}{\partial x})^2 + (-\frac{\partial p}{\partial y})^2\right]^{\frac{1}{2}} z}{\eta} \tag{4.209}$$

Substituting these into Eqs. (4.205) and (4.206) yields

$$\bar{v}_x = \frac{-\frac{\partial p}{\partial x}}{\frac{H}{2}}\int_0^{\frac{H}{2}}\int_z^{\frac{H}{2}} \frac{z'}{\eta} dz' dz \tag{4.210}$$

$$\bar{v}_y = \frac{-\frac{\partial p}{\partial y}}{\frac{H}{2}}\int_0^{\frac{H}{2}}\int_z^{\frac{H}{2}} \frac{z'}{\eta} dz' dz \tag{4.211}$$

Integrating Eqs. (4.210) and (4.211) by parts, we get the average velocities through the thickness:

$$\bar{v}_x = \frac{S}{\frac{H}{2}}\left(-\frac{\partial p}{\partial x}\right) \tag{4.212}$$

$$\bar{v}_y = \frac{S}{\frac{H}{2}}\left(-\frac{\partial p}{\partial y}\right) \tag{4.213}$$

where S is the flow conductance that is defined as

$$S = \int_0^{\frac{H}{2}} \frac{z^2}{\eta} dz \tag{4.214}$$

Substituting Eqs. (4.212) and (4.213) into the thickness-averaged continuity (Eq. (4.204)) gives the governing equation for the pressure distribution when H is known or for the thickness distribution when p is known. This is the well-known Reynolds equation:

$$\frac{\partial}{\partial x}\left[S\frac{\partial p}{\partial x}\right]+\frac{\partial}{\partial y}\left[S\frac{\partial p}{\partial y}\right]=0 \tag{4.215}$$

In particular, for the Newtonian case with uniform thickness throughout the domain, the above equations reduce to

$$S=\frac{H^3}{24\eta} \tag{4.216}$$

$$v_x=\frac{-\dfrac{\partial p}{\partial x}}{2\eta}\left(\left(\frac{H}{2}\right)^2-z^2\right) \tag{4.217}$$

$$v_y=\frac{-\dfrac{\partial p}{\partial y}}{2\eta}\left(\left(\frac{H}{2}\right)^2-z^2\right) \tag{4.218}$$

$$\overline{v}_x=\frac{H^2}{12\eta}\left(-\frac{\partial p}{\partial x}\right) \tag{4.219}$$

$$\overline{v}_y=\frac{H^2}{12\eta}\left(-\frac{\partial p}{\partial y}\right) \tag{4.220}$$

which are well-known results for the classical Hele-Shaw flow.

4.2.2 Drag Flows

The **drag flow** is the flow that occurs between two surfaces, one of which is mobile and the other immobile. These surfaces can be flat or curved. An important feature of this flow is that it is not necessary to have a pressure gradient in the flow direction for its implementation. Examples of drag flows are shown in Figure 2.1.

In the further part of this section, the basic drag flow is considered, i.e. the flow between two parallel plates.

To analyze this, the following basic assumptions are made:

- the flow is steady, laminar, and isothermal,
- the flow is fully developed,
- there is no slip on the channel wall,
- gravity and inertia forces are neglected,
- the fluid is incompressible and its viscosity does not depend on the pressure.

4.2.2.1 Isothermal Flow between Parallel Plates

The schematic of drag flow between two parallel plates, one of which is moving at the velocity V_z, is shown in the Cartesian system *xyz* in Figure 4.12.

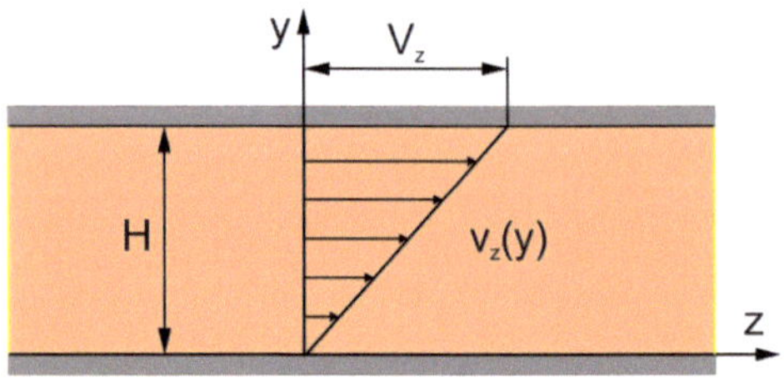

Figure 4.12 Schematic of drag flow in a flat channel (between parallel plates): $v_z(y)$ - velocity distribution, V_z - velocity of moving plate

It is assumed, that the velocity field has the form

$$v_x = 0, \quad v_y = 0, \quad v_z = v_z(y) \tag{4.221}$$

Under such conditions, the equation of motion in the flow direction, i.e. in the *z*-axis (Table 1.3), is expressed as follows:

$$0 = -\frac{\partial p}{\partial z} + \frac{\partial \tau_{yz}}{\partial y} \tag{4.222}$$

Since there is no pressure gradient in the drag flow, the equation simplifies to

$$0 = \frac{\mathrm{d}\tau_{yz}}{\mathrm{d}y} \tag{4.223}$$

After integrating Eq. (4.223) we get

$$\tau_{yz} = \mathrm{const} \tag{4.224}$$

Thus, the stress is constant here, and does not depend on the *y*-coordinate.

In the case of the Newtonian fluid, under assumed flow conditions, the stress component τ_{yz} (Table 1.7) is expressed by Eq. (4.7).

After substituting Eq. (4.7) into Eq. (4.224) we get

$$\mu \frac{\mathrm{d}v_z}{\mathrm{d}y} = \mathrm{const} \tag{4.225}$$

and this means that also the shear rate is constant in this flow:

$$\frac{\mathrm{d}v_z}{\mathrm{d}y} = \mathrm{const} \tag{4.226}$$

After integrating Eq. (4.226) we have

$$v_z = C_1 y + C_2 \tag{4.227}$$

where the constants C_1 and C_2 can be determined from the no slip condition on the walls of the channel, i. e. for $v_z(y = 0) = 0$ and $v_z(y = H) = V_z$.

According to these conditions, we have

$$C_2 = 0, \quad C_1 = \frac{V_z}{H} \tag{4.228}$$

and, as a result, we have a linear **velocity distribution** in the form

$$v_z = \frac{V_z}{H} y \tag{4.229}$$

On the basis of this equation the **shear rate** (Eq. (4.226)) can be expressed as

$$\frac{dv_z}{dy} = \frac{V_z}{H} = \text{const} \tag{4.230}$$

and the **shear stress** (4.224) is equal to

$$\tau_{yz} = \mu \frac{V_z}{H} = \text{const} \tag{4.231}$$

The **flow rate** can be determined by considering an elementary flow dQ through a flat channel of a width W and a thickness dy, occurring at the local velocity $v_z(y)$. Then, the flow rate with reference to the entire cross-section will be equal to

$$Q = W \int_0^H v_z(y) \, dy \tag{4.232}$$

and after substituting Eq. (4.229) we get

$$Q = WH \frac{V_z}{2} \tag{4.233}$$

When analyzing the flow of the non-Newtonian fluid of Ostwald-de Waele under assumed flow conditions, the stress component τ_{yz} (Table 1.7) for this model will be determined by Eq. (4.29). After substituting it into Eq. (4.224) we get

$$m \left| \frac{dv_z}{dy} \right|^{n-1} \frac{dv_z}{dy} = \text{const} \tag{4.234}$$

and this means that the **shear rate** is also constant in this case, i. e.

$$\frac{dv_z}{dy} = \text{const} \tag{4.235}$$

and the further solutions will be the same as for the Newtonian fluid.

The shear rate will be determined by Eq. (4.230). The **shear stress** (Eq. (4.29)), although also constant, will be determined according to a different formula, that is

$$\tau_{yz} = m\left(\frac{V_z}{H}\right)^n = \text{const} \tag{4.236}$$

Thus, the velocity distribution, the shear rate, and the flow rate in the drag flow between two parallel plates do not depend on the assumed rheological model of the fluid, and the shear rate is constant here. The shear stress is also constant, but it depends on the rheological properties of the fluid. For any implementation of the drag flow, under the same geometrical and kinematic conditions (at the same velocity V_z) for two different fluids (e.g. of different viscosity), it is necessary to apply a different force for the movement of one of the surfaces.

4.2.2.2 Non-Isothermal Flow between Parallel Plates

In order to consider the non-isothermal drag flow between two parallel plates, the case will be considered in which the material with an initial temperature T_0 flows into a channel whose walls have a constant temperature equal to T_0. It is also assumed that the viscosity of the polymer does not depend on the temperature.

The velocity field has the form of Eq. (4.221), and the temperature field is determined by

$$T = T(y, z) \tag{4.237}$$

Since these conditions are the same as for the pressure flow between parallel plates, the energy equation takes the form of Eq. (4.45).

In the case of the Newtonian fluid, under assumed flow conditions, the stress component τ_{yz} is determined by Eq. (4.7), so that after substituting it into Eq. (4.45) we get

$$\rho c_p v_z \frac{\partial T}{\partial z} = k\frac{\partial^2 T}{\partial y^2} + \mu\left(\frac{\partial v_z}{\partial y}\right)^2 \tag{4.238}$$

and after using Eq. (4.230)

$$\rho c_p v_z \frac{\partial T}{\partial z} = k\frac{\partial^2 T}{\partial y^2} + \mu\left(\frac{V_z}{H}\right)^2 \tag{4.239}$$

This equation can be considered in two limiting cases, i.e. in the equilibrium conditions and in the adiabatic flow conditions.

In the equilibrium conditions, the convective part of Eq. (4.239) is equal to zero and the equation, after transformation, takes the form

$$\frac{\mathrm{d}^2 T}{\mathrm{d}y^2} = -\frac{\mu V_z^2}{kH^2} \tag{4.240}$$

After double integration, we get

$$T = -\frac{\mu V_z^2}{kH^2}\frac{y^2}{2} + C_1 y + C_2 \tag{4.241}$$

where the constants C_1 and C_2 can be determined from the flow conditions on the channel walls, i.e. for $T(y=0) = T_0$ and $T(y=H) = T_0$.

According to these conditions we have

$$C_2 = T_0, \quad C_1 = \frac{\mu V_z^2}{2kH} \tag{4.242}$$

and, as a result, we have a parabolic **temperature distribution** in the form

$$T = T_0 + \frac{\mu}{2k}\left(\frac{V_z}{H}\right)^2 y\left(H-y\right) \tag{4.243}$$

where the maximum temperature $T_{\max}$ occurs for $y = H/2$, that is, in the plane of symmetry of the channel, and expresses as

$$T_{max} = T_0 + \frac{\mu V_z^2}{8k} \tag{4.244}$$

so that the maximum temperature increase is equal to

$$\Delta T_{max} = \frac{\mu V_z^2}{8k} \tag{4.245}$$

Under adiabatic conditions, the part representing the heat transfer in the energy equation (Eq. (4.239)) is negligibly small, so that the equation takes the form

$$\rho c_p v_z \frac{\mathrm{d}T}{\mathrm{d}z} = \mu\left(\frac{V_z}{H}\right)^2 \tag{4.246}$$

After substituting the velocity v_z determined by Eq. (4.229) into Eq. (4.246) we get

$$\rho c_p \left(\frac{V_z}{H}\right) y \frac{\mathrm{d}T}{\mathrm{d}z} = \mu\left(\frac{V_z}{H}\right)^2 \tag{4.247}$$

This equation can be solved by seeking the average temperature T^*. To do this, we need to write it in a global form relative to the volume $WH\mathrm{d}z$. Then, after division both sides of it by dz, the following relationship is obtained:

$$\rho c_p v_z \frac{\mathrm{d}T^*}{\mathrm{d}z} W \int_0^H \left(\frac{V_z}{H}\right) y \mathrm{d}y = W \int_0^H \mu \left(\frac{V_z}{H}\right)^2 \mathrm{d}y \tag{4.248}$$

After integration we get

$$\rho c_p WH \frac{V_z}{2}\frac{dT^*}{dz} = WH\mu\left(\frac{V_z}{H}\right)^2 \tag{4.249}$$

and after transformation

$$\frac{\mathrm{d}T^*}{\mathrm{d}z} = \frac{2\mu V_z}{\rho c_p H^2} \tag{4.250}$$

Next, the expression in Eq. (4.250) must be integrated with the condition $T(z = 0) = T_0$, and as a result a linear expression is obtained that determines the **average temperature**:

$$T^* = T_0 + \frac{2\mu V_z}{\rho c_p H^2} z \tag{4.251}$$

4.2.3 Pressure-Drag Flow between Parallel Plates

The pressure-drag flow is the flow between two surfaces, one of which is mobile and the other stationary. These surfaces can be flat or curved. In this flow, in a contrast to the drag flow, there is a pressure gradient.

The schematic of pressure-drag flow between two parallel plates is shown in Figure 4.13. This flow will be considered as isothermal and Newtonian, assuming the flow conditions as in Section 4.2.1.1 and Section 4.2.2.1.

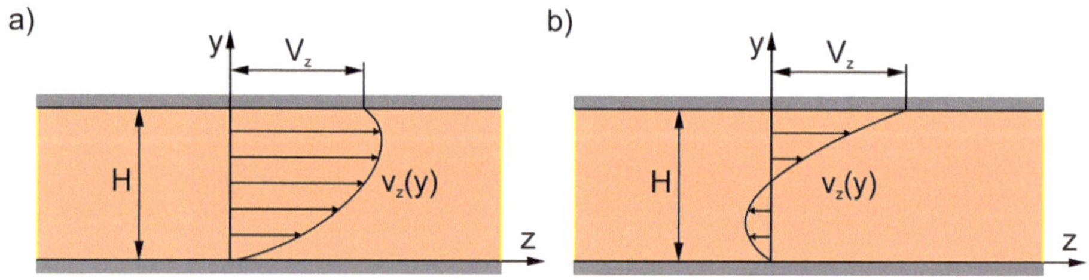

Figure 4.13 Schematic of pressure-drag flow in a flat channel (between two parallel plates): a) $p_0 > p_L (\Delta p > 0,\ dp < 0)$, b) $p_0 < p_L (\Delta p < 0,\ dp > 0)$; $v_z(y)$ - velocity distribution, V_z - velocity of a moving plate

It is assumed that the velocity field has the form of Eq. (4.221). Under these conditions, the equation of motion in the flow direction, i.e. in the z-axis (Table 1.3), expresses as

$$0 = -\frac{\partial p}{\partial z} + \frac{\partial \tau_{yz}}{\partial y} \tag{4.252}$$

Since the pressure is only the z-coordinate function, and the stress component τ_{yz} depends only on the y-coordinate, Eq. (4.252) can be written as

$$0 = -\frac{\mathrm{d}p}{\mathrm{d}z} + \frac{\mathrm{d}\tau_{yz}}{\mathrm{d}y} \tag{4.253}$$

The **pressure gradient** in this flow is constant and equals

$$-\frac{dp}{dz}=\frac{\Delta p}{L} \tag{4.254}$$

where $\Delta p = p_0 - p_L$ is the pressure drop along the channel (along z-axis); p_0 is the pressure at the beginning of the channel; p_L is the pressure at the end of the channel; L is the channel length.

This gradient can be positive or negative. When $p_0 > p_L$, the pressure gradient is negative ($dp/dz < 0$), which means that the pressure will drop ($\Delta p > 0$). However, when $p_0 < p_L$, the pressure gradient is positive ($dp/dz > 0$), which means that the pressure will increase ($\Delta p < 0$).

After substituting Eq. (4.254) into Eq. (4.253) we get

$$\frac{d\tau_{yz}}{dy}=-\frac{\Delta p}{L} \tag{4.255}$$

In the case of the Newtonian fluid, under the flow conditions assumed here, the stress component τ_{yz} (Table 1.7) is expressed by Eq. (4.7), so that the relationship in Eq. (4.255) can be written as

$$\frac{d^2 v_z}{dy^2}=-\frac{1}{\mu}\frac{\Delta p}{L} \tag{4.256}$$

After double integration of Eq. (4.256) we get

$$v_z=-\frac{1}{\mu}\frac{\Delta p}{L}\frac{y^2}{2}+C_1 y+C_2 \tag{4.257}$$

where the constants C_1 and C_2 can be determined from the no slip condition on the channel walls, i. e. for $v_z(y=0)=0$ and $v_z(y=H)=V_z$.

According to these conditions, it is obtained

$$C_2=0,\quad C_1=\frac{V_z}{H}+\frac{1}{\mu}\frac{\Delta p}{L}\frac{H}{2} \tag{4.258}$$

and, as a result, we have a **velocity distribution** in the form

$$v_z=\frac{V_z}{H}y+\frac{1}{2\mu}\frac{\Delta p}{L}y(H-y) \tag{4.259}$$

On the basis of this equation the **shear rate** may be expressed as

$$\frac{dv_z}{dy}=\frac{V_z}{H}+\frac{1}{2\mu}\frac{\Delta p}{L}(H-2y) \tag{4.260}$$

and then the **shear stress** (Eq. (4.7)) in the form

$$\tau_{yz} = \mu \frac{V_z}{H} + \frac{\Delta p}{2L}(H - 2y) \tag{4.261}$$

The **flow rate** can be determined on the basis of Eq. (4.232). After substituting Eq. (4.259) into this equation we get

$$Q = W \int_0^H \left[\frac{V_z}{H} y + \frac{1}{2\mu} \frac{\Delta p}{L} y(H - y) \right] \mathrm{d}y \tag{4.262}$$

and after integration

$$Q = WH \frac{V_z}{2} + \frac{WH^3}{12\mu} \frac{\Delta p}{L} \tag{4.263}$$

The first part of this expression $WH\frac{V_z}{2}$ represents the **drag flow,** and the second one $\frac{WH^3}{12\mu}\frac{\Delta p}{L}$ expresses the **pressure flow.** When $\Delta p > 0$, the pressure flow is added to the drag flow, whereas when $\Delta p < 0$, the pressure flow is subtracted from the drag flow.

It can be seen from Eq. (4.263) that the flow rate of the Newtonian pressure-drag flow is a simple **superposition** of the drag flow (Eq. (4.233)) and the pressure flow (Eq. (4.17)). These can be calculated independently of each other on the basis of the equations of motion with appropriate boundary conditions as in Section 4.2.1.1 and Section 4.2.2.1, and then they can be summed up. **Such a procedure is not acceptable in the case of the non-Newtonian flow,** when the solution of the equation of motion (Eq. (4.252)), at the boundary conditions of the pressure-drag flow, does not lead to two separate expressions representing the drag flow and the pressure flow. This solution is quite complicated [23] and will not be presented here.

■ 4.3 Numerical Methods

4.3.1 Introduction

In this chapter, several analytical solutions for some simple flow problems were presented. These simple solutions, formulas, and equations can be used to solve a lot of engineering issues. However, most of the polymer processing models are represented by sets of non-linear partial differential equations, which do not have analytical solutions. These problems can be solved using methods of numerical analysis.

Solving the equations numerically requires a discretization of the flow domain to convert the partial differential equations into a set of algebraic equations. These equations are then solved using the technigues of linear algebra, and provide the values of flow field parameters at discrete points within the domain.

Thus, the basic idea of numerical analysis is to **discretize the flow problem.** Instead of solving a continuous problem, the differential equations for a discrete problem are solved. This can be done in two basic ways. One way is selecting the finite number of points, and approximating the derivatives directly by finite differences. This is the basic idea of the **finite difference method** (FDM). The other way is selecting some functions, e.g. polynomials, and approximating the exact solution by combination of these functions over small interconnected elements. This is the basic idea of the **finite element method** (FEM).

In the finite difference method, a regular grid is necessary over the flow domain. In the finite element method, elements of different sizes can be applied, which means geometrical flexibility. Both methods can be used to solve the partial differential equations of fluid mechanics. The finite difference method is easier to program. The finite element method is more difficult to program but can much better handle irregular flow geometries and unusual boundary conditions. The question whether finite differences or finite elements should be used for a given problem is difficult to answer. Both methods have advantages and disadvantages, and the decision of using one of them may simply depend on the experience and background of the problem solver.

4.3.2 Finite Difference Method

4.3.2.1 Basic Formulations

The finite difference method is a very useful tool for the solution of partial differential equations. In this method the differential equations are approximated at a finite number of points (nodes) of the flow domain, i. e. the domain of integration.

The following steps of the computing procedure can be distinguished [34]:

- discretizing the domain of integration using a one-, two-, or three-dimensional grid depending on the flow problem,
- writing the appropriate finite difference approximations of the derivatives,
- approximating the partial differential equations with its finite difference approximation at each node of the discretized domain,
- solving the resulting difference equations to obtain approximate values of the unknown variables at each node.

Finite difference approximations of the derivatives can be obtained by using the expansion of a function in Taylor series.

We will consider the flow domain discretized using the two-dimensional grid (mesh) depicted in Figure 4.14.

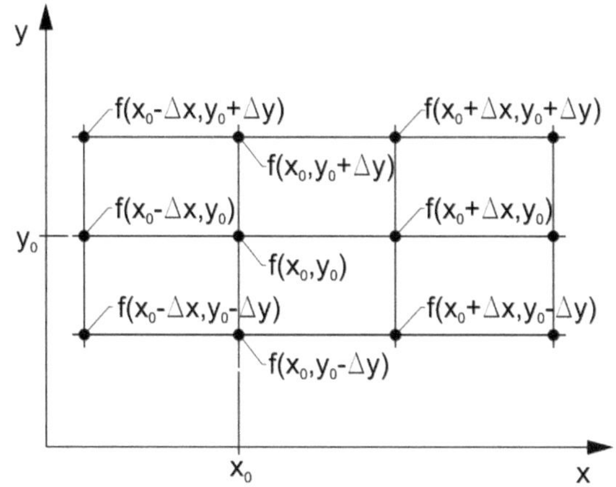

Figure 4.14 Finite difference mesh (grid): Δx, Δy – finite differences; x_0, y_0, $x_0 + \Delta x$, $y_0 + \Delta y$ – coordinates of nodal points; $f(x_0,y_0)$, $f(x_0 + \Delta x, y_0 + \Delta y)$ – functions $f(x, y)$ at nodal points

If $f(x, y)$ is a continuous differentiable function it may be written

$$f(x_0+\Delta x)=f(x_0)+\Delta x\frac{df(x_0)}{dx}+\frac{(\Delta x)^2}{2!}\frac{d^2f(x_0)}{dx}+\ldots \tag{4.264}$$

or

$$f(x_0-\Delta x)=f(x_0)-\Delta x\frac{df(x_0)}{dx}+\frac{(\Delta x)^2}{2!}\frac{d^2f(x_0)}{dx}-\ldots \tag{4.265}$$

Thus, the first order derivative can be approximated by a forward difference

$$\frac{df(x_0)}{dx}=\frac{f(x_0+\Delta x)-f(x_0)}{\Delta x}+0(\Delta x) \tag{4.266}$$

or a backward difference

$$\frac{df(x_0)}{dx}=\frac{f(x_0)-f(x_0-\Delta x)}{\Delta x}+0(\Delta x) \tag{4.267}$$

The error in both cases is of order (Δx).

Subtracting Eq. (4.265) from Eq. (4.264), a central difference approximation is obtained:

$$\frac{df(x_0)}{dx}=\frac{f(x_0+\Delta x)-f(x_0-\Delta x)}{2\Delta x}+0(\Delta x)^2 \tag{4.268}$$

The error is of order $(\Delta x)^2$.

Adding Eq. (4.265) and (4.264), an approximation of the second order derivative is obtained:

$$\frac{d^2 f(x_0)}{dx^2} = \frac{f(x_0+\Delta x) - 2f(x_0) + f(x_0 - \Delta x)}{(\Delta x)^2} + 0(\Delta x)^2 \tag{4.269}$$

All the above steps can be repeated for the y-variable. The expressions corresponding to Eqs. (4.266), (4.267), (4.268), and (4.269) are found to be as follows:

- first order, forward difference approximation for the first derivative:

$$\frac{df(y_0)}{dy} = \frac{f(y_0+\Delta y) - f(y_0)}{\Delta y} + 0(\Delta y) \tag{4.270}$$

- first order, backward difference approximation for the first derivative:

$$\frac{df(y_0)}{dy} = \frac{f(y_0) - f(y_0 - \Delta y)}{\Delta y} + 0(\Delta y) \tag{4.271}$$

- second order, central difference approximation for the first derivative:

$$\frac{df(y_0)}{dy} = \frac{f(y_0+\Delta y) - f(y_0 - \Delta y)}{2\Delta y} + 0(\Delta y)^2 \tag{4.272}$$

- second order, central difference approximation for the second derivative:

$$\frac{d^2 f(y_0)}{dy^2} = \frac{f(y_0+\Delta y) - 2f(y_0) + f(y_0 - \Delta y)}{(\Delta y)^2} + 0(\Delta y)^2 \tag{4.273}$$

This procedure can be used for the determination of finite difference approximations of derivatives of any order. These expressions are further used to approximate the partial differential equations at each node of the finite difference grid as explained in the example below.

4.3.2.2 Example of Computations

As an example of using the method of finite differences the pressure flow between two parallel plates will be considered [34], and the results will be compared to the analytical solution (see Section 4.2.1.1).

The equation for pressure driven flow of the Newtonian fluid between two parallel plates may be written as

$$\mu \frac{\partial^2 v_x}{\partial y^2} = \frac{dp}{dx} \tag{4.274}$$

with the boundary conditions of no slip on the channel wall, i. e. with $v_x(y = H/2) = 0$ and $v_x(y = -H/2) = 0$.

We simplify the solution procedure using a one-dimensional grid of only five points (Figure 4.15).

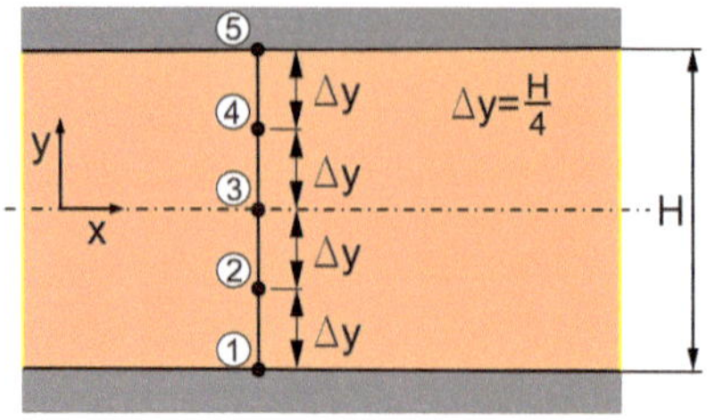

Figure 4.15
One-dimensional grid: Δy - finite difference, H - channel height

Defining $v_x = V$ and $\left(\frac{1}{\mu}\right)\frac{\mathrm{d}p}{\mathrm{d}x} = -A$ (minus because the pressure gradient is negative in the direction of flow), we have

$$A = -\frac{\mathrm{d}p}{\mathrm{d}x}\frac{1}{\mu} \tag{4.275}$$

and

$$\frac{\mathrm{d}^2 V}{\mathrm{d}y^2} = -A \tag{4.276}$$

The finite difference approximation using Eq (4.273) has the form

$$\frac{\mathrm{d}^2 V}{\mathrm{d}y^2} = \frac{V(y+\Delta y) - 2V(y) + V(y-\Delta y)}{(\Delta y)^2} \tag{4.277}$$

The differential equations can be approximated at the subsequent nodes as follows.

node 2:

$$\frac{V_3 - 2V_2 + V_1}{(\Delta y)^2} = -A \tag{4.278}$$

node 3:

$$\frac{V_4 - 2V_3 + V_2}{(\Delta y)^2} = -A \tag{4.279}$$

node 4:

$$\frac{V_5 - 2V_4 + V_3}{(\Delta y)^2} = -A \tag{4.280}$$

Noting that $V_1 = V_5 = 0$ and introducing a new variable V^* defined as

$$V^* = \frac{V}{A(\Delta y)^2} \tag{4.281}$$

we have

$$V_3^* - 2V_2^* = -1 \tag{4.282}$$

$$V_4^* - 2V_3^* + V_2^* = -1 \tag{4.283}$$

$$V_5^* - 2V_4^* = -1 \tag{4.284}$$

Solving these algebraic equations we get

$$V_2^* = 1.5$$

$$V_3^* = 2$$

$$V_4^* = 1.5 \tag{4.285}$$

Rearranging Eq. (4.281), we have

$$V = V^* A(\Delta y)^2 \tag{4.286}$$

and substituting A defined by Eq. (4.275), we get

$$V = -V^* \frac{\mathrm{d}p}{\mathrm{d}x}\frac{1}{\mu}(\Delta y)^2 \tag{4.287}$$

Thus, the values of velocity at the subsequent nodes are as follows:

node 2:

$$V_2 = -1.5\frac{\mathrm{d}p}{\mathrm{d}x}\frac{1}{\mu}\left(\frac{H}{4}\right)^2 \tag{4.288}$$

node 3:

$$V_2 = -2\frac{\mathrm{d}p}{\mathrm{d}x}\frac{1}{\mu}\left(\frac{H}{4}\right)^2 \tag{4.289}$$

node 4:

$$V_2 = -1.5\frac{\mathrm{d}p}{\mathrm{d}x}\frac{1}{\mu}\left(\frac{H}{4}\right)^2 \tag{4.290}$$

These results can be compared to Eqs.(4.12) and (4.13), representing a parabolic velocity distribution obtained from analytical calculations (Figure 4.16). Obviously, if more nodes were used the profile could be made virtually identical to an analytical parabola.

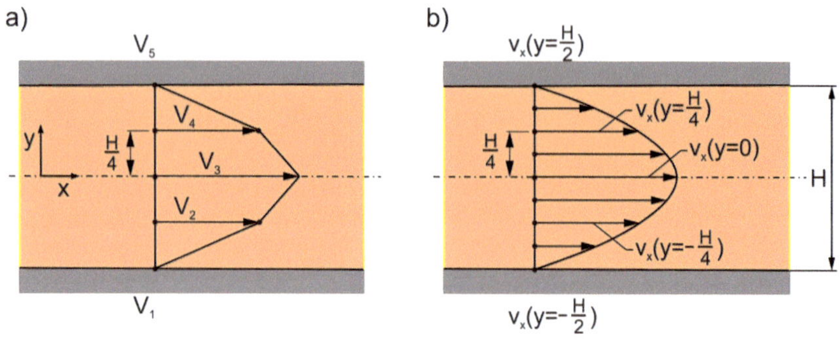

Figure 4.16 Pressure flow between parallel plates: a) finite difference computations, b) analytical calculations

References

1. Bird, R.B., Stewart, W.E., Lightfoot, E.N., *Transport Phenomena* (1960) John Wiley & Sons, New York
2. Wilkinson, W.L., *Non-Newtonian Fluids* (1960) Pergamon Press Ltd., London
3. McKelvey, J.M., *Polymer Processing* (1962) John Wiley & Sons, New York
4. Pearson, J.R. A., *Mechanical Principles of Polymer Melt Processing* (1966) Pergamon Press, Oxford
5. Schenkel, G., *Plastics Extrusion. Technology and Theory* (1966) Illife Books Ltd., London
6. Fenner, R.T., *Extruder Screw Design* (1966) Illife-Butterworth, London
7. Astarita, G., Marucci, G., *Principles of Non-Newtonian Fluid Mechanics* (1974) McGraw Hill, London
8. Torner, R.W., *Grundprocesse der Verarbeitung von Polymeren* (1974) VEB Deutscher Verlag, Leipzig
9. Middleman, S., *Fundamentals of Polymer Processing* (1977) McGraw-Hill, New York
10. Pearson, J.R. A., Richardson, S.M., *Computational Analysis of Polymer Processing* (1983) Elsevier Publisher, London
11. Bird, R.B., Armstrong, R.C., Hassager, O., *Dynamic of Polymeric Liquids* (1987) John Wiley & Sons, New York
12. Isayev, A.I., *Injection and Compression Molding Fundamentals* (1987) Marcel Dekker, New York
13. Tucker, C.L., *Fundamentals of Computer Modeling for Polymer Processing* (1987) Carl Hanser Verlag, Munich
14. Hensen, F., Knappe, W., Potente, H., *Handbuch der Kunststoff-Extrusionstechnik. Grundlagen* (1989) Carl Hanser Verlag, Munich
15. White J. L, *Principles of Polymer Engineering Rheology* (1990) John Wiley & Sons, New York
16. Collyer A.A., Utracki L.A., *Polymer Rheology and Processing* (1990) Elsevier, London
17. Isayev, A.I. (Ed.), *Modeling of Polymer Processing. Recent Developments* (1991) Carl Hanser Verlag, Munich
18. Baird, D.G., Collias, D. I ., *Polymer Processing. Principles and Design* (1995) Butterworth-Heinemann, Oxford
19. White, J.L., *Rubber Processing* (1995) Carl Hanser Verlag, Munich
20. Griskey, R.G., *Polymer Process Engineering* (1995) Chapman and Hall, New York
21. Kennedy P., *Flow Analysis of Injection Molds* (1995) Carl Hanser Verlag, Munich
22. White, J.L., Potente, H., *Screw Extrusion* (2002) Carl Hanser Verlag, Munich
23. Tadmor, Z., Gogos, C.G., *Principles of Polymer Processing* (2006) John Wiley & Sons, New Jersey

24. Osswald, T., Hernandez-Ortiz, J.P., *Polymer Processing. Modeling and Simulation* (2006) Hanser Publishers, Munich

25. Denn, M.M., *Polymer Melt Processing. Foundations in Fluid Mechanics and Heat Transfer* (2008) Cambridge University Press, Cambridge

26. Manas-Zloczower, I. (Ed.), *Mixing and Compounding of Polymers. Theory and Practice* (2009) Carl Hanser Verlag, Munich

27. Osswald, T., Turng, L.-S., Gramann, P., *Injection Molding Handbook* (2009) Carl Hanser Verlag, Munich

28. Kamal, M.R., Isayev, A.I., Liu, S.-J., *Injection Molding. Technology and Fundamentals* (2009) Carl Hanser Verlag, Munich

29. White, J.L., Kim E.K., *Twin Screw Extrusion* (2010) Carl Hanser Verlag, Munich

30. Vlachopoulos, J., *Polymer Rheology and Extrusion* (2011) McMaster University, Hamilton

31. Rao, S.N., Schott, N.R., *Plastics Engineering Calculations* (2012) Carl Hanser Verlag, Munich

32. Rauwendaal, Ch., *Polymer Extrusion* (2014) Hanser Publishers, Munich

33. Campbell, G.A., Spalding, M.A., *Analyzing and Troubleshooting Single-Screw Extruders* (2014) Carl Hanser Verlag, Munich

34. Vlachopoulos, J., *Fundamentals of Fluid Mechanics* (2016) Polydynamics, Hamilton

35. Agassant, J.-F., Avenas, P., Carreau, P.J., Vergnes, B., Vincent, M., *Polymer Processing: Principles and Modeling* (2017) Carl Hanser Verlag, Munich

36. Hopmann, Ch., Michaeli, W., *Extrusion Dies for Plastics and Rubber* (2017) Carl Hanser Verlag, Munich

37. Vlachopoulos, J., Polychronopoulos, N.D., *Understanding Rheology and Technology of Polymer Extrusion* (2019) Polydynamics Inc., Dundas

5 Modeling of Extrusion

5.1 Introduction

Extrusion is the most important technology in the polymer processing industry since more than half of all polymers are extruded. Moreover, extrusion is fundamental for compounding, i. e. mixing, pelletizing, filling, reinforcing, etc.

In general, extrusion can be divided into single screw extrusion and twin screw extrusion (see Section 3.1). Single screw extruders may have smooth or grooved barrels and can be equipped with conventional screws or non-conventional ones with elements intensifying melting or mixing action. Twin screw extruders can be co-rotating or counter-rotating, with varying degrees of screw meshing.

Extruders are usually fed with solid polymers. This means these are plasticating extruders. Extrusion can be performed with gravitational feeding or feeding with metering. Single screw extruders are usually fed by gravity (flood fed extruders), while twin screw extruders are fed by metering (starve fed extruders).

The first and fundamental models of polymer processing were developed for single screw extrusion, and these early concepts are still highly important in modeling of polymer processing. In this book, the discussion on modeling of polymer processing is limited to this fundamental technique.

We will start with a physical model of extrusion which is a basis for modeling. Then, we will discuss the basic assumptions of extrusion theory, and next the polymer flow in an extruder, i. e. solid conveying, melting, and metering, as well as the flow in extrusion dies. Finally, co-operation of the extruder/die system and global modeling will be discussed.

■ 5.2 Single Screw Extrusion

5.2.1 Physical Model of Extrusion

Extrusion is the process of continuous forming of polymer products. It consists in melting a polymer in the plasticating unit of the extruder and then pushing it, under the pressure developed in this unit, through the die forming the product. Melting occurs as a result of the material being heated up by the heating unit as well as by the energy dissipation inside the barrel.

Thus, in the extrusion process, the physical state of the material changes from the solid to the melt. According to these different physical states, different process zones are distinguished in the extruder (Figure 5.1).

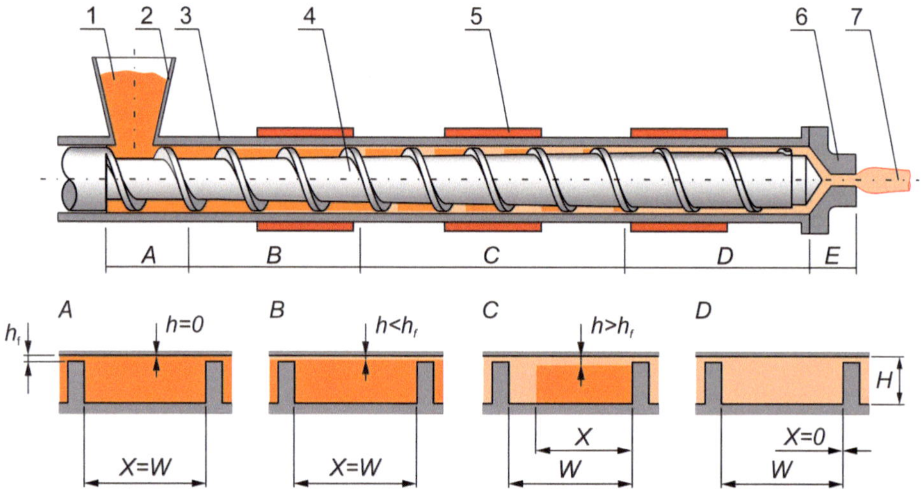

Figure 5.1 Extrusion process: 1 - polymer 2 - hopper, 3 - barrel, 4 - screw, 5 - heaters, 6 - die, 7 - extrudate; *A* - solid conveying zone, *B* - pre-melting zone (delay zone), *C* - melting zone, *D* - melt conveying zone in the barrel, *E* - melt flow zone in the die; *X* - width of the solid bed, *W* - width of the screw channel, h_f - clearance between the screw flights and the barrel, *h* - polymer melt thickness.

The solid polymer (1), in the form of granulate or powder, is fed into the extruder barrel through the hopper (2). Movement of the polymer in this zone usually takes place under the influence of gravity.

The first physical zone of the process (*A*) is in an initial part of the barrel (3), which is filled with the solid polymer, which is compacted, and takes the form of a uniform layer, called the **solid bed.** This is the **solid conveying** zone. The movement of the solid is caused by the relative motion of the screw (4) and the barrel, as well as by the friction between the polymer and the screw and barrel.

The second zone of the process (*B*), the so-called **pre-melting** zone (*delay* zone), starts when the polymer melt film forms on the surface of contact between the polymer and the heated wall of the barrel, the thickness of which increases as it moves towards the die (*6*). When this thickness is large enough (larger than the gap between the tips of the screw flight and the barrel), the material is scraped off by the flight of the screw. The molten polymer then accumulates in the rear, active part of the channel, while the front passive part is filled with the solid material. The place corresponding to this occurrence is the beginning of the **melting** zone (*C*). During the melting, due to high pressure in the melting region, the width of the solid bed decreases. In the final stage of melting, the solid disintegrates *(breaks up)* to small pieces, which are melted independently until the material is completely melted.

The melting mechanism presented here is valid for materials that show adhesion to the barrel wall, and therefore for most polymers. In the case, however, when there is a slip on the barrel wall, the melting course is different. Then, the polymer melt film formed at the barrel wall is not scraped off by the screw flight, but flows through the gap between the flight and the barrel, and accumulates in the passive part of the channel.

In the place where melting ends, the **melt conveying** zone begins (*D*). The polymer melt flow is the result of the relative motion of the screw and the barrel, and the pressure gradient in the extruder screw.

The last process zone is the **melt flow** (*E*) in the die, through which the polymer is pushed under the pressure developed in the barrel, and where the product is formed.

5.2.2 Basic Assumptions of Extrusion Theory

The geometrical characteristics of the classical screw, in relation to the extruder barrel, are shown in Figure 5.2 (the clearance between the screw flight and the barrel is omitted).

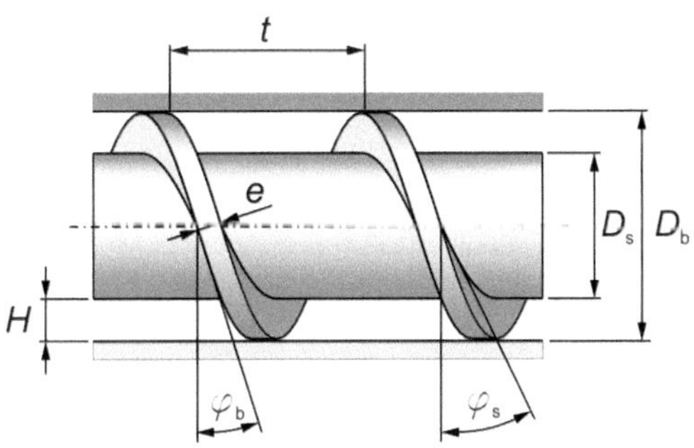

Figure 5.2
Screw geometry

The basic dimensions of this system are as follows:

- the internal barrel diameter D_b,
- the screw root diameter D_s,
- the screw channel depth H,
- the screw flight width e,
- the screw pitch (lead) t; this is usually equal to the diameter D_b, and then the screw flight angle (helix angle) is equal to $\varphi = 17°40'$,
- the screw flight angle (helix angle) φ; this is dependent on the radius and is largest at the screw root, and smallest at the tips of screw flight; it is expressed by the following formula:

$$\mathrm{tg}\varphi(r) = \frac{t}{2\pi r} \tag{5.1}$$

where $\varphi(r)$ is the screw flight angle (helix angle); r is the cylindrical coordinate; t is the pitch of the screw.

At the screw root surface, this angle equals

$$\tan\varphi_s = \frac{t}{\pi D_s} \tag{5.2}$$

and, at the barrel wall

$$\tan\varphi_b = \frac{t}{\pi D_b} \tag{5.3}$$

The other geometrical parameters of the screw can be determined based on the unwinding of the screw channel on the plane (Figure 5.3).

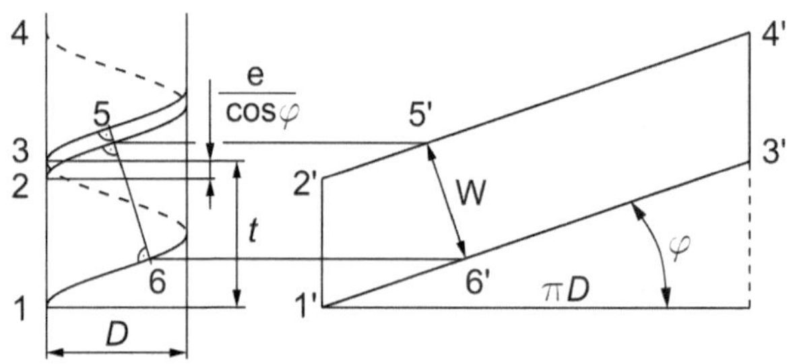

Figure 5.3 Unwinding of the screw channel

Then, the screw channel width expresses as

$$W(r) = t\cos\varphi(r) - e \tag{5.4}$$

where $W(r)$ varies in the radial direction, as $\cos\varphi(r)$.

The length of the screw channel corresponding to the pitch can be determined from

$$L = \frac{\pi D}{\cos \varphi} \tag{5.5}$$

In the basic analysis of the process, the gap between the screw flight and the barrel is omitted, but the existence of which is however included in selected issues.

The basic assumption of the classical extrusion theory concerning the geometry is an approximation of the screw channel with a rectangular channel (Figure 5.4). Such an approximation is justified when the ratio of the depth of the channel to its width is small.

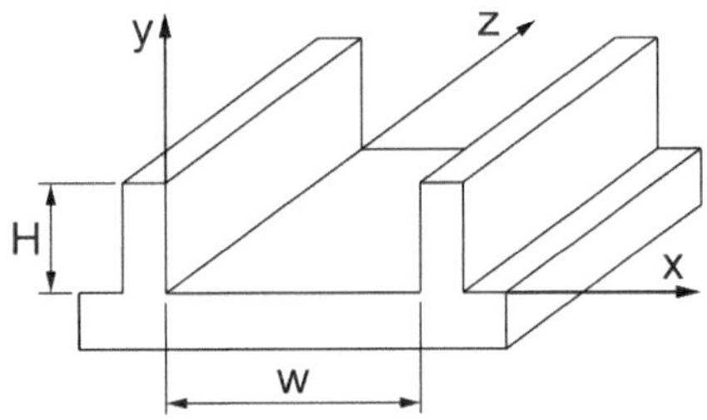

Figure 5.4
Screw geometry approximation with a rectangular channel

The next important assumption concerns the relative motion of the screw and the barrel. It is assumed that the screw is stationary and the barrel rotates, and the coordinate system is located on the screw surface.

This assumption is justified because the only slight difference between the two systems, the rotating barrel and the rotating screw, manifests itself in the stress distribution. It results from the occurrence of various centrifugal forces, which when extruding polymers are not significant.

This assumption, in relation to the rectangular channel, leads to the case presented in Figure 5.5. The barrel, represented by the plate, moves in relation to the channel at the angle φ with the linear velocity V_b equal to

$$V_b = \pi D_b N \tag{5.6}$$

where N is the screw rotational speed.

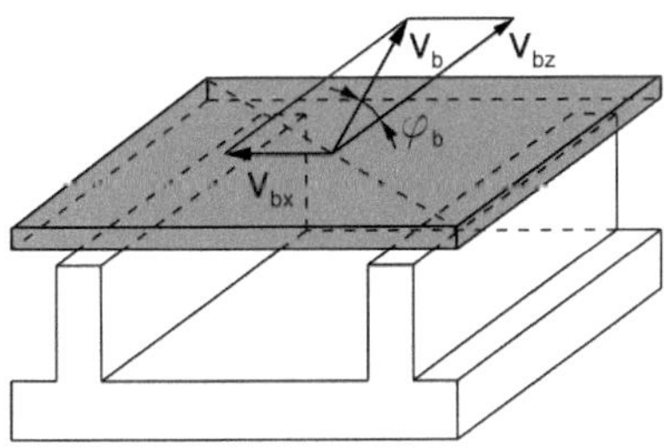

Figure 5.5
Assumption of relative motion of the barrel and the screw

This velocity can be decomposed into the components in the direction of the channel length (down-channel), i.e. in the *z*-direction, and in the direction of the channel width (cross-channel), i.e. in the *x*-direction. These components are expressed as

$$V_{bz} = V_b \cos\varphi \tag{5.7}$$

$$V_{bx} = V_b \sin\varphi \tag{5.8}$$

where V_{bz} is the longitudinal (down-channel) velocity component; V_{bx} is the transverse (cross-channel) velocity component.

As a result of this decomposition, the flow in the extruder can be considered as the sum of two flows: longitudinal and transverse.

The basic and common assumptions for the extrusion process are as follows:

- the flow is steady,
- the flow is laminar,
- there is no slip on the screw and the barrel,
- gravity and inertia forces are neglected.

5.2.3 Solid Conveying

5.2.3.1 Solid Conveying Mechanism

The material is supplied to the extruder in the form of granulate or powder, which is compressed in the initial part of the barrel and takes the form of a non-deformable solid layer, called solid bed, that slides between the barrel and the screw. The basic model of transport of this layer was proposed by Darnell and Mol [6, 29].

In order to analyze the solid bed transport, the model presented in Figure 5.6 is considered. The solid bed moves here with respect to the barrel with the velocity $\underset{\sim}{V}_j$, equal to the vector difference of the barrel velocity $\underset{\sim}{V}_b$ and the velocity of the solid bed in the *z*-direction, i. e. $\underset{\sim}{V}_{sz}$, that is

$$\underset{\sim}{V}_j = \underset{\sim}{V}_b - \underset{\sim}{V}_{sz} \tag{5.9}$$

where $\underset{\sim}{V}_b$ is the vector of the barrel velocity; $\underset{\sim}{V}_{sz}$ is the vector of the solid bed velocity.

The velocity vectors $\underset{\sim}{V}_j$ and $\underset{\sim}{V}_{sz}$ can be determined on the basis of geometrical relations, resulting from Figure 5.6, as

$$\underset{\sim}{V}_j = \underset{\sim}{V}_b \frac{\sin\varphi_b}{\sin\left(\theta + \varphi_b\right)} \tag{5.10}$$

$$V_{sz} = V_b \frac{\sin\theta}{\sin(\theta + \varphi_b)} \tag{5.11}$$

where φ_b is the helix angle at the barrel wall; θ is the angle between the direction of the barrel movement and the direction of the solid bed movement, i.e. the angle of the solid bed transport.

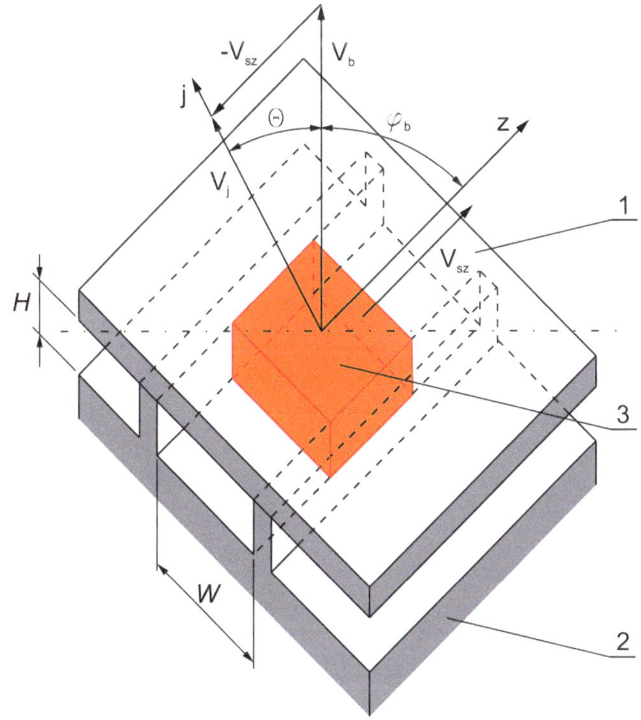

Figure 5.6 Mechanism of solid bed transport: 1 - barrel, 2 - screw, 3 - solid bed

Analysis of Eq. (5.11) leads to the following conclusions:

- when the angle $\theta = 0°$, the velocity $V_{sz} = 0$; in this case, there is no transport of the solid bed in the channel, i.e. (in the real system) it rotates with the screw, and this is the case when the friction coefficient on the screw is much larger than on the barrel, i.e. when $f_s >> f_b$,
- when the angle $\theta = 90° - \varphi_b$, the velocity $V_{sz} = V_b \sin(90° - \varphi_b) = V_b \cos\varphi_b = V_{bz}$; that is, the solid bed has the velocity equal to the longitudinal component of the barrel velocity V_{bz}, and this is the case when the coefficient of friction on the barrel is much larger than on the screw, i.e. when $f_b >> f_s$,
- when the angle $\theta = 90°$, the velocity $V_{sz} = V_b/\sin(90° + \varphi_b) = V_b/\cos\varphi_b$; then, the solid bed has the maximum velocity equal to $V_b/\cos\varphi_b$, and this may occur

when the barrel has a very high coefficient of friction f_{bt} in the circumferential direction (tangential), i.e. when it is much larger than the coefficient of friction on the screw f_s and the coefficient of friction f_{bl} on the barrel in the longitudinal direction, i.e. when $f_{bt} >> f_s$ and $f_{bt} >> f_{bl}$.

These conclusions can be related to the movement of a nut placed on the rotating screw. Then, in the first case, when the angle $\theta = 0°$, the nut will rotate with the screw without a translational movement. In the third case, when $\theta = 90°$, the nut will be fixed relative to the rotational movement of the screw and will move in the axial direction. The analysis of Eq. (5.11) is the basis of the concept of grooved barrels, in which the longitudinal grooves form a high coefficient of friction f_{bt} on the barrel surface in the tangential direction and force an intense axial transport.

5.2.3.2 Flow Rate

The volume flow rate of the solid bed can be determined as a product of the solid axial velocity and the solid cross-section, that is

$$Q = V_{sl} A \tag{5.12}$$

where V_{sl} is the velocity of the solid bed in the direction of the barrel axis; A is the cross-section of the layer.

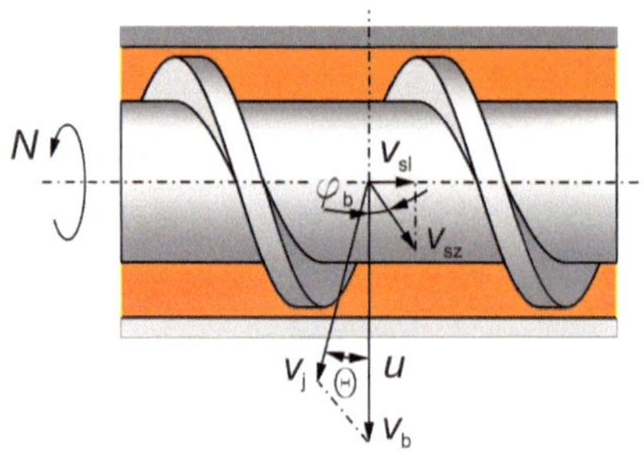

Figure 5.7
Schematic of movement of the solid bed

The axial velocity V_{sl} can be determined from Figure 5.7 as

$$V_{sl} = V_{sz} \sin \varphi_b \tag{5.13}$$

and, after using Eq. (5.11), it can be presented in the form

$$V_{sl} = V_b \frac{\tan\theta \tan\varphi_b}{\tan\theta + \tan\varphi_b} \tag{5.14}$$

The cross-section, on the other hand, is expressed in accordance with Figure 5.2 as

$$A = \pi \left(D_b - H\right) H - \frac{eH}{\sin\varphi_a} \tag{5.15}$$

where φ_a is the helix angle of the screw, halfway up the channel.

As a result, the volume flow rate of the solid bed can be represented by the formula

$$Q = V_b \frac{\tan\theta \tan\varphi_b}{\tan\theta + \tan\varphi_b} \left[\pi \left(D_b - H \right) H - \frac{eH}{\sin\varphi_a} \right] \tag{5.16}$$

5.2.3.3 Pressure Development

The schematic of loading the solid bed with thickness dz is shown in Figure 5.8.

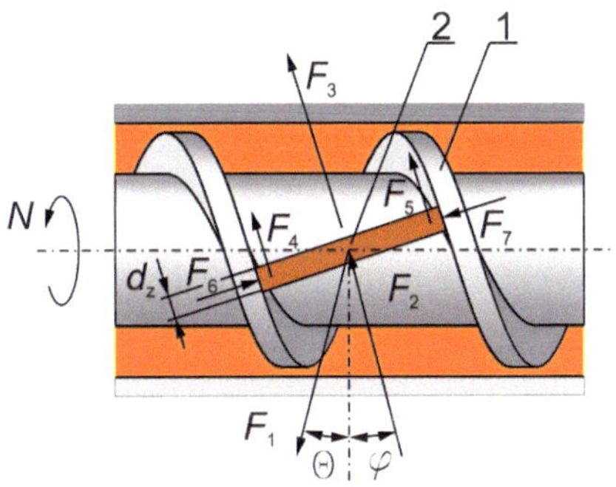

Figure 5.8
Schematic of loading the solid bed:
1 - screw, 2 - solid bed

There are forces on this layer that express themselves as follows:

- the friction force F_1 between the barrel and the solid bed:

$$F_1 = f_b p W_b \mathrm{d}z_b \tag{5.17}$$

- the force F_2 resulting from the pressure gradient along the solid bed thickness dz:

$$F_2 = H W_a \mathrm{d}p \tag{5.18}$$

- the friction force F_3 betwen the screw root and the solid bed:

$$F_3 = f_s p W_s \mathrm{d}z_s \tag{5.19}$$

- the friction force F_4 between the active flight of the screw and the solid bed:

$$F_4 = f_s F_6 \tag{5.20}$$

- the friction force F_5 between the passive flight of the screw and the solid bed:

$$F_5 = f_s F_7 \tag{5.21}$$

- the force F_6 exerted by the active flight of the screw on the solid bed:

$$F_6 = pH\mathrm{d}z_a + F_d \tag{5.22}$$

- the force F_7 exerted by the passive flight of the screw on the solid bed:

$$F_7 = pH\mathrm{d}z_a \tag{5.23}$$

where f_b is the friction coefficient on the barrel; f_s is the friction coefficient on the screw; dp is the elementary pressure increse along the thickness of the solid bed dz; H is the height of the screw channel; W_b, W_a, W_s is the width of the screw channel, repectively, on the barrel surface, halfway up the channel, and on the screw surface; dz_b, dz_a, dz_s is the thickness of the solid bed, respectively, on the barrel surface, halfway up the channel, and on the screw surface; F_d is the additional force, normal to the active flight of the screw.

Each of these forces can be decomposed into the components in the direction of the screw axis and in the direction perpendicular to it. As a result, we can make the balance of forces acting in the direction of the screw axis and the balance of moments of forces relative to this axis.

The equation of the balance of forces is as follows:

$$F_1 \sin\theta + (F_2 + F_4 + F_5)\sin\varphi_a + F_3 \sin\varphi_s - (F_6 - F_7)\cos\varphi_a = 0 \tag{5.24}$$

and the equation of the balance of moments of forces is

$$F_1 \cos\theta\left(\frac{D_b}{2}\right) - (F_2 + F_4 + F_5)\cos\varphi_a\left(\frac{D_a}{2}\right) - F_3 \cos\varphi_s\left(\frac{D_s}{2}\right) - \\ +(F_6 - F_7)\sin\varphi_a\left(\frac{D_a}{2}\right) = 0 \tag{5.25}$$

where θ is the angle of the solid bed transport; D_b, D_a, D_s is the diameter, respectively, on the barrel surface (barrel diameter), halfway up the screw channel, and on the screw surface; φ_a, φ_s is the helix angle, respectively, halfway up the channel and on the screw surface.

After substituting Eqs. (5.17) to (5.23) into Eqs. (5.24) and (5.25), and eliminating the unknown F_d, the following expression is obtained:

$$(B_1 - E_1 K)\mathrm{d}z_b = (B_2 + E_2 K)\frac{\mathrm{d}p}{p} \tag{5.26}$$

where

$$B_1 = f_b W_b \cos\theta - 2 f_s H \sin\varphi_b \cot\varphi_a\left(\frac{D_a}{D_b}\right) - f_s W_s \sin\varphi_b \cot\varphi_s\left(\frac{D_s}{D_b}\right) \tag{5.27}$$

$$B_2 = W_a H \cos\varphi_a\left(\frac{D_a}{D_b}\right) \tag{5.28}$$

$$E_1 = f_b W_b \sin\theta + 2 f_s H \sin\varphi_b + f_s W_s \sin\varphi_b \tag{5.29}$$

$$E_2 = W_a H \sin\varphi_a \tag{5.30}$$

$$K = \frac{D_a \sin\varphi_a + f_s \cos\varphi_a}{D_b \cos\varphi_a - f_s \sin\varphi_a} \tag{5.31}$$

After integrating Eq. (5.26), in the range from $z_b = 0$ where $p = p_1$ to z_b where $p = p_2$, we get

$$(B_1 - E_1 K) z_b = (B_2 + E_2 K) \ln\left(\frac{p_2}{p_1}\right) \tag{5.32}$$

and next, after transformation, we obtain the equation defining the pressure increase in the solid bed zone in the form

$$p_2 = p_1 \exp\left(\frac{B_1 - E_1 K}{(B_2 + E_2 K)} z_b\right) \tag{5.33}$$

As it results from Eq. (5.33), the pressure increases exponentially, but a certain initial value is required to determine it, or otherwise only the pressure ratio p_2/p_1 can be determined. This initial value can be the pressure at the outlet of the hopper, which can be determined on the basis of an appropriate analysis.

Equation (5.32) can be transformed into the form that allows determining the angle of transport θ. This equation can be written as

$$\cos\theta = K \sin\theta + M \tag{5.34}$$

where

$$M = 2\frac{f_s}{f_b}\frac{H}{W_b}\sin\varphi_b\left(K + \frac{D_a}{D_b}\cot\varphi_a\right) + \frac{f_s}{f_b}\frac{W_s}{W_b}\sin\varphi_b\left(K + \frac{D_a}{D_b}\cot\varphi_s\right) + \\ + \frac{1}{f_b}\frac{W_a}{W_b}\frac{H}{z_b}\sin\varphi_a\left(K + \frac{D_a}{D_b}\cot\varphi_s\right)\ln\left(\frac{p_2}{p_1}\right) \tag{5.35}$$

and then, transforming to the form [6]

$$\sin\theta = \frac{\sqrt{1 + K^2 - M^2} - KM}{1 + K^2} \tag{5.36}$$

where K, M are the constants determined by Eqs. (5.31) and (5.35).

Knowing the angle of transport θ allows us to calculate the flow rate according to Eq. (5.16).

Considering Eqs. (5.16) and (5.34), and assuming (in order to simplify) the lack of pressure gradient and the equality of friction coefficients $f_b = f_s = f$, we obtain the formulas for the relationship between the material flow rate and the helix angle φ_b and the coefficient of friction f (Figure 5.9).

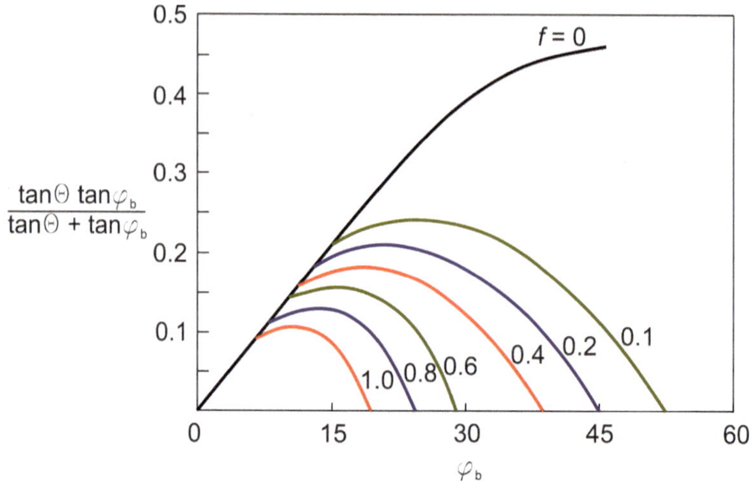

Figure 5.9 Expression $\tan\theta\tan\varphi_b/(\tan\theta + \tan\varphi_b)$, defining the flow rate of solid bed, as a function of the screw helix angle φ_b and the friction coefficient f, adapted from [29]

It is seen that the optimal angle φ_b that ensures a maximum efficiency is a decreasing function of friction coefficient f. Since for most polymers this coefficient is in the range of 0.25–0.5, the optimal value of angle φ_b will be approximately equal to $\varphi_b = 20°$. This explains why the screw pitch is usually equal to the screw diameter (then, $\varphi_b = 17°40'$).

5.2.4 Plasticating

5.2.4.1 Pre-Melting

Melting starts when the polymer melt film is formed on the surface of contact the polymer with the heated barrel. In the initial phase of this, called **pre-melting** (also *delay* zone), the thickness of the film h gradually increases (Figure 5.1). When this thickness becomes greater than the gap h_f between the tips of the screw flight and the barrel, a considerable pressure is created in it, according to the Rayleigh mechanism. It deforms the solid bed, creating conditions for accumulation of the melted material near the active flight of the screw (Figure 5.10). From that moment, melting proceeds according to a different mechanism, the so-called **melting** mechanism.

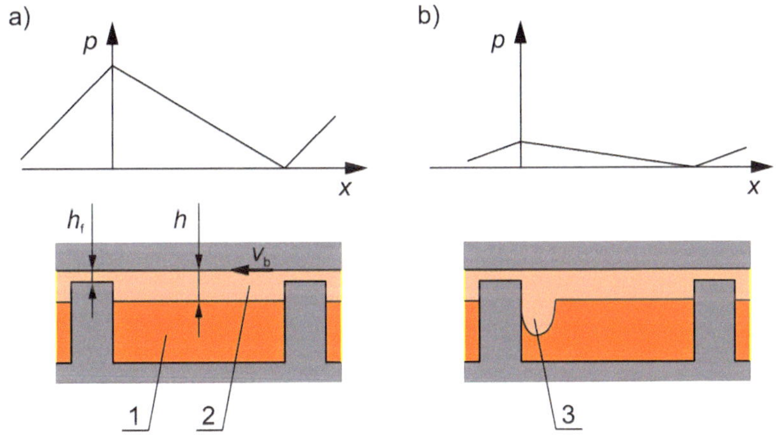

Figure 5.10 Initiation of melting mechanism: 1 - solid bed, 2 - melt film, 3 - melt pocket, *p* - pressure, adapted from [29]

Experimental research shows [6] that the deformation of the solid bed begins when the thickness of melted material is equal to

$$h_d \approx 5.64 h_f \tag{5.37}$$

where h_d is the thickness of melted material in the final phase of pre-melting (maximum thickness); h_f is the gap between the tips of the screw flight and the barrel.

The length of pre-melting zone is a function of the dimensionless parameter Ψ [6] that characterizes the melting process, which is defined by

$$z_d = f\left(\Psi^{-1}\right) \tag{5.38}$$

where z_d is the length of pre-melting zone (along the screw channel); Ψ is the dimensionless parameter which is proportional to the polymer melting rate and inversely proportional to the polymer flow rate.

Function (5.38) can be linear or exponential.

The motion mechanism of solid bed in the pre-melting zone is similar to the motion mechanism in the feed zone. The main difference is that there is no dry friction at the interface between the solid bed and the barrel, but there is the melt film where shearing occurs.

This results in changing the balance of forces and moments of forces acting on the solid bed, defined by Eqs. (5.24) and (5.25). Namely, the dry friction force F_1, defined by Eq. (5.17), is replaced by the viscous friction force F_1^* defined as

$$F_1^* = \tau W_b \mathrm{d}z_b \tag{5.39}$$

where τ is the shear stress in the melt film; W_b is the width of the screw channel at the barrel surface; $\mathrm{d}z_b$ is the thickness of the elementary solid bed at the barrel surface.

As a result, after some rearrangements, as in the case of the feed zone, the formula determining the pressure increase in the pre-melting zone can be obtained in the form

$$p_2 = p_1 \exp\left(\frac{B_1^* - E_1^* K}{B_2 + E_2} z_b\right) + \\ + \frac{\tau W_b \left(\cos\theta - K \sin\theta\right)}{B_1^* - E_1^* K}\left[\exp\left(\frac{B_1^* - E_1^* K}{B_1' + E_1' K} z_b\right) - 1\right] \tag{5.40}$$

where

$$B_1^* = B_1 - f_b W_b \cos\theta \tag{5.41}$$

$$E_1^* = E_1 - f_b W_b \sin\theta \tag{5.42}$$

and the other parameters are defined in Section 5.2.3.

5.2.4.2 Melting

Mechansim of Melting Initiation

Melting starts when the pressure in the melt film is so large that it can deform the solid bed (Figure 5.10). Then, in the place of this deformation (in the rear, active part of the channel), a cavity is formed in which melted material accumulates. As melting progresses, the width of the solid bed decreases, which is shown in Figure 5.11 and Figure 5.12. In the final phase of this, the solid breaks into small pieces (*solid bed break-up*), which will become melted independently, until the polymer is completely molten.

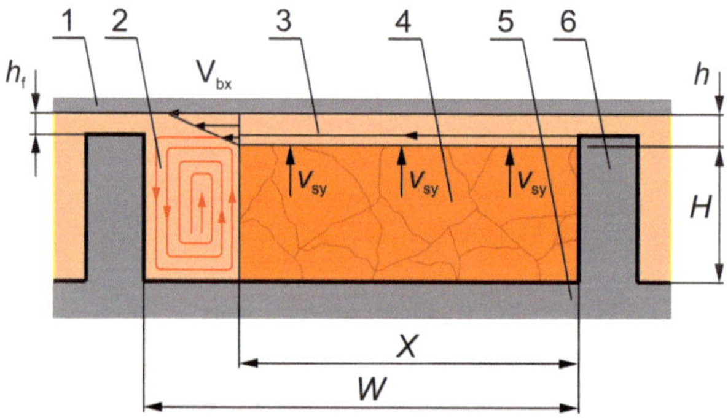

Figure 5.11 Cross-section of screw channel in the melting zone: 1 - barrel, 2 - melt, 3 - melt film, 4 - solid bed, 5 - screw, 6 - flight

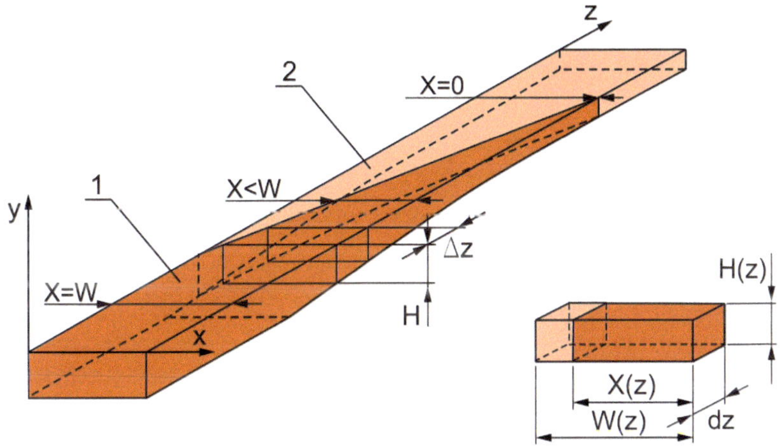

Figure 5.12 Melting course (unwinding the screw channel): 1 - solid bed, 2 - melt

Pressure in the melt film is developed in accordance with the Rayleigh mechanism. In the plane perpendicular to the screw axis (Figure 5.13), the pressure increase Δp can be expressed by the formula [29]

$$\Delta p = \frac{6\eta V_b}{\sin\varphi_b} \frac{h - h_f}{\left(\frac{h^3}{W} + \frac{h_f^3}{e}\right)} \tag{5.43}$$

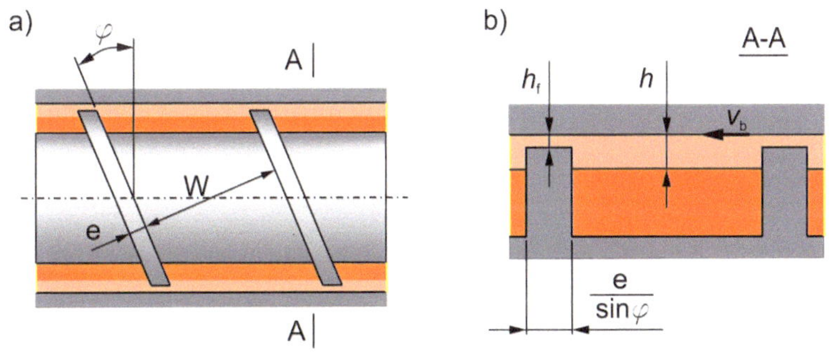

Figure 5.13 Geometrical characteristics of pre-melting zone: a) screw, b) cross-section A-A

The relationship in Eq. (5.43) is illustrated in Figure 5.14, from which it is apparent that the pressure increase Δp strongly depends on the ratio h/h_f and reaches the maximum for $h/h_f \approx 5$. This corresponds to the experimental observation, expressed by Eq. (5.37). It follows from Figure 5.14 that with increase of the gap h_f the pressure increase drops, so that it may be less than a certain threshold value p_c which is necessary to deform the solid bed. So there may be conditions under which the melting mechanism will not be initiated.

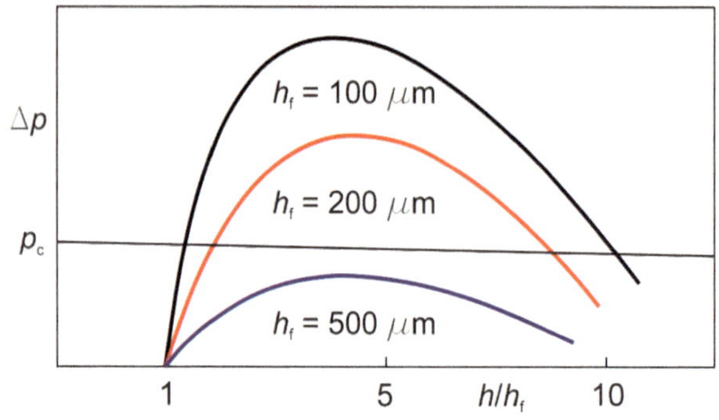

Figure 5.14 Pressure increase Δp in the melt film in the pre-melting zone in dependence on the gap thickness h_f, and the ratio h/h_f; p_c - critical pressure, adapted from [29]

Mass and Heat Balance in the Melting Zone

The melting of polymer is characterized by the ratio of the width of solid bed X to the width of screw channel W, represented as a function of the channel length z, i.e. as

$$X / W = f(z) \tag{5.44}$$

The fundamental model of melting was proposed by Tadmor [6]. The main assumptions of this model, in addition to the simplifications given in Section 5.2.2, are as follows:

- the cross-section of the screw channel is rectangular,
- melting takes place from the barrel side on the surface of the phase separation, between a thin melt film and a solid bed,
- the interfaces are sharp boundaries,
- heat exchange takes place in the direction of the height of the screw channel and by conduction only,
- the solid bed is continuous and homogeneous,
- the velocity of solid bed along the screw channel length is constant, i.e. $V_{sz} =$ const.

The basis of the melting model is the mass and heat balance in the melting region. To form such a balance it is necessary to determine the velocity and temperature profiles in the melt film and the temperature profile in the solid bed.

The flow conditions in the melt film are illustrated in Figure 5.15. The rectangular coordinate system is assigned to the phase separation surface, so that the flow in the melt film will be one-dimensional drag flow between two plates, the stationary interface of the phase separation and the movable wall of the barrel. The barrel wall moves at the velocity $\underset{\sim}{V}_j$, equal to the vector difference in the barrel velocity $\underset{\sim}{V}_b$ and the velocity of the solid in the z-direction, i.e. $\underset{\sim}{V}_{sz}$ (see Section 5.2.3.1).

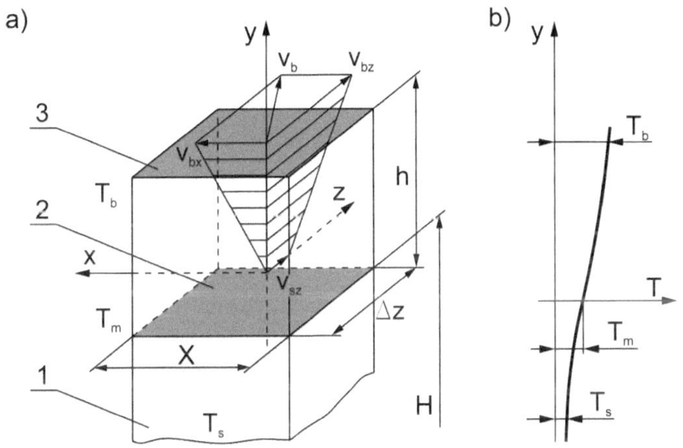

Figure 5.15 Flow conditions in the elementary material volume in the melting zone: a) velocity profile, b) temperature profile; 1 - solid bed, 2 - phase separation surface, 3 - barrel wall, adapted from [6]

The value of the vector $\underset{\sim}{V}_j$ can be determined from geometrical relations, according to the formula

$$V_j = \left(V_b^2 + V_{sz}^2 - 2V_b V_{sz} \cos\varphi_b\right)^{1/2} \tag{5.45}$$

The velocity V_{sz} can be determined in this case using the equation defining the mass flow rate (see Section 5.2.3.2):

$$G = \rho_s V_{sz} WH \tag{5.46}$$

where G is the mass flow rate; ρ_s is the solid density.

Under assumed flow conditions, the equations of motion (Table 1.8) and energy (Table 1.9), with respect to the melt film, take the form

$$\eta \frac{d^2 v_j}{dy^2} = 0 \tag{5.47}$$

$$k_m \frac{d^2 T}{dy^2} + \eta \left(\frac{dv_j}{dy}\right)^2 = 0 \tag{5.48}$$

where η is the viscosity; k_m is the melt thermal conductivity; T is the temperature; y is the coordinate in the direction of the melt film height; v_j is the velocity in the j-direction.

As a result of integrating Eq. (5.47), with boundary conditions $v_j(y = 0) = 0$ and $v_j(y = h) = V_j$, a linear velocity distribution is obtained:

$$v_j = V_j \frac{y}{h} \tag{5.49}$$

After substituting Eq. (5.49) into the energy equation (Eq. (5.48)), and integrating it with boundary conditions $T(y=0)=T_m$ and $T(y=h)=T_b$, a temperature distribution is obtained in the form

$$\frac{T-T_m}{T_b-T_m}=\frac{\eta V_j^2}{2k_m\left(T_b-T_m\right)}\frac{y}{h}\left(1-\frac{y}{h}\right)+\frac{y}{h} \tag{5.50}$$

where T_m is the temperature at the interface of phase separation (melting temperature of the material); T_b is the barrel temperature.

The energy equation in relation to the solid bed (Table 1.9), under assumed flow conditions, is expressed as

$$\rho_s c_s V_{sy}\frac{\mathrm{d}T}{\mathrm{d}y}=k_s\frac{\mathrm{d}^2T}{\mathrm{d}y^2} \tag{5.51}$$

where c_s is the specific heat of the solid material; k_s is the thermal conductivity of the solid; V_{sy} is the velocity of the solid bed in the channel height direction, i.e. in the y-direction.

After integrating Eq. (5.51), with boundary conditions $T(y=0)=T_m$ and $T(y=-\infty)=T_s$, we have a temperature distribution in the form

$$\frac{T-T_s}{T_m-T_s}=\exp\left(V_{sy}\frac{\rho_s c_s}{k_s}y\right) \tag{5.52}$$

where T_s is the initial temperature of the solid bed (temperature of the polymer feeding the extruder).

Determination of the velocity and temperature profiles in the melt film and the temperature profile in the solid bed makes it possible to determine the heat balance on the phase separation surface and the mass balance in the melt film and in the solid bed (Figure 5.16).

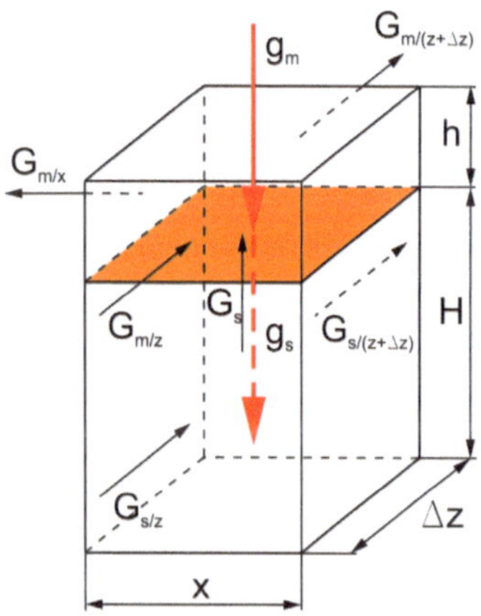

Figure 5.16
Mass balance and heat balance in the elementary material volume in the melting zone

According to the heat balance, the heat flux density used for material melting q_M is equal to the difference between the heat flux density q_M fed to the phase separation surface from the melt film and the heat flux density q_s discharged from this surface to the solid bed, that is

$$q_M = q_m - q_s \tag{5.53}$$

The components of Eq. (5.53) are expressed as follows

$$q_M = \rho_s V_{sy} \lambda \tag{5.54}$$

$$q_m = k_m \left(\frac{dT}{dy} \right)_{y=0} \tag{5.55}$$

$$q_s = k_s \left(\frac{dT}{dy} \right)_{y=0} \tag{5.56}$$

where λ is the heat of fusion.

The temperature gradient in the melt film and in the solid bed can be obtained by differentiating Eqs. (5.50) and (5.52). As a result, the heat balance equation (Eq. (5.53)) takes the form

$$\rho_s V_{sy} \lambda = \left[\frac{k_m}{h} \left(T_b - T_m \right) + \frac{\eta V_j^2}{2h} \right] - \left[\rho_s c_s V_{sy} \left(T_m - T_s \right) \right] \tag{5.57}$$

and after transformation

$$\rho_s V_{sy} \left[c_s \left(T_m - T_s \right) + \lambda \right] = \frac{k_m}{h} \left(T_b - T_m \right) + \frac{\eta V_j^2}{2h} \tag{5.58}$$

The expression on the left side of Eq. (5.58) defines a heat necessary to increase the temperature of the material from T_s to T_m, and then melt it. The expressions on the right side of the equation, in turn, determine a heat conduction through the melt film and a heat released in this film as a result of viscous energy dissipation.

The mass balance in the melt film with reference to the elementary volume with the dimensions Δz, h, X (Figure 5.16), is expressed as

$$G_{m/z} + G_s = G_{m/(z+\Delta z)} + G_{m/x} \tag{5.59}$$

where $G_{m/z}$ is the mass flow rate of the material flowing into the elementary volume in the z-direction; G_s is the mass flow rate within the elementary volume; $G_{m/(z+\Delta z)}$ is the mass flow rate of the material flowing out the elementary volume in the z-direction; $G_{m/x}$ is the mass flow rate of the material flowing out the elementary volume in the x-direction.

The components of Eq. (5.59) are expressed by

$$G_{m/z} = \rho_m \frac{V_{bz} + V_{sz}}{2} (Xh)_z \tag{5.60}$$

$$G_s = \rho_s V_{sy} X \Delta z \tag{5.61}$$

$$G_{m/(z+\Delta z)} = \rho_m \frac{V_{bz} + V_{sz}}{2} (Xh)_{z+\Delta z} \tag{5.62}$$

$$G_{m/x} = \rho_m \frac{V_{bz}}{2} h \Delta z \tag{5.63}$$

After substituting Eqs. (5.60) to (5.63) into the equation of mass (Eq. (5.59)), and then dividing this equation on both sides by Δz, and considering infinitesimal increments, we obtain

$$\rho_m \frac{V_{bz} + V_{sz}}{2} \frac{\mathrm{d}(Xh)}{\mathrm{d}z} + \rho_s V_{sy} X - \rho_m \frac{V_{bz}}{2} h = 0 \tag{5.64}$$

Since the first component of this relationship is negligible in relation to the other two, the mass balance equation can be written in the form

$$\rho_s V_{sy} X = \rho_m \frac{V_{bx}}{2} h = \omega \tag{5.65}$$

where ω is the melting rate of the material per unit of the screw channel length, (kg·s^{-1})/m.

The mass balance in the solid bed with reference to the elementary volume with the dimensions Δz, H, X (Figure 5.16), is expressed by

$$G_{s/z} = G_s + G_{s/(z+\Delta z)} \tag{5.66}$$

where $G_{s/z}$ is the mass flow rate of the material flowing into the elementary volume in the z-direction; G_s is the mass flow rate within the elementary volume; $G_{s/(z+\Delta z)}$ is the mass flow rate of the material flowing out the elementary volume in the z-direction.

The component G_s of Eq. (5.66) is determined by Eq. (5.61), and the next components are

$$G_{s/z} = \rho_s V_{sz} (HX)_z \tag{5.67}$$

$$G_{s/(z+\Delta z)} = \rho_s V_{sz} (HX)_{(z+\Delta z)} \tag{5.68}$$

After substituting Eqs. (5.61), (5.67), and (5.68) into the equation of mass (Eq. (5.66)), and then dividing this equation on both sides by Δz, and considering infinitesimal increments, we obtain the relationship

$$-\frac{\mathrm{d}(HX)}{\mathrm{d}z}=\frac{\omega}{\rho_s V_{sz}} \tag{5.69}$$

where ω is defined by Eq. (5.65).

Melting Process

The equations of mass balance and heat balance (Eqs. (5.58), (5.65), and (5.69)) make it possible to determine the melting of the material, i. e. to define the function in Eq. (5.44).

On the basis of Eqs. (5.58) and (5.65), by determining the velocity V_{sy}, it is possible to calculate the thickness h of the melt film and the melting rate ω.

These parameters are defined as

$$h=\left\{\frac{\left[2k_m\left(T_b-T_m\right)+\eta V_j^2\right]}{\rho_s V_{bx}\left[c_s\left(T_m-T_s\right)+\lambda\right]}X\right\}^{1/2} \tag{5.70}$$

$$\omega=\Phi X^{1/2} \tag{5.71}$$

where Φ is defined by

$$\Phi=\left\{\frac{\rho_m V_{bx}\left[k_m\left(T_b-T_m\right)+\dfrac{\eta V_j^2}{2}\right]}{2\left[c_s\left(T_m-T_s\right)+\lambda\right]}\right\}^{1/2} \tag{5.72}$$

The width of the solid bed can be determined by substituting Eq. (5.71) into Eq. (5.69). Then we get

$$-\frac{\mathrm{d}(HX)}{\mathrm{d}z}=\frac{\Phi X^{1/2}}{\rho_s V_{sz}} \tag{5.73}$$

The solution of Eq. (5.73) depends on whether the depth of channel H is constant or whether it varies in the direction of the z-axis.

In the case of a channel with a constant depth, the equation can be written as

$$\frac{\mathrm{d}X}{\mathrm{d}z}=-\frac{\Phi X^{1/2}}{\rho_s V_{sz}H} \tag{5.74}$$

Integrating Eq. (5.74), with the boundary condition $X(z=z_1)=H_1$, leads to the formula

$$\frac{X}{W}=\frac{X_1}{W}\left[1-\frac{\psi}{2H}\left(z-z_1\right)\right]^2 \tag{5.75}$$

where X, X_1 is the width of the solid bed as a function of a position in the direction of the screw channel length, respectively z, z_1; ψ is the dimensionless parameter defined as

$$\psi = \frac{\Phi}{V_{sz}\rho_s X_1^{1/2}} \tag{5.76}$$

Eq. (5.75) in the case when $z_1 = 0$ and $X_1 = W$ expresses as

$$\frac{X}{W} = \left(1 - \frac{\psi}{2H}z\right)^2 \tag{5.77}$$

and, on the basis of this formula, the length of melting zone can be determined.

Melting ends when $X = 0$. After substituting this condition into Eq. (5.77), we obtain the expression determining the length of melting:

$$L_M = \frac{2H}{\psi} \tag{5.78}$$

where L_M is the total length of the screw channel which is neccesary for complete melting of the material.

In the case of a tapered channel, the channel depth decreases linearly in the z-direction, according to the dependence

$$H = H_1 - A_t z \tag{5.79}$$

where H_1 is the channel depth for $z = 0$; A_t is the degree of tapering.

After inserting this relationship into Eq. (5.73), we obtain

$$\left(H_1 - A_t z\right)\frac{dX}{dz} - XA_t = -\frac{\Phi X^{1/2}}{\rho_s V_{sz}} \tag{5.80}$$

Integrating this equation, with the boundary condition $X(z = z_1) = X_1$, leads to the formula

$$\frac{X}{W} = \frac{X_1}{W}\left[\frac{\psi}{A_t} - \left(\frac{\psi}{A_t} - 1\right)\left(\frac{H_1}{H}\right)^{1/2}\right]^2 \tag{5.81}$$

This relationship, in the case when $z_1 = 0$ and $X_1 = W$, has the form

$$\frac{X}{W} = \left[\frac{\psi}{A_t} - \left(\frac{\psi}{A_t} - 1\right)\left(\frac{H_1}{H}\right)^{1/2}\right]^2 \tag{5.82}$$

and, on the basis of Eq. (5.82), using Eq. (5.79), the length of melting zone can be determined, which in this case is

$$L_M = \frac{H_1}{\psi}\left(2 - \frac{A_t}{\psi}\right) \tag{5.83}$$

Figure 5.17 shows the melting process defined by the *X/W* distribution in the case of a channel with a constant depth and in the case of a tapered channel. The channel tapering causes the thickness of the solid layer to decrease, but its width is increased. As a result, the melting rate increases per unit of the channel length and the total melting length decreases.

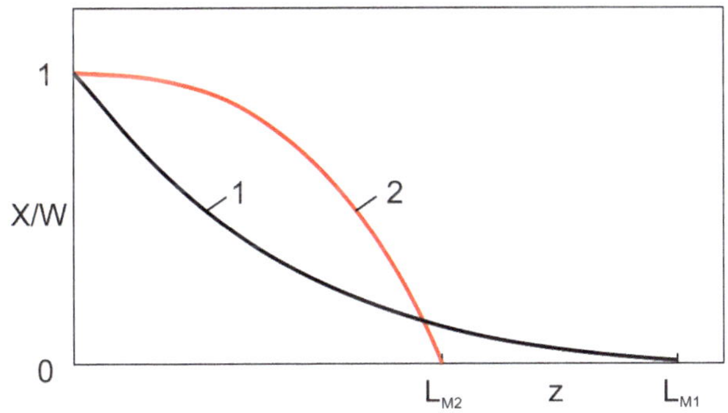

Figure 5.17 Melting process: 1 – the channel of constant depth ($A_t = 0$), 2 – the convergent channel ($A_t > 0$), L_M – length of melting

5.2.5 Melt Conveying in Conventional Screws

5.2.5.1 Classification of Models

The flow of molten polymer occurs in the extruder in the compression zone (with a tapered channel) and in the metering zone (with a constant depth channel). In the compression zone this flow usually takes place in the presence of a solid bed, while in the metering zone it is most often the flow of the polymer that is completely melted. The basic extrusion model in this range refers to the flow of polymer in the metering zone.

Extrusion models are usually classified [28] in dependence on the model assumptions in relation to the five most important elements of the flow description, as follows:

- model of the material
  - Newtonian
  - non-Newtonian
- number of coordinates defining the flow
  - one-dimensional model: longitudinal flow (down-channel)
  - two-dimensional model: longitudinal and transverse flow (down and cross-channel)
  - three-dimensional model: longitudinal, transverse, and radial flow

- effect of the temperature
  - viscosity not dependent on the temperature
  - viscosity dependent on the temperature
- effect of the flight flanks
  - negligible, as in the flow in a channel of infinite width; it is when $W/H > 10$
  - not negligible, as in the flow in a channel of finite width
- flight clearance
  - negligible
  - not negligible

5.2.5.2 Newtonian Model

The Basic Model

A schematic of geometry of the basic model of flow in the extruder screw channel is depicted in Figure 5.18. It is assumed in this model, next to the assumptions given in Section 5.2.2, that:

- the polymer is a Newtonian fluid,
- the polymer viscosity does not depend on the temperature,
- the screw channel is flat (curvature of the channel is neglected),
- the channel width is infinite (effect of flights is neglected),
- the flight clearance is neglected (between the screw flight and the barrel),
- the flow is fully developed.

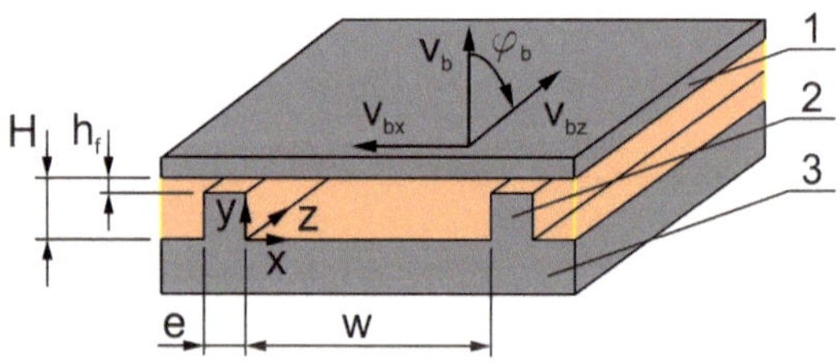

Figure 5.18 Screw channel geometry: 1 – barrel, 2 – flight, 3 – screw root

The velocity field in this flow is defined as

$$v_x = v_x(y), \quad v_y = 0, \quad v_z = v_z(y) \tag{5.84}$$

so that, for the assumed-here Newtonian model of the fluid, the equations of motion in the cross-channel direction x and in the down-channel direction z have the following form (Table 1.8):

$$\frac{\partial p}{\partial x} = \eta \frac{d^2 v_x}{dy^2} \tag{5.85}$$

$$\frac{\partial p}{\partial z} = \eta \frac{d^2 v_z}{dy^2} \tag{5.86}$$

After integrating the expression in Eq. (5.85), with boundary conditions $v_x(y = 0) = 0$ and $v_x(y = H) = -V_{bx}$, the velocity profile in the cross-channel direction x is obtained in the form

$$v_x = -V_{bx} \frac{y}{H} + \frac{y(y-H)}{2\eta} \frac{\partial p}{\partial x} \tag{5.87}$$

The pressure gradient $\partial p/\partial x$ can be determined on the basis of the condition of zero cross-channel flow, according to which

$$Q = W \int_0^H v_x(y) dy = 0 \tag{5.88}$$

Then, after substituting here Eq. (5.87) and integrating it, we have

$$\frac{\partial p}{\partial x} = -6\eta \frac{V_{bx}}{H^2} \tag{5.89}$$

After inserting this relationship into Eq. (5.87), we get the velocity profile in the cross-channel direction (Figure 5.19)

$$v_x = V_{bx} \frac{y}{H} \left(2 - 3\frac{y}{H} \right) \tag{5.90}$$

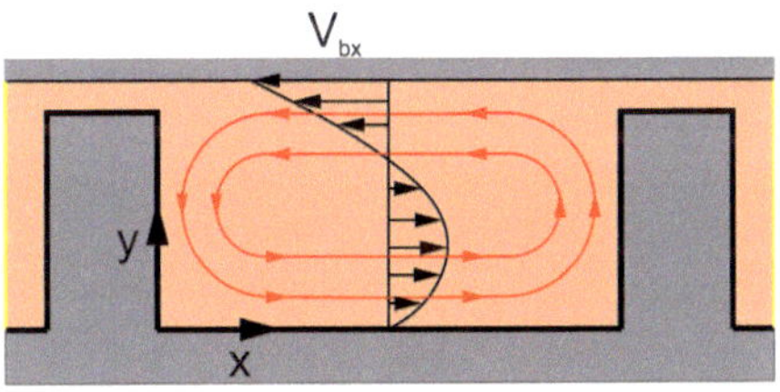

Figure 5.19
Cross-channel velocity profile $v_x(y)$

It is worth noting that the velocity v_x zeroes at the level of 2/3 of the channel depth, which corresponds to experimental observations.

The flow rate in the extruder can be determined on the basis of the analysis of flow in the direction z.

After integrating Eq. (5.86), with boundary conditions $v_z(y = 0) = 0$ and $v_z(y = H) = v_{bz}$, the velocity profile in the down-channel direction is obtained in the form

$$v_z = V_{bz}\frac{y}{H} + \frac{y(y-H)}{2\eta}\frac{\partial p}{\partial z} \tag{5.91}$$

In this relationship, two components of the velocity v_z can be distinguished, the drag flow component v_{zd}, represented by the first term of Eq. (5.91), and the pressure flow component v_{zp}, determined by its second term (see Section 4.2.3).

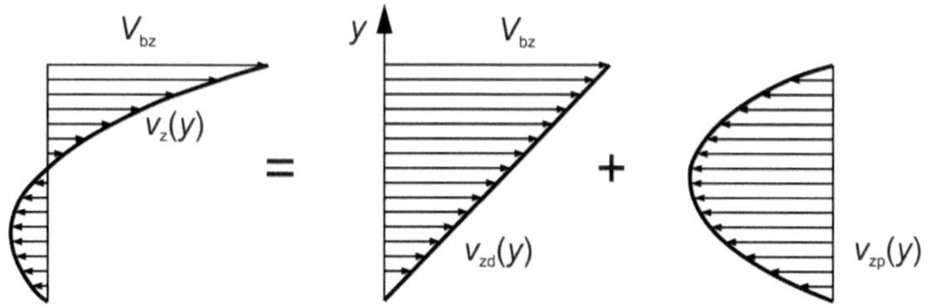

Figure 5.20 Down-channel velocity profile $v_z(y)$

The velocity profile is illustrated in Figure 5.20. In the general case, three situations can be distinguished depending on the pressure gradient $\partial p/\partial z$ sign, that is,

- when $\partial p/\partial z > 0$, the pressure component is subtracted from the drag component; this occurs when the pressure increases in the metering zone (Figure 5.20),
- when $\partial p/\partial z = 0$, the pressure component does not exist; this is the case with constant pressure in the metering zone,
- when $\partial p/\partial z < 0$, the pressure component is added to the drag component; this occurs when the pressure drops in the metering zone.

The flow rate in the extruder can be determined on the basis of the relationship

$$Q = \int_0^H v_z(y)\,\mathrm{d}y \tag{5.92}$$

After substituting Eq. (5.91) into Eq. (5.92) and integrating it, we have

$$Q = \frac{V_{bz}WH}{2} - \frac{WH^3}{12\eta}\frac{\partial p}{\partial z} \tag{5.93}$$

The first term of Eq. (5.93) $\frac{V_{bz}WH}{2}$ determines the **drag flow** Q_d, and the second term $\frac{WH^3}{12\eta}\frac{\partial p}{\partial z}$ represents the **pressure flow** Q_p.

The viscosity η in this equation is a function of the temperature and the shear rate in the screw channel. It can be approximated on the basis of the so-called characteristic shear rate $\dot{\gamma}_c$, which can be determined by

$$\dot{\gamma}_c = \frac{V_b}{H} \tag{5.94}$$

The flow of polymer in the extruder is often described by means of the ratio of pressure flow to drag flow, i. e. by the **throttle coefficient** a, which is equal to

$$a = \frac{Q_p}{Q_d} = \frac{H^2}{6\eta V_{bz}} \frac{\partial p}{\partial z} \tag{5.95}$$

After inserting it into Eqs. (5.93) and (5.91) we have, respectively

$$\frac{Q}{Q_d} = 1 - a \tag{5.96}$$

$$\frac{v_z}{V_{bz}} = \frac{y}{H}\left(1 - 3a + 3a\frac{y}{H}\right) \tag{5.97}$$

On this basis, the following flow conditions in the extruder can be distinguished:

- when $a = 1$, there is a total throttling of the flow; the pressure gradient is then positive, $\partial p/\partial z > 0$, and it has the maximum value,
- when $0 < a < 1$, the pressure gradient is still positive, $\partial p/\partial z > 0$, so that the direction of the pressure flow is still opposite to the drag flow; this is the case during extrusion when the pressure at the beginning of the metering zone is lower than at the end of this zone,
- when $a = 0$, there is no throttling; the pressure gradient is zero, $\partial p/\partial z = 0$, and there is no pressure flow,
- when $a < 0$, the pressure gradient changes the sign to negative, $\partial p/\partial z < 0$, so that the direction of the pressure flow changes; this is a common case during extrusion, when the pressure at the beginning of the metering zone is greater than at the end of this zone.

The velocity profiles in the screw channel, in the down-channel, cross-channel, and axial directions, for various values of the throttle coefficient a, are depicted in Figure 5.21. The axial velocity component v_1 (Figure 5.18) is expressed by

$$v_l = v_z \sin\varphi + v_x \cos\varphi \tag{5.98}$$

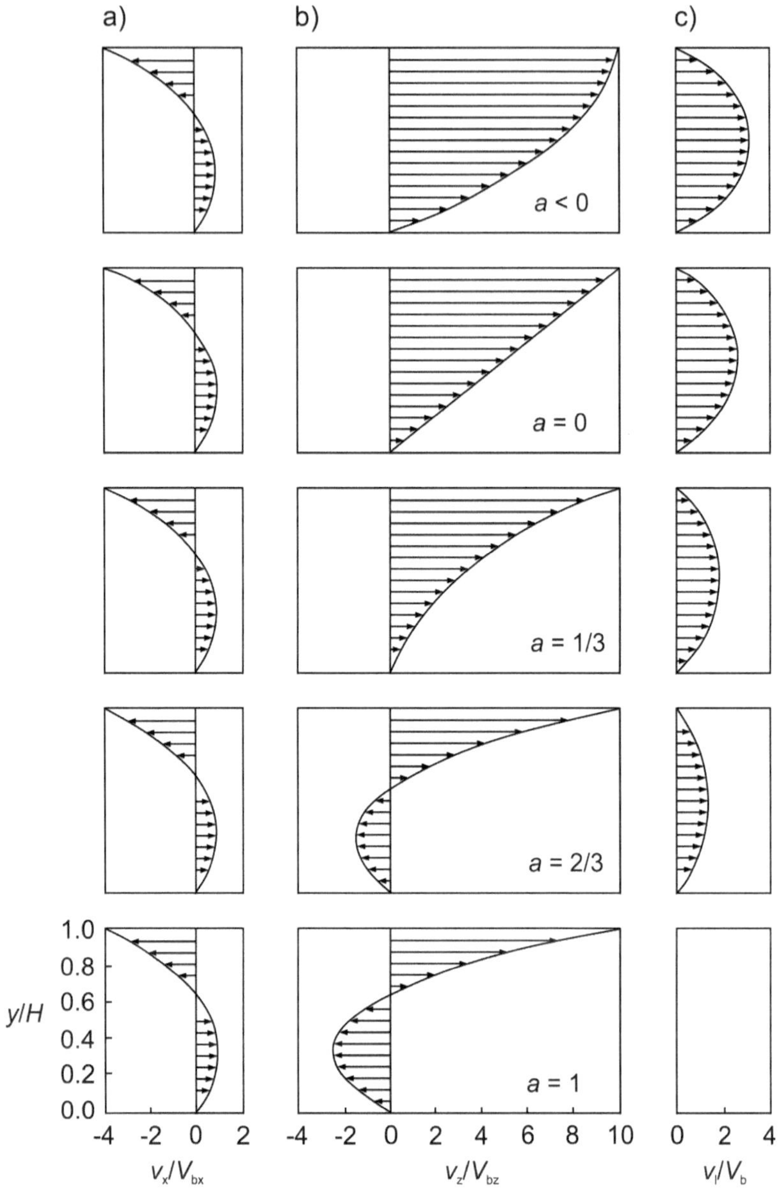

Figure 5.21 Velocity profile in the screw channel: a) cross-channel profile v_x/V_{bx}, b) down-channel profile v_z/V_{bz}, c) axial profile v_l/V_b, for different values of the throttle coefficient *a*, adapted from [15]

Effect of Flight Flanks

If the channel depth is not negligible in relation to its width, i.e. when $W/H < 10$, the influence of the flight flanks on the *z*-axis flow must be taken into account.

Then, the velocity component in this direction is described by

$$v_z = v_z(x,y) \tag{5.99}$$

and the equation of motion takes the form

$$\frac{\partial p}{\partial z} = \eta\left(\frac{\partial^2 v_z}{\partial x^2} + \frac{\partial^2 v_z}{\partial y^2}\right) \tag{5.100}$$

After integrating this equation, with boundary conditions $v_z(x,y=0)=0$, $v_z(x,y=H)=V_{bz}$, $v_z(x=0,y)=0$, $v_z(x=W,y)=0$, the velocity profile is obtained in the following form [6]:

$$v_z = \frac{4V_{bz}}{\pi}\sum_{i=1,3,5}^{\infty}\frac{\sinh\left(\frac{i\pi y}{W}\right)}{i\sinh\left(\frac{i\pi H}{W}\right)}\sin\left(\frac{i\pi x}{W}\right) - \frac{H^2}{2\eta}\frac{\partial p}{\partial z}\cdot$$
$$\left[\left(\frac{y}{H}\right)^2 - \frac{y}{H} + \frac{8}{\pi^3}\sum_{i=1,3,5}^{\infty}\frac{\cosh\left[i\pi\frac{W}{H}\left(\frac{x}{W}-\frac{1}{2}\right)\right]}{i^3\cosh\left(\frac{i\pi W}{2H}\right)}\sin\left(\frac{i\pi y}{H}\right)\right] \tag{5.101}$$

The flow rate can now be determined from the equation

$$Q = \int_0^H\int_0^W v_z(x,y)\,dx\,dy \tag{5.102}$$

After substituting Eq. (5.101) into this equation and integrating it, we have the equation

$$Q = \frac{V_{bz}WH}{2}F_d - \frac{WH^3}{12\eta}\frac{\partial p}{\partial z}F_p \tag{5.103}$$

where F_d and F_p are the shape coefficients that determine the reducing effect of the flight flanks on the drag and pressure flows. These coefficients are defined by

$$F_d = \frac{16W}{\pi^3 H}\sum_{i=1,3,5}^{\infty}\frac{1}{i^3}\tanh\left(\frac{i\pi H}{2W}\right) \tag{5.104}$$

$$F_p = 1 - \frac{192H}{\pi^5 W}\sum_{i=1,3,5}^{\infty}\frac{1}{i^5}\tanh\left(\frac{i\pi H}{2H}\right) \tag{5.105}$$

The coefficients F_d and F_p depend on the H/W ratio as shown in Figure 5.22. This ratio, for most screws is approximately equal to $H/W \approx 0.1$, and thus omitting the influence of flight flanks on the flow rate means an error of approximately 5%.

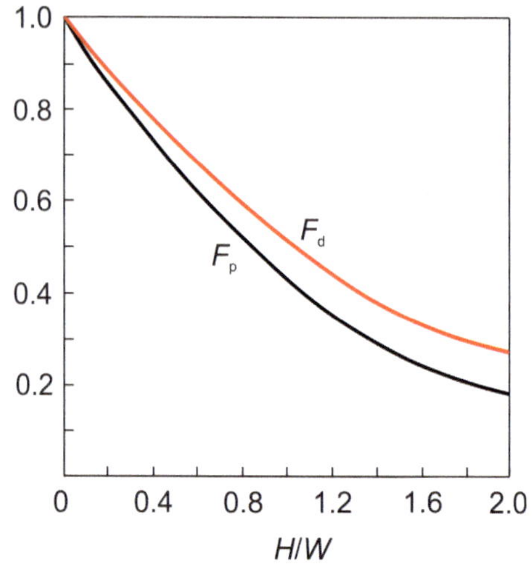

Figure 5.22 Shape coefficients F_d and F_p in dependence on the ratio H/W: H - screw channel depth, W - screw channel width

Equations (5.104) and (5.105), in relation to almost all screws ($H/W < 0.6$), can be approximated with the following simple equations [28]:

$$F_d = 1 - 0.571\frac{H}{W} \tag{5.106}$$

$$F_p = 1 - 0.625\frac{H}{W} \tag{5.107}$$

Leakage Flow

The basic flow model assumes that there is no gap between the tips of the screw and the barrel. In fact, there is such a gap, as well as the so-called **leakage flow.** It affects the flow rate, expressed by Eq. (5.93), in the following way [6]:

$$Q = \frac{V_{bz}WH}{2}\left(1 - \frac{h_f}{H}\right) - \frac{WH^3}{12\eta}\frac{\partial p}{\partial z}\left(1 + f_l\right) \tag{5.108}$$

where coefficient f_l is determined by the formula

$$f_l = \left(\frac{h_f}{H}\right)^3 \frac{e}{W}\frac{\eta}{\eta_f} + \left(1 + \frac{e}{W}\right)\frac{\left(\frac{6\eta V_{bz}\left(H - h_f\right)}{H^3\left(\partial p/\partial z\right)} + \frac{1 + e/W}{\tan^2\varphi}\right)}{\left(1 + \left(\frac{H}{h_f}\right)^3 \frac{e}{W}\frac{\eta_f}{\eta}\right)} \tag{5.109}$$

where f_l is the pressure flow correction coefficient; η is the viscosity in the screw channel; η_f is the viscosity in the gap.

It is worth noting that the viscosity in the gap will be much lower in the case of polymers than the viscosity in the channel, since the shear rate in the gap is much higher than in the channel. This viscosity can be determined on the basis of the characteristic shear rate $\dot{\gamma}_c$, which is expressed by

$$\dot{\gamma}_c = \frac{V_b}{h_f} \tag{5.110}$$

The leakage flow reduces the drag flow and increases the pressure flow. This flow does not matter if the gap is small. However, when the thickness of the gap increases, its role increases, with a particularly significant influence on the pressure flow.

Power Balance

The mechanical power e_m, introduced into the extruder as a result of the screw rotation, is dissipated and converted into heat in the screw channel and in the gap, and is used to develop a pressure gradient in the channel. So, we can write

$$e_m = e_s + e_f + e_p \tag{5.111}$$

where

$$e_s + e_p = e_{ch} \tag{5.112}$$

and e_m is the mechanical power introduced into the extruder as a result of the screw rotation; e_s is the power dissipated in the screw channel; e_f is the power dissipated in the gap between the screw flights and the barrel; e_p is the power used to develop a pressure gradient in the channel; e_{ch} is the power consumed in the screw channel.

The power consumed in the screw channel e_{ch}, in relation to the elementary volume of the channel $WH\mathrm{d}z$, can be determined as a product of the shear stress on the barrel surface and the velocity of this surface. This product can be integrated over the $W\mathrm{d}z$ area according to

$$\mathrm{d}e_{ch} = \left[\int_0^W \tau_{yz}(H) V_{bz} \mathrm{d}x + \int_0^W \tau_{yx}(H) V_{bx} \mathrm{d}x\right] \mathrm{d}z \tag{5.113}$$

where the shear stresses on the barrel surface $\tau_{yz}(H)$ and $\tau_{yx}(H)$ are equal to

$$\tau_{yz}(H) = \eta \left(\frac{\mathrm{d}v_z}{\mathrm{d}y}\right)_{y=H} \tag{5.114}$$

$$\tau_{yx}(H) = \eta \left(\frac{\mathrm{d}v_x}{\mathrm{d}y}\right)_{y=H} \tag{5.115}$$

The shear rates can be determined using Eqs. (5.91) and (5.90) as

$$\left(\frac{dv_z}{dy}\right)_{y=H} = \frac{V_{bz}}{H} + \frac{H}{2\eta}\frac{\partial p}{\partial z} \tag{5.116}$$

$$\left(\frac{dv_x}{dy}\right)_{y=H} = -\frac{4V_{bx}}{H} \tag{5.117}$$

As a result, after substituting Eqs. (5.114) to (5.117) into Eq. (5.113), and integrating it, the expression is obtained that determines the power consumed in the screw channel in the form

$$de_{ch} = \frac{\eta V_{bz}^2 W dz}{H}\left(1 + 4\tan^2\varphi + 3a\right) \tag{5.118}$$

where a is the throttle coefficient defined by Eq. (5.95).

The first term of Eq. (5.118) determines the fraction of the down-channel drag flow in the total power, the second term expresses the fraction of the cross-channel drag flow, and the third one defines the fraction of the pressure flow.

The power e_f dissipated in the gap between the screw flight and the barrel, by omitting the pressure that is not very important in this case, can be determined in a similar way, based on the relationship

$$de_f = \tau V_b e dz \tag{5.119}$$

where the shear stress τ can be expressed as

$$\tau = \eta_f \frac{V_b}{h_f} \tag{5.120}$$

After substituting Eq. (5.120) into Eq. (5.119) we have

$$de_f = \eta_f \frac{V_b^2}{h_f} e dz \tag{5.121}$$

The power de_p used to generate the pressure gradient in the channel can be determined based on a product of the volume flow rate Q and the pressure gradient, as

$$de_p = Q\frac{\partial p}{\partial z}dz \tag{5.122}$$

or

$$de_p = Q\,dp \tag{5.123}$$

It is worth noting that, depending on the sign of the pressure gradient, the de_p component of the power (in an elementary segment of the channel) can be positive or negative, so it can be consumed or generated.

The power dissipated in the extruder, i.e. the sum of $e_s + e_f$, turns into heat and contributes to an increase of the polymer temperature. Usually, it is about 90% of power e_m. The power e_m, in this case, is the power effectively supplied to the barrel and does not include various types of losses, e.g. in the drive system of the extruder.

5.2.5.3 Non-Newtonian Model

The flow in the extruder is a complex flow, and the non-Newtonian analytical solutions are known in relation to one-dimensional flow of the power fluid [22, 28].

These solutions are quite complicated and will be given in a graphical form. They are usually presented as a function of dimensionless quantities, the so-called dimensionless flow rate Q^* and the dimensionless pressure gradient Q_p^*. They have some reference to Eq. (5.93), which defines the flow rate in the Newtonian solution. These quantities are defined as

$$Q^* = \frac{2Q}{WHV_{bz}} \tag{5.124}$$

$$Q_p^* = \frac{H^{n+1}}{6mV_{bz}^n}\frac{\mathrm{d}p}{\mathrm{d}z} \tag{5.125}$$

With respect to the Newtonian solution, they express the ratio of total flow rate to drag flow rate and the ratio of pressure flow rate to drag flow rate, i.e.

$$Q^* = \frac{Q}{Q_d} \tag{5.126}$$

$$Q_p^* = \frac{Q_p}{Q_d} = a \tag{5.127}$$

and then, the flow rate equation (Eq. (5.93)) expresses as

$$Q^* = 1 - Q_p^* \tag{5.128}$$

The non-Newtonian solutions of one-dimensional flow of the power law fluid, for various flow exponents n, are depicted in Figure 5.23 in the form

$$Q^* = f\left(Q_p^*\right) \tag{5.129}$$

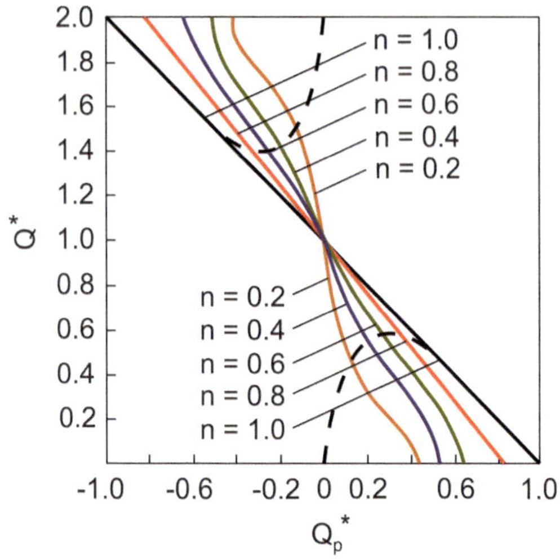

Figure 5.23 Non-Newtonian solutions of one-dimensional flow of a power law fluid for various flow exponents *n*, adapted from [15]

Since this solution is inconvenient to use, some approximate equations were developed [13], e. g. for one-dimensional flow:

$$Q^* = 1 - n^{-0.94} Q_p^* \tag{5.130}$$

whose scope of applicability is shown in Figure 5.23; for example, for $n = 0.4$ the equation is valid within the range $0.6 \leq Q^* \leq 1$.

With respect to two-dimensional flow of the power law fluid the numerical solutions are known. Also in this case, some approximations were developed [13], e. g. in the form

$$Q^* = A_1 - A_2 n^{-0.94} Q_p^* \tag{5.131}$$

where A_1, A_2 are the functions of the helix angle and the flow exponent.

The other approximation developed for two-dimensional flow of the power law fluid [28] is valid in a wider range of dimensionless flow rates $0 \leq Q^* \leq 1$. However, it is limited to helix angles within the range 15–25°. This is determined by the formula

$$Q^* = \frac{8+2n}{10} - \frac{3}{1+2n} Q_p^* \tag{5.132}$$

or

$$Q^* = \frac{4+n}{10} WHV_{bz} - \frac{1}{1+2n} \frac{WH^3}{4\eta} \frac{\partial p}{\partial z} \tag{5.133}$$

It is worth noting (see Section 4.2.3) that the non-Newtonian flow rate in the single screw extruder is not superposition of the drag flow and the pressure flow, as in the case of Newtonian solution (Eq. (5.93)).

5.2.6 Melt Conveying in Non-Conventional Screws

In order to increase the screw performance, non-conventional designs of screws are used. Such screws may have special elements (sections) that intensify shearing and mixing of the material. Although both of these processes occur simultaneously, there are elements in which shearing dominates and elements in which mixing dominates.

5.2.6.1 Dispersive Mixing Elements

The elements with predominance of shearing process, which are called elements of intensive shearing, have gaps in which the flowing polymer undergoes intensive shearing. The barrier sections of torpedo type (Figure 5.24) and the grooved sections (Figure 5.25) are of basic importance. The grooved elements may have the rectilinear grooves of Maddock type or the helical grooves of Egan type.

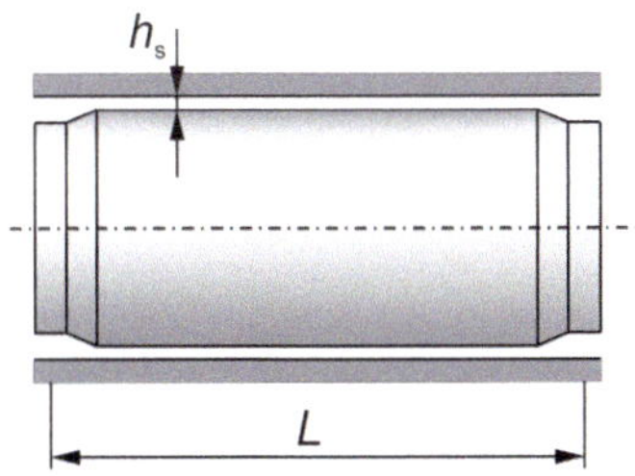

Figure 5.24
Barrier section of intensive shearing

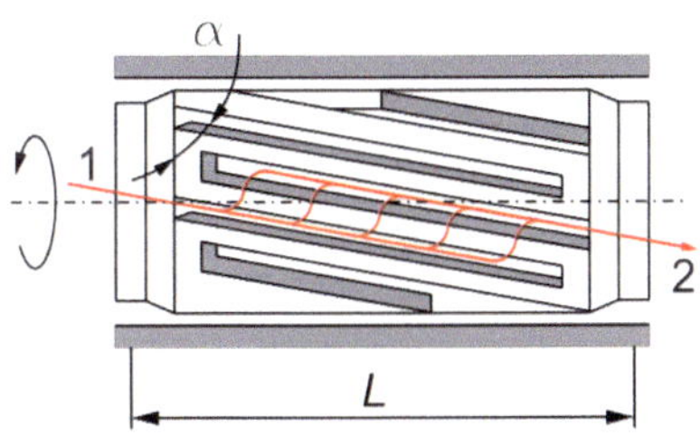

Figure 5.25
Grooved section of intensive shearing:
1 - inlet groove, 2 - outlet groove

Barrier Elements

The barrier sections of torpedo type (Figure 5.24) are cylindrical (smooth) sections of the screw in which intense shearing occurs in a narrow gap between the surface of section and the surface of barrel. The polymer flow in the gap results from the pressure gradient.

The pressure drop in this barrier section can be determined on the basis of Eq. (4.40), which is valid for the pressure flow in an annulus, according to the formula

$$\Delta p = \frac{2mL}{h_s}\left[\frac{2(2+1/n)Q}{\pi(D_s+h_s)h_s^2}\right]^n \tag{5.134}$$

where Δp is the pressure drop; L is the shearing section length; h_s is the gap thickness; D_s is the section diameter; Q is the flow rate; m, n are the parameters of the power law fluid.

Equation (5.134) does not take into account the drag flow in the gap, but the impact of this flow on the pressure drop is small.

Grooved Elements

The grooved sections (Figure 5.25) are systems of pairs of grooves, separated from each other (there is no polymer flow between them), one of which is an inlet groove (open at the inlet and closed at the outlet) and the other is an outlet groove (closed at the inlet and open at the outlet). The inlet groove is separated from the outlet groove by a barrier flight over which the material flows, undergoing intensive shearing in the gap between the flight and the barrel. The grooves can be rectilinear (parallel to the screw axis) or helical (inclined at a certain angle), with different cross-sections.

Numerical computations are necessary to describe the non-Newtonian flow in grooved sections. Therefore, for simplicity, we present here the Newtonian case [28].

To determine the pressure drop in the grooved section, one pair of grooves is considered with the assumption that the grooves have a rectangular cross-section.

The volume flow rate of the material at the entrance to the inlet channel is equal to

$$Q_1(0) = \frac{Q}{i} \tag{5.135}$$

where $Q_1(0)$ is the volume flow rate of the material at the entrance to the inlet channel; Q is the total flow rate; i is the number of pairs of grooves.

The flow rate in the inlet groove decreases in the down-channel direction due to the leakage flow over the barrier flight into the outlet groove, in which the flow rate increases. The flow rate in the inlet groove (as a function of channel length) can be determined based on Eqs. (5.103) and (5.108) as

$$Q_1(z) = \frac{V_b \cos\alpha}{2} WHF_d \left(1 - \frac{h_s}{H}\right) - \frac{WH^3}{12\eta}\frac{\mathrm{d}p_1}{\mathrm{d}z} F_p \tag{5.136}$$

where $Q_1(z)$ is the flow rate in the inlet channel; z is the down-channel direction; W, H are the width and the height, respectively, of the channel; h_s is the shearing gap; F_d, F_p are the shape coefficients (Eqs. (5.104) and (5.105), respectively); V_b is the circumferential velocity of the barrel; α is the groove angle; η is the viscosity; p_1 is the pressure in the inlet channel.

In contrast, the flow rate in the outlet channel can be determined as

$$Q_2(z) = \frac{V_b \cos\alpha}{2} WHF_d - \frac{WH^3}{12\eta} \frac{dp_2}{dz} F_p \tag{5.137}$$

where $Q_2(z)$ is the flow rate in the outlet channel; p_2 is the pressure in the outlet channel.

The unit leakage flow rate in the gap above the barrier flight (per unit of channel length) can be determined from

$$Q_s(z) = \frac{V_b \sin\alpha}{2} h_s + \frac{h_s^3}{12\eta_s} \frac{p_1 - p_2}{w_s} \tag{5.138}$$

where $Q_s(z)$ is the leakage flow rate; h_s, w_s are the height and the width, respectively, of the shearing gap; η_s is the viscosity in the gap.

This flow rate is equal to the local variation of the flow rate in the inlet channel according to

$$\frac{dQ_1(z)}{dz} = -Q_s(z) \tag{5.139}$$

From the equation of continuity it follows that the flow rate at a given place of the outlet channel (the local flow rate) is equal to the flow rate at the entrance to the inlet channel $Q_1(0)$ decreased by the local flow rate in this channel, i.e.

$$Q_2(z) = Q_1(0) - Q_1(z) \tag{5.140}$$

Assuming that the dimensions of inlet and outlet channels are the same, the sum of the pressure gradients in both channels will be constant along the section length. Thus, on the basis of Eqs. (5.140), (5.136), and (5.137), the following relationship is obtained:

$$\frac{dp_1(z)}{dz} + \frac{dp_2(z)}{dz} = -A \tag{5.141}$$

where

$$A = \frac{12\eta}{WH^3 F_p} \left[Q_1(0) - \frac{V_b \cos\alpha}{2} WHF_d \left(2 - \frac{h_s}{H} \right) \right] \tag{5.142}$$

As a result, after integrating the expression in Eq. (5.141), the relationship is obtained that connects the pressure in the outlet channel with the pressure in the inlet channel:

$$p_2 = -Az + B - p_1 \tag{5.143}$$

where B is the integration constant.

The next relationship can be obtained on the basis of Eq. (5.139) using Eqs. (5.136), (5.138), and (5.143), that is

$$\frac{d^2 p_1}{dz^2} - Cp_1 = \frac{1}{2} ACz + \frac{6\eta h_s V_b \sin\alpha}{WH^3 F_p} - \frac{1}{2} BC \tag{5.144}$$

where

$$C = \frac{2\eta h_s^3}{WH^3 F_p \eta_s w_s} \tag{5.145}$$

The solution of this heterogeneous second order differential equation makes it possible to determine the distribution of pressure in the inlet channel, and then, using Eq. (5.143), the distribution of pressure in the outlet channel. To determine the integration constants, it can be assumed that the pressure gradient at the beginning of the inlet channel is equal to the pressure gradient at the end of the outlet channel, which results from Eqs. (5.136) and (5.137).

The total pressure drop in the grooved section is equal to the difference between the pressure at the beginning of the inlet channel and the pressure at the end of the outlet channel, and expresses as

$$\Delta p = p_1(0) - p_2(z_g) \tag{5.146}$$

where Δp is the total pressure drop in the grooved section; $p_1(0)$ is the pressure at the beginning of the inlet channel; $p_2(z_g)$ is the pressure at the end of the outlet channel; z_g is the groove length.

After applying the solution of Eq. (5.144) and appropriate transformations [28], Eq. (5.146) takes the form

$$\Delta p = \frac{2E\left[1 + \exp(Dz_g)\right]}{1 - \exp(Dz_g)} - \frac{6\eta_s w_s V_b \sin\alpha}{h_s^2} + \frac{Az_g}{2} \tag{5.147}$$

where

$$D = C^{1/2} \tag{5.148}$$

$$E = -\frac{6\eta Q_1(0)}{WH^3 F_p D} \tag{5.149}$$

The presented solution applies to the grooves with rectangular cross-sections. In relation to the other geometries, the shape correction method may be used (see Section 5.3.6.2).

5.2.6.2 Distributive Mixing Elements

The elements with predominance of mixing process, called elements of intensive mixing, have regions in which the flowing material undergoes multiple subdivisions and combining, with a large amount of rotational motion which causes intensive mixing of the material.

The basic section of intensive mixing is the section with mixing rings (Figure 5.26). It is a system of successive mixing rings and smooth cylindrical sections of the screw. The rings may have working spaces (i.e. spaces through which the material flows) with different cross-sections.

The pressure drop across the mixing element is the sum of the pressure drop in the mixing rings and the gaps of the cylindrical sections, that is

$$\Delta p = i_m \Delta p_m + i_c \Delta p_c \tag{5.150}$$

where Δp is the total pressure drop in the mixing section; i_m is the number of mixing rings; Δp_m is the pressure drop in the mixing ring; i_c is the number of cylindrical sections (annular gaps); Δp_c is the pressure drop in the cylindrical section.

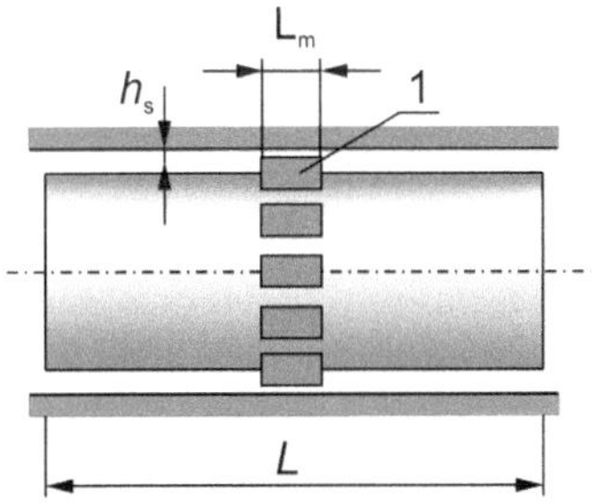

Figure 5.26
Section of intensive mixing with mixing ring: 1 – mixing ring

The pressure drop in the gap of the cylindrical section can be determined on the basis of Eq. (5.134) as

$$\Delta p_c = \frac{2mL_c}{h_s}\left[\frac{2(2+1/n)Q}{\pi(D_s+h_s)h_s^2}\right]^n \tag{5.151}$$

where Δp_c is the pressure drop; L_c is the cylindrical section length; h_s is the gap thickness; D_s is the section diameter; Q is the flow rate; m, n are the parameters of the power law model.

The pressure drop in the mixing ring can be determined using the general cross-section method and the concept of equivalent hydraulic radius (see Section 5.3.6.2). According to Eq. (4.98), in this case we have

$$\Delta p_m = \frac{2mL_m}{R_h}\left[\frac{(3+1/n)Q_o}{\pi R_h^3}\right]^n \tag{5.152}$$

where Δp_m is the pressure drop in the mixing ring; L_m is the mixing ring width; Q_o is the flow rate in the working space of the ring; R_h is the equivalent hydraulic radius, defined by Eq. (5.270).

The flow rate in the working space is equal to

$$Q_o = \frac{Q}{i_o} \tag{5.153}$$

and the equivalent radius can be expressed as

$$R_h = \left(\frac{A}{\pi}\right)^{1/2} \tag{5.154}$$

where Q is the total flow rate; i_o is the number of working spaces in the ring; A is the cross-section of working space.

5.2.7 Characteristics of Extruder Operation

The extrusion process is the result of cooperation between the screw and the die.

The equation of the polymer flow rate in the extruder, in reference to the Newtonian model, is expressed by Eq. (5.93). This equation, after taking into account the leakage flow (Eq. (5.108)) and the effect of the flight flanks on the flow rate (Eq. (5.103)), and after replacing the pressure gradient along the length of the screw channel with the gradient along the screw axis, can be presented as

$$Q = \frac{\pi D_b N \cos\varphi_b WH}{2}\left(1 - \frac{h_f}{H}\right)F_d - \frac{WH^3}{12\eta}\frac{\Delta p}{L}\sin\varphi_a\left(1 + f_1\right)F_p \tag{5.155}$$

or in the form

$$Q = \alpha N - \beta\frac{\Delta p}{\eta} \tag{5.156}$$

where α, β are the geometrical constants defined as

$$\alpha = \frac{\pi D_b N \cos\varphi_b WH}{2}\left(1 - \frac{h_f}{H}\right)F_d \tag{5.157}$$

$$\beta = \frac{WH^3}{12L}\sin\varphi_a\left(1 + f_1\right)F_p \tag{5.158}$$

where N is the screw rotational speed; Δp is the pressure change in the metering zone, the difference between the pressure at the end of the screw and the pressure at the beginning of the metering zone, i.e. $\Delta p = p_L - p_0$; L is the length of the metering zone; φ_b, φ_a is the helix angle, on the barrel surface and halfway up the channel, respectively.

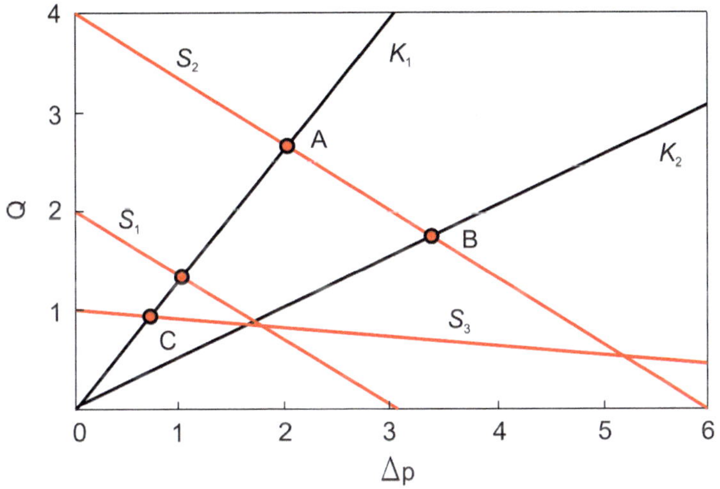

Figure 5.27 Extruder operating characteristics (Newtonian flow): Q - volume flow rate; Δp - pressure change in the extruder (increase) and in the die (drop); $S_1(H, N)$, $S_2(H, 2N)$, $S_3(H/2, N)$ - screw characteristics; H - screw channel depth; N - screw rotational screw; K_1, K_2 die characteristics, K - die conductivity, $K_1 > K_2$; A, B, C - extruder operating points

The flow rate equation for the screw (Eq. (5.156)), in the coordinate system of volume flow rate and pressure change, is represented by a straight line with a negative angle coefficient, which is the **screw characteristics** (Figure 5.27).

The flow rate equation for the die, with reference to the Newtonian model, can be expressed as

$$Q = K\frac{\Delta p}{\eta} \tag{5.159}$$

where Δp is the pressure drop in the die; K is the geometrical constant (conductivity).

This equation, in the coordinate system of volume flow rate and pressure change, is represented by a straight line with a positive angle coefficient, which is the **die characteristics** (Figure 5.27).

The extruder, from the point of view of flow conditions, is a serial connection of the barrel (screw) and the die, so that the formulas in Eq. (5.165) apply here.

For the sake of simplicity, a further analysis of the screw/die interactions are made with respect to the melt extrusion, i.e. assuming that the material is fed into the extruder as a melt.

Because, omitting the hopper, the pressure at the extruder inlet and the pressure at the die outlet are equal to atmospheric pressure, the pressure increase in the screw Δp_{screw} will be equal to the pressure drop in the die Δp_{die}, i.e.

$$\Delta p_{screw} = \Delta p_{die} \tag{5.160}$$

If we assume that the polymer density is the same in the screw and in the die, the continuity condition indicates that the flow rate in the screw and the flow rate in the die will be the same, i.e.

$$Q_{screw} = Q_{die} \tag{5.161}$$

After further assuming that the viscosity is the same in the screw and in the die, and using Eqs. (5.156) and (5.159), we obtain

$$\Delta p = \frac{\alpha N}{K+\beta}\eta \tag{5.162}$$

$$Q = K\frac{\alpha N}{K+\beta} \tag{5.163}$$

The Δp and Q values, expressed by Eqs. (5.162) and (5.163), constitute the solution of system of Eqs. (5.156) and (5.159), and determine the so-called **extruder operating point**. Figure 5.27 shows, for example, how the location of this point changes in dependence on the screw speed and the screw channel depth, as well as the die geometry.

With reference to the non-Newtonian model, the screw characteristics are presented in the coordinate system of dimensionless flow rate Q^* (Eq. (5.124)) and dimensionless pressure gradient Q_p^* (Eq. (5.125)). These characteristics are determined by the relationship in Eq. (5.129), which is illustrated in Figure 5.28.

The **die characteristics**, in general case, is here expressed by Eq. (5.178) as

$$Q = \frac{K}{m^{1/n}}\Delta p^{1/n} \tag{5.164}$$

where K is the geometrical constant (conductivity); m, n are the parameters of the power law model.

This equation, in the coordinate system of volume flow rate and pressure change, is represented by a parabola. This equation can also be represented in the coordinate system of dimensionless flow rate Q^* and dimensionless pressure gradient Q_p^*, as shown in Figure 5.28.

Figure 5.28 also shows how the extruder operating point moves as a function of the flow exponent n.

It is worth noting that the extruder operating characteristics are only demonstrative and are difficult to relate to the actual extrusion conditions with plasticizing the material. In this case, numerical iterative computations are needed to find the operating point.

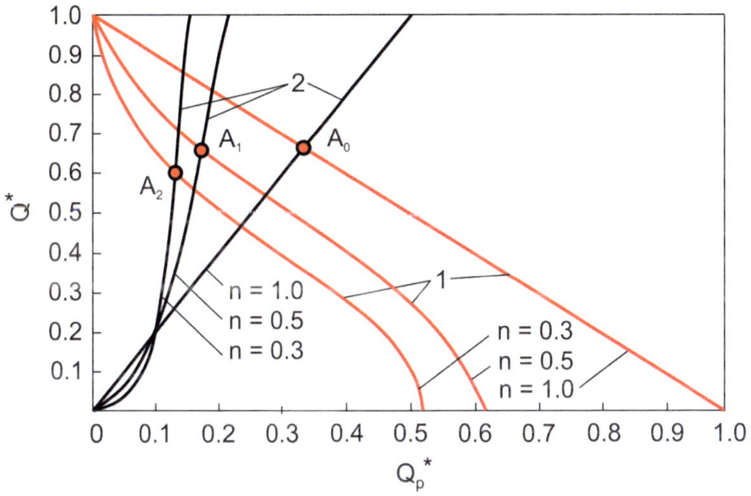

Figure 5.28 Extruder operating characteristics (non-Newtonian flow) at various flow exponents n: 1 - screw characteristics, 2 - die characteristics; Q^* - dimensionless flow rate, Q_p^* - dimensionless pressure gradient, A_0, A_1, A_2 - extruder operating points, adapted from [15]

■ 5.3 Extrusion Dies

5.3.1 Classification of Dies

The task of extrusion dies is to form the extrudate with desired shape and dimensions, ensuring thermal and mechanical homogeneity in the whole volume of the material.

Dies can be divided according to the geometry of forming section, i. e. the cross-section of the channel at the die exit. And thus, the following dies can be distinguished:

- dies with a circular exit cross-section,
- dies with a flat slit exit cross-section,
- dies with an annular exit cross-section,
- profile dies with an irregular exit cross-section.

The basic task of the die, that is, transformation of the annular stream of the polymer flowing out of the extruder barrel into the stream of desired shape and dimensions, is performed by gradually changing the cross-section of the flow channel or by using special distribution channels.

Dies can therefore be considered, from the point of view of modeling the polymer flow, as a series of (sometimes parallel) sets of connected channels with specific cross-sections. The basic die design calculations are based on determining the relationships between the polymer flow rate and the pressure gradient (drop) in the die. These calculations also include determining the residence time of the polymer in the die, due to possible thermal degradation (e.g. of polyvinyl chloride), and evaluating the global temperature increase of the polymer.

5.3.2 Methodology of Modeling

5.3.2.1 General Assumptions

The flow in extrusion dies is pressure flow (see Section 4.2.1). It occurs due to the pressure difference between the pressure at the extruder barrel exit and the atmospheric pressure at the die exit. When modeling the flow in extrusion dies, the following basic assumptions are accepted [30], which correspond to the real flow conditions:

- the flow is steady,
- the flow is laminar; in the case of polymer flows, the Reynolds number is very small (usually $Re < 1$), so that it is always smaller than the limit value that determines the laminar flow, i.e. $Re < 2300$,
- the flow is isothermal, that is, all particles of the flowing material have the same temperature; also, the die has the same temperature, so that there is no heat transfer between the material and the die,
- the flow is fully developed (the end effects are neglected); the entrance section L_e is proportional to the Reynolds number, which is very small, so that this section is very short,
- the fluid is incompressible and its viscosity does not depend on the pressure,
- there is no slip on the channel wall; this assumption is not valid when the shear stress at the wall exceeds a certain threshold value (see Section 2.4.3), which often occurs, e.g. in the case of polyethylene (at higher shear rates), hard poly(vinyl chloride), and elastomers,
- gravity and inertia forces are neglected.

Modeling (computations) of pressure flows can be carried out on the basis of an analogy to electric current flow. According to this analogy, the flow rate corresponds to the electric current and the pressure drop to the voltage drop (electric potential difference). Then, with a serial connection of the channels, the total pressure drop in the die will be the sum of the pressure drops in the individual channels, and the flow rate will be the same in each of them. However, with a parallel connection of the channels (e.g. in a breaker plate), the flow rate is divided into individual channels, and the pressure drop is the same (Figure 5.29).

Thus, for the serial connection of the channels we have

$$Q = Q_i = \text{const} \quad \text{and} \quad \Delta p = \sum_{i=1}^{k} \Delta p_i \tag{5.165}$$

and, for the parallel connection

$$\Delta p = \Delta p_i = \text{const} \quad \text{and} \quad Q = \sum_{i=1}^{k} Q_i \tag{5.166}$$

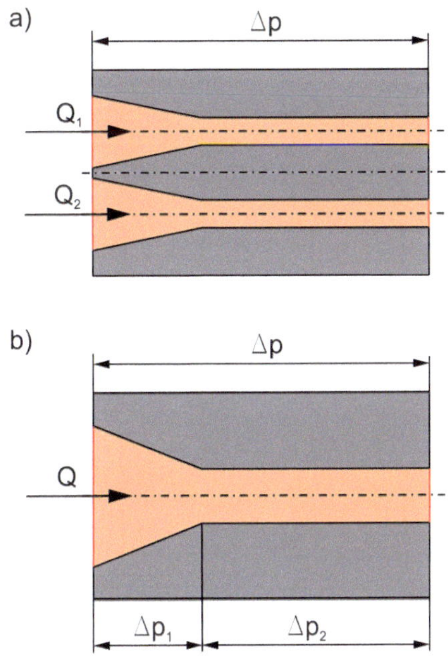

Figure 5.29 Relationship between flow rate Q and pressure drop Δp in connected channels: a) in series, b) in parallel; Q - total flow rate, Q_i - flow rate in the ith channel, Δp - total pressure drop; Δp_i - pressure drop in the ith channel

This approach can be also applied for flow computations in the convergent and divergent channels. These can be considered as a system (serial and parallel connections) of elementary channels (cylindrical or flat) with elementary dimensions, e.g. Δz and Δr (Figure 5.30). This applies, among others, in the computer calculations.

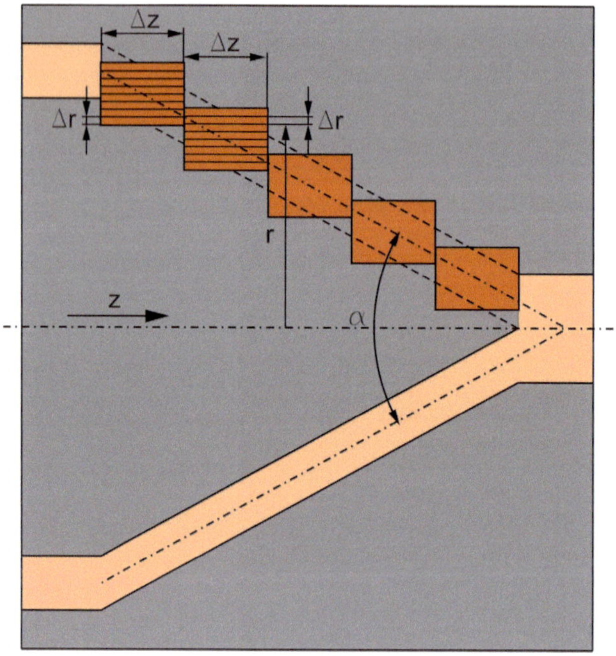

Figure 5.30 Geometrical approximation of conical channel with a set of elementary channels with constant geometry

Calculations of dies also include determining the time of residence of the material in the die and the total temperature increase of the material.

The mean residence time can be determined as the ratio of the volume of die channels to the volume flow rate:

$$t_r = \frac{V_{ch}}{Q} \tag{5.167}$$

where t_r is the mean residence time; V_{ch} is the volume of die channels; Q is the volume flow rate.

The polymer temperature increase in the die can be determined assuming adiabatic flow conditions, according to Eq. (4.67).

$$\Delta T = \frac{\Delta p}{\rho c_p} \tag{5.168}$$

where ΔT is the temperature increase; Δp is the pressure drop.

5.3.2.2 Newtonian Model

The basic calculations of extrusion dies consist in determining the relationship between the flow rate and the pressure drop in the die.

Such relationships, in the case of Newtonian fluids, are expressed for the cylindrical channel by Eqs. (4.78) and (4.79), and with reference to the flat channel by Eqs. (4.17) and (4.18).

These equations can be expressed, in general, in the form

$$Q = \frac{K}{\mu}\Delta p \tag{5.169}$$

and

$$\Delta p = K^{*}\mu Q \tag{5.170}$$

where K, K^* are the constants dependent on the channel geometry; K is the conductivity of the channel; K^* is the resistance of the channel, where

$$K = \frac{1}{K^{*}} \tag{5.171}$$

For example, in the case of the circular channel, using Eq. (4.78), the conductivity K is equal to

$$K = \frac{\pi R^4}{8L} \tag{5.172}$$

and, in the case of the flat channel, using Eq. (4.17), it equals

$$K = \frac{WH^3}{12L} \tag{5.173}$$

Assuming that the viscosity μ is constant in the entire die, Eqs. (5.165) and (5.166) can be expressed as

$$Q = Q_i = \text{const} \quad \text{and} \quad \Delta p = \left(\sum_{i=1}^{k} K_i^{*}\right)\mu Q \tag{5.174}$$

$$\Delta p = \Delta p_i = \text{const} \quad \text{and} \quad Q = \left(\sum_{i=1}^{k} K_i\right)\frac{\Delta p}{\mu} \tag{5.175}$$

where

$$\sum_{i=1}^{k} K_i^{*} = K_D^{*} \tag{5.176}$$

$$\sum_{i=1}^{k} K_i = K_D \tag{5.177}$$

and, where K_D^{*} and K_D are the total resistance and the total conductivity.

In the case of polymers, Eqs. (5.174) and (5.175), defined on the basis of the total resistance and the total conductivity, are essentially of an illustrative nature, since the viscosity is not constant in the entire die.

Therefore, when using the Newtonian model, Eqs. (5.165) and (5.166) are applied, calculating the viscosity separately in each die segment. It is determined on the basis of the viscosity curve, in relation to the specific temperature and the shear rate on the channel wall in this segment. Usually, instead of the viscosity curve, the appropriate rheological equation, e.g. the Klein model (see Section 1.2.4.2) is used.

The geometrical constants (conductivity or resistance) of the basic types of die channels, with reference to the Newtonian model, can be determined on the basis of the formulas presented in Table 4.1. There are also the formulas defining the shear rate on the channel walls.

5.3.2.3 Non-Newtonian Model

The approach used for the analysis made with reference to the Newtonian model can also be applied in the case of the non-Newtonian fluid, e.g. described by the power law model of Ostwald de Waele.

Then, the relations between the volume flow rate and the pressure drop in the die are expressed for the cylindrical channel by Eqs. (4.97) and (4.98), and in the case of the flat channel have the form of Eqs (4.39) and (4.40). These formulas can also be represented generally by

$$Q = \frac{K}{m^{1/n}} \Delta p^{1/n} \tag{5.178}$$

and

$$\Delta p = \left(K^*\right)^n mQ^n \tag{5.179}$$

where K, K^* are the constants dependent on the channel geometry and the flow exponent n; K is the conductivity of the channel; K^* is the resistance of the channel (the conductivity is an inverse of the resistance).

In the case of the cylindrical channel, on the basis of Eq. (4.97), the conductivity K is equal to

$$K = \frac{\pi R^{3+1/n}}{2^{1/n}\left(3+1/n\right)} \frac{1}{L^{1/n}} \tag{5.180}$$

and with reference to the flat channel, on the basis of Eq. (4.39), it is expressed by

$$K = \frac{WH^{2+1/n}}{2^{1+1/n}\left(2+1/n\right)} \frac{1}{L^{1/n}} \tag{5.181}$$

The rheological parameters of the power law equation, i.e. the consistency coefficient m and the flow exponent n, are practically constant in the wide range of the shear rate for polymers, so that the relationships in Eqs. (5.165) and (5.166) can be represented as follows:

$$Q = Q_i = \text{const} \quad \text{and} \quad \Delta p = \left(\sum_{i=1}^{k} K_i^{*n} \right) m Q^n \tag{5.182}$$

$$\Delta p = \Delta p_i = \text{const} \quad \text{and} \quad Q = \left(\sum_{i=1}^{k} K_i \right) \frac{\Delta p^{1/n}}{m^{1/n}} \tag{5.183}$$

where the expressions in brackets define the total resistance and the total conductivity.

The geometrical constants (conductivity or resistance) of the basic types of die channels, with reference to the non-Newtonian model, can be determined on the basis of the formulas presented in Table 4.1. There are also given the formulas defining the shear rate on the channel walls.

5.3.2.4 The Concept of Representative Viscosity

The modeling of flow in extrusion dies, as well as other tools and machines in polymer processing, using Newtonian solutions is convenient but not very accurate, as it does not take into account the non-Newtonian character of polymers. Non-Newtonian solutions, for example, using the power law model, are accurate, but more complex and sometimes uncomfortable to use.

Therefore, a reasonable compromise in the modeling of pressure flows, e.g. in extrusion dies, is using the concept of **representative viscosity**.

The basis of this concept [30] is the fact that in the pressure flow one can indicate the place where the shear rate determined according to the Newtonian model is equal to the shear rate determined according to the non-Newtonian model described by the power law (Figure 5.31). This place is referred to as the representative point, and the corresponding shear stress τ^*, shear rate $\dot{\gamma}^*$, and viscosity η^* are also referred to as representative.

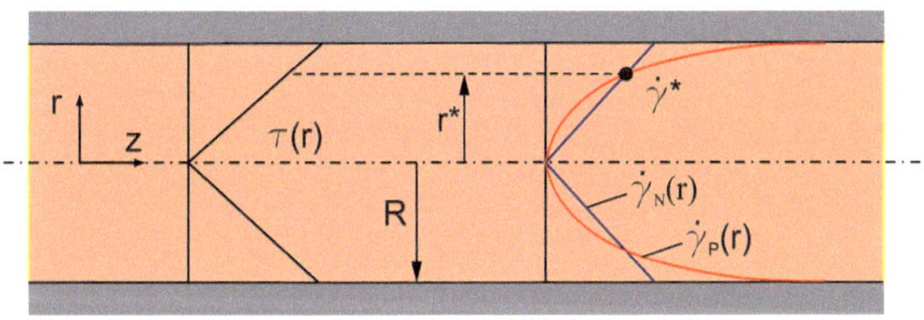

Figure 5.31 The concept of representative viscosity: r^* - representative radius, τ - shear stress, $\dot{\gamma}$ - shear rate, $\dot{\gamma}^*$ - representative shear rate, $\dot{\gamma}_N$ - Newtonian shear rate, $\dot{\gamma}_P$ - non-Newtonian (pseudoplastic) shear rate

The location of the representative point can be determined by comparing the functions of the shear rate distribution (in channels of a particular shape) for the Newtonian and non-Newtonian model.

In the case of the cylindrical channel, on the basis of Eqs. (4.80) and (4.99) we have

$$\frac{4Q}{\pi R^4} r = \frac{(3+1/n)Q}{\pi R^3}\left(\frac{r}{R}\right)^{1/n} \tag{5.184}$$

from which it follows that the shear rate will be the same in the position determined by the formula

$$e_r = \frac{r^*}{R} = \left(\frac{4}{3+1/n}\right)^{n/(1-n)} \tag{5.185}$$

where e_r is the relative representative radius; r^* is the representative radius; R is the radius of the channel; n is the flow exponent.

In the case of the flat channel, on the basis of Eqs.(4.19) and (4.41), we obtain

$$\frac{12Q}{WH^3} y = \frac{2(2+1/n)Q}{WH^2}\left(\frac{2y}{H}\right)^{1/n} \tag{5.186}$$

which results in the formula

$$e_h = \frac{h^*}{H/2} = \left(\frac{3}{2+1/n}\right)^{n/(1-n)} \tag{5.187}$$

where e_h is the relative representative height; h^* is the representative height; W is the channel width; H is the channel height.

The position of the representative point is a function of the flow exponent n. However, with respect to typical materials, i.e. with flow exponents in the range $0.25 < n < 0.5$, the fixed values of relative representative dimensions can be assumed equal to

$e_r = 0.815$ and $e_h = 0.772$

Then, the relative error of these values does not exceed 1.8% in the case of a circular channel, and 2.5% in the case of a rectangular channel [30].

The use of representative viscosity η^* in modeling of polymer processing consists in using Newtonian solutions of the equations of motion and energy (calculation formulas), and then introducing to these solutions the value of the representative viscosity corresponding to the process conditions regarding the shear rate and the temperature. This viscosity is determined on the basis of the **representative shear rate,** i.e. the shear rate obtained as a result of the Newtonian solution, but corrected by the value of the relative representative dimension, i.e.

$$\eta^* = \eta\left(\dot{\gamma}^*\right) \tag{5.188}$$

where $\dot{\gamma}^* = e_r \dot{\gamma}_w$ for the circular channel and $\dot{\gamma}^* = e_h \dot{\gamma}_w$ for the flat channel, and where $\dot{\gamma}_w$ is the Newtonian shear rate (otherwise the apparent shear rate $\dot{\gamma}_a$) expressed by Eqs. (4.81) and (4.21), respectively.

Calculations based on the concept of representative viscosity can be made on the basis of the formulas appropriate to the Newtonian solutions which are given in Table 4.1.

5.3.3 Circular Dies

Circular dies, i.e. dies with a circular exit cross-section (forming section) are used, among other things, for the production of rods, strands, and fibers. Usually they consist of segments with a circular or annular cross-section of the channel.

The problem of flow modeling and calculations in the channels of this type is presented in Section 5.3.2.

5.3.4 Slit Dies

Slit dies, i.e. dies with a flat slit exit cross-section (forming section), also called flat dies, are used to extrude films and sheets. The criterion according to which the films differ from the sheets is their thickness and winding capacity (without permanent deformation and damage). It is conventionally assumed that the limiting value is a thickness of 0.5 to 0.7 mm.

During the extrusion process, the material leaving the die should be thermally and mechanically homogeneous throughout its volume. Since it is usually fed to the die in the form of a cylindrical stream (a circular channel), it needs to be transformed into a stream with a rectangular section. This is done by means of a properly designed distribution channel (the manifold).

The task of the manifold is to distribute the material evenly along the width of the die, so that the outflow velocity will be independent of the position and its properties will be uniform.

The condition for obtaining a constant average outflow velocity is the linear change in the volume flow rate in the distribution channel (as a function of the die width) and the same flow resistance along all the polymer flow paths in the forming gap (the land). These conditions will be met when the total pressure drop along all the flow paths will be constant (Figure 5.32).

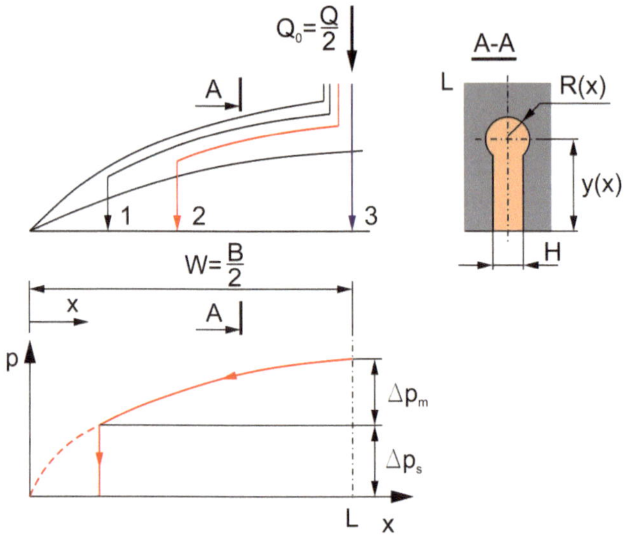

Figure 5.32 Flow conditions in a slit die: 1, 2, 3 – polymer flow paths, Q – volume flow rate, B – slit width, H – slit height, $y(x)$ – slit length, $R(x)$ – radius of distributive channel (manifold), x – coordinate along the die width, p – pressure, Δp_m – pressure drop in the distributive channel (manifold), Δp_s – pressure drop in the slit channel (land)

The homogeneity of properties of the extruded material can be obtained to a large extent by preserving the conditions of the same residence time of all polymer particles in the die and the same actions shear stresses, and thus the same shear rates.

The purpose of design calculations for the slit dies [30] is to define the geometrical conditions of the flow, i. e. to determine the geometry of the die at which the previously stated requirements will be met. They can be formulated in the following form:

a) the condition of constant mean velocity of the flowing out material

$$V(x) = \text{const} \tag{5.189}$$

b) the condition of constant total pressure drop along all flow paths

$$\Delta p(x) = \text{const} \tag{5.190}$$

c) the condition of constant shear stress, which is equivalent to the constant shear rate condition

$$\dot{\gamma}(x) = \text{const} \tag{5.191}$$

where x means the coordinate in the direction of the die width.

The calculations will be carried out using the representative viscosity method [30].

To meet the condition (a), the flow rate in the distribution channel should vary linearly $Q_m(x)$, i.e. according to the dependence

$$Q_m\left(x\right)=Q_0\frac{x}{W} \tag{5.192}$$

where $Q_m(x)$ is the flow rate in the manifold; $Q_0 = Q/2$ is one-half of the extruder output; x is the coordinate in the direction of the die width; $W = B/2$ is one-half of the land width B.

Based on the condition (c), the geometry of the manifold can be determined. The radius $R(x)$ of the manifold can be determined from the constant shear rate condition at the channel wall. While maintaining this condition, based on Eq. (4.81), the following formula is obtained:

$$\frac{4Q_m\left(x\right)}{\pi R^3\left(x\right)}=\frac{4Q_0}{\pi R_0^3} \tag{5.193}$$

from which, after taking into account Eq. (5.192), it follows that

$$R\left(x\right)=R_0\left(\frac{x}{W}\right)^{1/3} \tag{5.194}$$

The unknown initial radius R_0 can be determined from the condition of uniform shear rate in the manifold and in the land. This condition is expressed by the corresponding equality of the representative shear rates:

$$\dot{\gamma}_m^* = \dot{\gamma}_s^* \tag{5.195}$$

where $\dot{\gamma}_m^*$ is the representative shear rate in the manifold; $\dot{\gamma}_s^*$ is the representative shear rate in the land.

These rates are determined, on the basis of Table 4.1, and Eqs. (5.188) and (5.193), by the following expressions

$$\dot{\gamma}_m^* = \frac{4Q_0}{\pi R_0^3}e_r \tag{5.196}$$

$$\dot{\gamma}_s^* = \frac{6Q_0}{WH^2}e_h \tag{5.197}$$

where $e_r = 0.815$; $e_h = 0.772$.

When applying the condition of Eq. (5.195), this leads to the relationship between the manifold radius R_0 and the dimensions of the extruded profile in the form

$$R_0=\left(\frac{1}{3\pi}\frac{e_r}{e_h}BH^2\right)^{1/3} \tag{5.198}$$

or, after substituting the values e_r and e_h, to the formula

$$R_0 = 0.482\left(BH^2\right)^{1/3} \tag{5.199}$$

Based on the condition (b), i.e. the constant total pressure drop along all flow paths, the length of the forming slit (the land) y as a function of the position along the channel width, i.e. $y(x)$, can be determined. This condition is expressed as

$$\mathrm{d}\Delta p(x)/\mathrm{d}x = 0 \tag{5.200}$$

The total pressure drop in the die (along the specific flow paths) is equal to the sum of the pressure drops in the manifold and the land, that is

$$\Delta p(x) = \Delta p_m(x) + \Delta p_s(x) \tag{5.201}$$

where $\Delta p(x)$ is the total pressure drop; $\Delta p_m(x)$ is the pressure drop in the manifold; $\Delta p_s(x)$ is the pressure drop in the land.

These pressure drops can be determined on the basis of the Pouiseille flow solutions in Eqs. (4.79) and (4.18), using the concept of representative viscosity, and considering due to symmetry of the system half of the die (Figure 5.33). According to these solutions

$$\Delta p_m(x) = \int_x^W \frac{8\overset{*}{\eta}_m(x)Q_m(x)}{\pi R^4(x)}\mathrm{d}x \tag{5.202}$$

$$\Delta p_s(x) = \frac{12\overset{*}{\eta}_s(x)Q_0}{WH^3}y(x) \tag{5.203}$$

where $\overset{*}{\eta}_m(x)$, $\overset{*}{\eta}_s(x)$ is the representative viscosity in the distribution channel (manifold) and the forming slit (land), respectively; $y(x)$ is the land length; H is the land height.

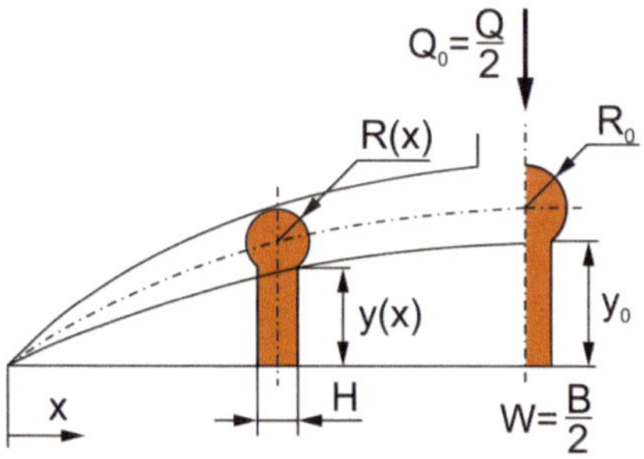

Figure 5.33 Geometry of a distributor: R_0 - maximum radius of manifold, y_0 - maximum height of land gap

Then, the condition in Eq. (5.200), using the relationship in Eq. (5.192), has the form

$$\frac{d\Delta p(x)}{dx} = \frac{d}{dx}\left[\int_{x}^{W} \frac{8\overset{*}{\eta}_m(x)Q_0}{\pi R^4(x)}\frac{x}{W}dx + \frac{12\overset{*}{\eta}_s(x)Q_0}{WH^3}y(x)\right] = 0 \tag{5.204}$$

The condition (c), the constant shear stress, indicates that the representative viscosity of the material does not depend on the x-position, such that

$$\overset{*}{\eta}_m = \text{const} \neq f(x) \tag{5.205}$$

and

$$\overset{*}{\eta}_s = \text{const} \neq f(x) \tag{5.206}$$

and then, the dependence in Eq. (5.204) can be transformed into the form

$$\frac{dy(x)}{dx} = \frac{\overset{*}{\eta}_m}{\overset{*}{\eta}_s}\frac{2H^3 x}{3\pi R^4(x)} \tag{5.207}$$

and next, after using the relationship in Eq. (5.194), can be written as

$$\frac{dy(x)}{dx} = \frac{\overset{*}{\eta}_m}{\overset{*}{\eta}_s}\frac{2H^3 x}{3\pi R_0^4\left(\frac{x}{W}\right)^{4/3}} \tag{5.208}$$

Integrating the expression in Eq. (5.208) leads to the formula defining the length of the forming land:

$$y(x) = \frac{\overset{*}{\eta}_m}{\overset{*}{\eta}_s}\frac{H^3 W^{4/3} x^{2/3}}{\pi R_0^4} \tag{5.209}$$

which, after taking into account the condition that $y(x = W) = y(x = B/2) = y_0$, can be expressed as

$$y(x) = y_0\left(\frac{x}{W}\right)^{2/3} \tag{5.210}$$

or

$$y_0 = \frac{\overset{*}{\eta}_m}{\overset{*}{\eta}_s}\frac{H^3 B^2}{4\pi R_0^4} \tag{5.211}$$

From the condition in Eq. (5.195), it follows that the representative viscosity in the distribution channel and the forming gap should be equal, i. e.

$$\overset{*}{\eta}_m = \overset{*}{\eta}_s \tag{5.212}$$

such that after taking Eq. (5.198) into account, the following relationship is obtained:

$$y_0 = \frac{1}{4}\left(3\frac{e_h}{e_r}\right)^{4/3}\left(\pi HB^2\right)^{1/3} \tag{5.213}$$

or

$$y_0 = 1.474\left(HB^2\right)^{1/3} \tag{5.214}$$

Equations (5.194), (5.199), (5.210), and (5.214) define the geometry of the slit die for extruding flat profiles. They provide a constant average flow velocity of the material flowing out of the die (independent of the position) and the same shear stress (representative shear rate) throughout the flow space.

The total pressure drop in the die can be determined on the basis of Eq. (5.203) for $y(x = W) = y_0$. Then we have the following relationship:

$$\Delta p = \frac{12\eta_s^* Q}{BH^3} y_0 \tag{5.215}$$

where $Q = 2Q_0$ is the total flow rate.

The polymer temperature increase in the die can be determined, assuming adiabatic conditions, using Eq. (5.168).

The residence time of the polymer in the die results from the flow velocity and the length of the flow path, and is expressed by the formula

$$t(x) = \int_x^W \frac{1}{v_m(x)} \mathrm{d}x + \frac{y(x)}{v_s(x)} \tag{5.216}$$

where $t(x)$ is the flow time; v_m, v_s is the velocity in the distribution channel and in the forming land, respectively.

The mean residence time can be determined based on the ratio of the average flow velocity at the die exit to the largest length of the forming land y_0, that is

$$t_{av} = \frac{BHy_0}{Q} \tag{5.217}$$

5.3.5 Annular Dies

Annular dies, i.e. dies with an annular exit cross-section (forming section), are used for the extrusion of pipes, hoses, tubular films, for extrusion blow molding and for cable coating.

These dies can be classified as:

- center fed, when the extruder axis coincides with the die axis (e.g. dies for extrusion of pipes),

- side fed, when the extruder axis forms a certain angle with the die (e.g. wire coating dies).

Generally, for design reasons, several types of tubular dies can be distinguished, that is, in the group of **center fed dies**:

- mandrel support dies (spider supported),
- screen pack dies,
- spiral mandrel dies,

and, in the group of **side fed dies**:

- dies with a cylindrical mandrel,
- dies with a tapered mandrel.

5.3.5.1 Center Fed Mandrel Dies

Center fed dies usually consist of elements with circular or annular channel cross-sections. The flow modeling and calculations in such channels are presented in Section 5.3.2. When calculating this type of dies, the method of dividing the flow channel and the method of the channel shape correction are used, which are presented in Section 5.3.6.

5.3.5.2 Side Fed Mandrel Dies

In the side fed dies, the even distribution of material (so that it is thermally and mechanically homogeneous in all its volume) is realized by means of the properly designed distributors (the mandrels), cylindrical (Figure 5.34) or tapered (Figure 5.35). The purpose of using the mandrel is to distribute the material around the circumference of the die so that the velocity of its outflow is independent on the position, and its properties are as uniform as possible. The conditions for fulfilling these requirements are the same as for the flat profile dies. They can therefore be determined on the basis of conditions given in Eqs. (5.189), (5.190), and (5.191), except that they will be in the function of the variable β, i.e. the cylindrical coordinate in the circumferential direction.

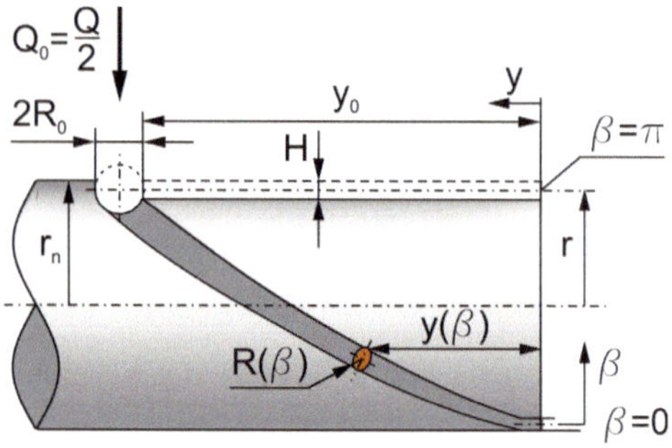

Figure 5.34 Geometry of a cylindrical mandrel: Q - volume flow rate, $R(\beta)$ - radius of distribution channel, R_0 - maximum radius of distribution channel, $y(\beta)$ - land length, y_0 - maximum land length, H - land height, r_n - radius of mandrel, r - mean radius of forming land, β - cylindrical coordinate in the direction of die circumference, y - coordinate along the mandrel axis

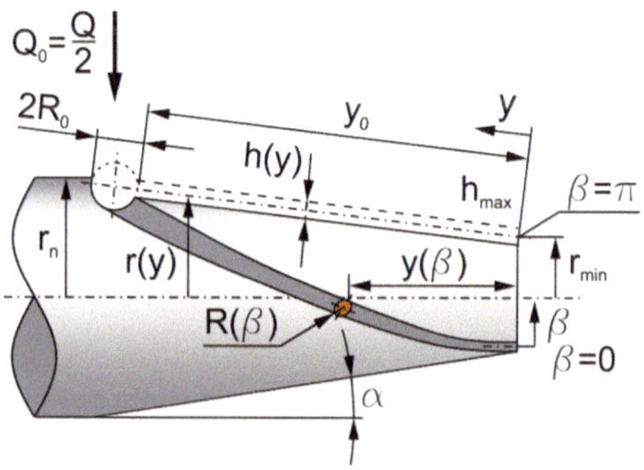

Figure 5.35 Geometry of a tapered mandrel: Q - volume flow rate, $R(\beta)$ - radius of distribution channel, R_0 - maximum radius of distribution channel, $y(\beta)$ - land length, y_0 - maximum land length, $h(y)$ - land height, h_{max} - maximum land height, $r(y)$ - mean radius of forming land, r_n - maximum mandrel radius, r_{min} - minmum radius of forming land, α - taper angle, β - cylindrical coordinate in the direction of die circumference, y - coordinate along the mandrel axis

Thus, these conditions have the following form:

a) the condition of constant mean velocity of the flowing out material

$$V(\beta) = \text{const} \tag{5.218}$$

b) the condition of constant total pressure drop along all flow paths

$$\Delta p(\beta) = \text{const} \tag{5.219}$$

c) the condition of constant shear stress, which is equivalent to the constant shear rate condition

$$\dot{\gamma}(\beta) = \text{const} \tag{5.220}$$

The calculation method [30] will be also the same as in the calculation of flat slit dies. As with these dies, the material distribution system (the mandrel) consists of the distribution channel (the manifold of circular cross section with variable radius) and the forming slit (the land). However, this time the land is in the form of an annulus with a constant thickness (at the cylindrical manifold) or with variable thickness (at the tapered mandrel).

In general, the solution of the problem of annular dies design consists in transforming the solutions obtained in relation to the slit dies in the Cartesian system to the solutions in the cylindrical system. It should be clearly stated that such a procedure is only permissible if the diameter of the annular slit is much greater than its thickness, i. e. $D/H >> 8$.

Side Fed Dies with a Cylindrical Mandrel

The geometry of the cylindrical mandrel (Figure 5.34), when unfolded onto the plane, is identical to the geometry of the flat die. This enables the use of the calculation method used in Section 5.3.4.

In order to meet the condition (a), expressed with the condition of Eq. (5.218), it is necessary to change the linear flow rate in the distribution channel $Q_m(\beta)$, i. e. according to the dependence

$$Q_m(\beta) = Q_0 \frac{\beta}{\pi} \tag{5.221}$$

where $Q_m(\beta)$ is the flow rate in the manifold; $Q_0 = Q/2$ is one-half of the extruder output; β is the coordinate in the direction of the die circumference.

On the basis of the condition (c), expressed by Eq. (5.220), the geometry of the manifold can be determined. The radius $R(\beta)$ of the manifold can be determined from the constant shear rate condition at the channel wall. While maintaining this condition, based on Eq. (4.81), the following formula is obtained:

$$\frac{4Q_m(\beta)}{\pi R^3(\beta)} = \frac{4Q_0}{\pi R_0^3} \tag{5.222}$$

from which, after taking into account Eq. (5.221), it follows that

$$R(\beta) = R_0 \left(\frac{\beta}{\pi}\right)^{1/3} \tag{5.223}$$

The unknown initial radius R_0 can be determined from the condition of uniform shear rate in the manifold and in the land. This condition is expressed by the corresponding equality of the representative shear rates:

$$\dot{\gamma}_m^* = \dot{\gamma}_s^* \tag{5.224}$$

where $\dot{\gamma}_m^*$ is the representative shear rate in the manifold; $\dot{\gamma}_s^*$ is the representative shear rate in the land.

These rates are determined, on the basis of Table 4.1 and Eqs. (5.188) and (5.222), by the following expressions:

$$\dot{\gamma}_m^* = \frac{4Q_0}{\pi R_0^3} e_r \tag{5.225}$$

$$\dot{\gamma}_s^* = \frac{6Q_0}{\pi r H^2} e_h \tag{5.226}$$

where r is the radius of the annular forming land; $e_r = 0.815$; $e_h = 0.772$.

When applying the condition in Eq. (5.224), this leads to the relationship between the manifold radius R_0 and the dimensions of the extruded profile in the form

$$R_0 = \left(\frac{2}{3} \frac{e_r}{e_h} r H^2 \right)^{1/3} \tag{5.227}$$

or, after substituting the values e_r and e_h, to the formula

$$R_0 = 0.889 \left(r H^2 \right)^{1/3} \tag{5.228}$$

Based on the condition (b), i.e. the constant total pressure drop along all flow paths, the length of the forming land y as a function of the position along the channel width, i.e. $y(\beta)$, can be determined. This condition is expressed as

$$\mathrm{d}\Delta p(\beta) / \mathrm{d}\beta = 0 \tag{5.229}$$

The total pressure drop in the die (along the specific flow paths) is equal to the sum of the pressure drops in the manifold and the land, that is

$$\Delta p(\beta) = \Delta p_m(\beta) + \Delta p_s(\beta) \tag{5.230}$$

where $\Delta p(\beta)$ is the total pressure drop; $\Delta p_m(\beta)$ is the pressure drop in the manifold; $\Delta p_s(\beta)$ is the pressure drop in the land.

These pressure drops can be determined on the basis of the Pouiseille flow solutions in Eqs. (4.79) and (4.18), using the concept of representative viscosity, and considering due to symmetry of the system half of the die (Figure 5.34). Then we have

$$\Delta p_m(\beta) = \int_{\beta}^{\pi} \frac{8\eta_m^*(\beta) Q_m(\beta)}{\pi R^4(\beta)} r \mathrm{d}\beta \tag{5.231}$$

$$\Delta p_s(\beta) = \frac{12\eta_s^*(\beta) Q_0}{\pi r H^3} y(\beta) \tag{5.232}$$

where $\overset{*}{\eta}_m(\beta)$, $\overset{*}{\eta}_s(\beta)$ is the representative viscosity in the manifold and in the forming land; $y(\beta)$ is the land length; r is the mean radius of the annular forming land; H is the land height.

Then, the condition in Eq. (5.229), using the relationship in Eq. (5.221), expresses as

$$\frac{\mathrm{d}\Delta p(\beta)}{\mathrm{d}\beta} = \frac{\mathrm{d}}{\mathrm{d}\beta}\left[\int_{\beta}^{\pi} \frac{8\overset{*}{\eta}_m(\beta) Q_0}{\pi R^4(\beta)} \frac{\beta}{\pi} r\mathrm{d}\beta + \frac{12\overset{*}{\eta}_s(\beta) Q_0}{\pi r H^3} y(\beta)\right] = 0 \tag{5.233}$$

The condition (c), the constant shear stress, indicates that the representative viscosity of the material does not depend on the β-position, such that

$$\overset{*}{\eta}_m = \text{const} \neq f(\beta) \tag{5.234}$$

$$\overset{*}{\eta}_s = \text{const} \neq f(\beta) \tag{5.235}$$

and then, Eq. (5.233) can be transformed into the form

$$\frac{\mathrm{d}y(\beta)}{\mathrm{d}\beta} = \frac{\overset{*}{\eta}_m}{\overset{*}{\eta}_s} \frac{2H^3 r^2}{3\pi R^4(\beta)} \beta \tag{5.236}$$

and next, after using Eq. (5.223), can be written as

$$\frac{\mathrm{d}y(\beta)}{\mathrm{d}\beta} = \frac{\overset{*}{\eta}_m}{\overset{*}{\eta}_s} \frac{2H^3 r^2}{3\pi R_0^4 \left(\frac{\beta}{\pi}\right)^{4/3}} \beta \tag{5.237}$$

Integrating the Eq. (5.237) leads to the formula defining the length of the forming land:

$$y(\beta) = \frac{\overset{*}{\eta}_m}{\overset{*}{\eta}_s} \frac{H^2 r^3 \pi^{1/3} \beta^{2/3}}{R_0^4} \tag{5.238}$$

which, after taking into account the condition that $y(\beta = \pi) = y_0$, can be expressed as

$$y(\beta) = y_0 \left(\frac{\beta}{\pi}\right)^{2/3} \tag{5.239}$$

or

$$y_0 = \frac{\overset{*}{\eta}_m}{\overset{*}{\eta}_s} \frac{\pi r^2 H^3}{R_0^4} \tag{5.240}$$

From the condition in Eq. (5.224), it follows that the representative viscosity in the distribution channel and the forming gap should be equal, i. e.

$$\overset{*}{\eta}_m = \overset{*}{\eta}_s \tag{5.241}$$

such that, after taking Eq. (5.227) into account, the following relationship is obtained

$$y_0 = \left(\frac{3}{2}\frac{e_h}{e_r}\right)^{4/3} \pi\left(Hr^2\right)^{1/3} \tag{5.242}$$

or

$$y_0 = 5.016\left(Hr^2\right)^{1/3} \tag{5.243}$$

Equations (5.223), (5.228), (5.239), and (5.243) define the geometry of the annular die with cylindrical mandrel. They provide a constant average flow velocity of the material flowing out of the die (independent of the position) and the same shear stress (representative shear rate) throughout the flow space.

The total pressure drop in the die can be determined on the basis of Eq. (5.232) for $y(\beta = \pi) = y_0$. Then we have the following relationship

$$\Delta p = \frac{12\eta_s^* Q}{2\pi r H^3} y_0 \tag{5.244}$$

where $Q = 2Q_0$ is the total flow rate.

The polymer temperature increase in the die can be determined, assuming adiabatic conditions, using Eq. (5.168).

The residence time of the polymer in the die results from the flow velocity and the length of the flow path, and is expressed by the formula

$$t(\beta) = \int_{\beta}^{\pi} \frac{1}{v_m(\beta)} \mathrm{r d}\beta + \frac{y(\beta)}{v_s(\beta)} \tag{5.245}$$

where $t(\beta)$ is the flow time; v_m, v_s is the velocity in the distribution channel and in the forming land, respectively.

The mean residence time can be determined based on the ratio of the average flow velocity at the die exit to the largest length of the forming land y_0, that is

$$t_{av} = \frac{2\pi r H y_0}{Q} \tag{5.246}$$

It is worth noting that the solutions obtained in this section, at the substitutions $x = r\beta$, $\mathrm{d}x = r\mathrm{d}\beta$, $W = \pi r$, correspond to the results obtained in the case of the slit die.

Side Fed Dies with a Tapered Mandrel

The geometry of the tapered mandrel (Figure 5.35), when unfolded onto the plane, is similar to the geometry of the cylindrical mandrel, differing only in the variable thickness of the forming slit. Since the radius of the cone $r(y)$ decreases in the direction of flow, the thickness of the slit $h(y)$ must increase to maintain the constant flow velocity of the material. It is worth noting that at the cone angle $\alpha = 0$, the tapered mandrel becomes the cylindrical one.

This geometrical similarity makes it possible to use for tapered dies the calculation method used previously for cylindrical and flat dies [30].

The variable dimensions of the forming land can be determined by applying the condition (c) to maintain the constant shear rate in this land. The basis will be Eq. (5.226), which in this case has the form

$$\dot{\gamma}_s^* = \frac{6Q_0}{\pi r(y)h^2(y)} e_h \tag{5.247}$$

where $r(y)$ is the cone radius; $h(y)$ is the slit thickness; y is the coordinate in the direction of the slit length.

On the basis of this equation and geometrical relations resulting from Figure 5.35, the following formulas describing the geometry of the forming land [30] are obtained:

$$h(y) = h_{max}\left(\frac{r_{min}}{r(y)}\right)^{1/2} \tag{5.248}$$

$$r(y) = r_{min} + y\sin\alpha \tag{5.249}$$

$$r_n = r_{min} + (2R_0 + y_0)\sin\alpha + \frac{h_{max}}{2}\cos\alpha \tag{5.250}$$

The remaining dimensions defining the geometry of the annular die with a tapered mandrel are expressed as follows [30]:

- the radius of the distribution channel (manifold)

$$R(\beta) = R_0\left(\frac{\beta}{\pi}\right)^{1/3} \tag{5.251}$$

- the maximum radius of the distribution channel (manifold)

$$R_0 = 0.889\left(r_{min}h_{max}^2\right)^{1/3} \tag{5.252}$$

- the length of the forming land

$$y(\beta) = y_0\left(\frac{\beta}{\pi}\right)^{2/3}\left(\frac{\pi^{1/3}\beta^{2/3}h_{max}^3 r_{min}\sin\alpha + 4R_0^4}{\pi h_{max}^3 r_{min}\sin\alpha + 4R_0^4}\right) \tag{5.253}$$

- the maximum length of the forming land

$$y_0 = h_{max}^{1/3} r_{min}^{1/3}\left(6.29h_{max}^{1/3}\sin\alpha + 5.016r_{min}^{1/3}\right) \tag{5.254}$$

It is worth noting that at the cone angle $\alpha = 0$, the tapered mandrel becomes cylindrical, and then the obtained formulas, after substitutions $r_{\min} = r$, $h_{\max} = H$, are transformed respectively into the solutions in Eqs. (5.223), (5.228), (5.239), and (5.243) obtained in the case of the cylindrical die.

The total pressure drop in the die and the mean residence time of the material can be determined on the basis of the following formulas [30]:

$$\Delta p = \frac{4\eta_s^* Q}{\pi h_{max}^3 r_{min}^{2/3} \sin\alpha}\left[\left(r_{min} + y_0 \sin\alpha\right)^{3/2} - r_{min}^{3/2}\right] \tag{5.255}$$

$$t_{av} = \frac{4\pi h_{max} r_{min}^{1/2}}{3\sin\alpha Q}\left[\left(r_{min} + y_0 \sin\alpha\right)^{3/2} - r_{min}^{3/2}\right] \tag{5.256}$$

where Q is the total flow rate.

The polymer temperature increase in the die can be determined, assuming adiabatic conditions, using Eq. (5.168).

5.3.6 Profile Dies

The computations presented in the previous sections concerned the dies with relatively simple cross-sections, i. e. with a circular, annular, or flat cross-section. And, although some of these computations, e. g. regarding annular or flat dies, are quite complex, there are well-known exact solutions of one-dimensional flow equations.

In polymer processing, however, dies with a much more complex geometry are often used. These dies with irregular cross-sections are called profile dies.

In this case, the only accurate method of design is flow modeling using numerical methods, the finite difference method (FDM) or finite element method (FEM). However, such modeling is difficult and time-consuming (expensive), and therefore reluctantly used in industrial practice. For this reason, substitute methods are used.

There are two basic substitute methods, that is, the method of dividing the channel cross-section and the method of correcting the channel shape. Sometimes, a simple general cross-section method is used.

5.3.6.1 Computation by Cross-Section Division

The method of dividing the cross-section of the channel is derived from the basic principles of the flow in serial and parallel channel connections. These are expressed by Eqs. (5.165) and (5.166).

This method consists in the division of the complex cross-sections of flow channels into simple-shaped elements. Then, such channels can be considered, with some approximation (e. g. end effects are omitted), as the parallel connection of elementary channels with a simple shape. This is described in the assumptions of Eq. (5.166).

An example of using this method is the calculation of ribbed profiles with thin, flat or curved ribs (Figure 5.36). These profiles, with some approximation, can be considered as a parallel arrangement of elementary flat channels of the same height.

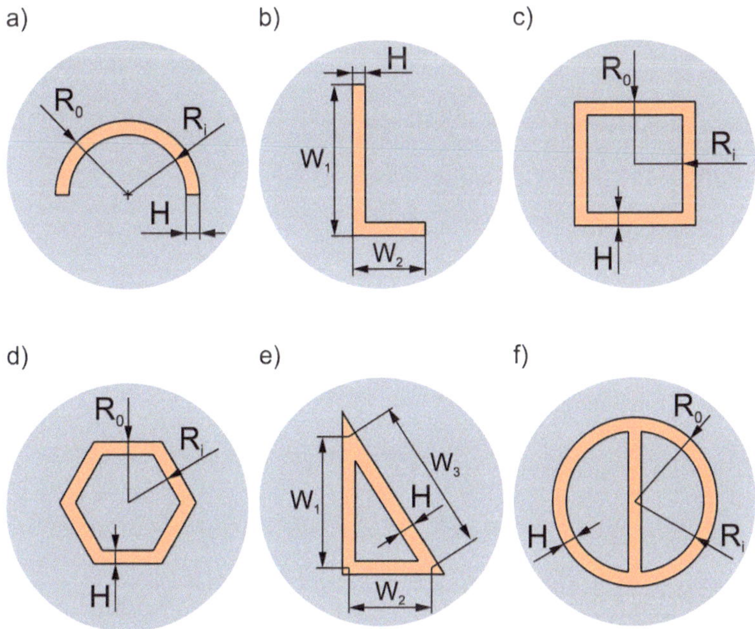

Figure 5.36 Ribbed profiles which can be calculated based on the flow approximation in a flat channel with height H and equivalent widths W^*: a) $W^* = 0.5\pi(R_0 + R_i)$, b) $W^* = W_1 + W_2$, c) $W^* = 4(R_0 + R_i)$, d) $W^* = 3.46(R_0 + R_i)$, e) $W^* = W_1 + W_2 + W_3$, f) $W^* = \pi R_0 + (2 + \pi)R_i$

In such channels, for example using the representative viscosity method, the following relationship (Eq. (4.17)) between the flow rate and the pressure gradient is valid:

$$Q = \frac{WH^3}{12\eta_s^*}\frac{\Delta p}{L} \tag{5.257}$$

where η_s^* is the representative viscosity in the flat channel.

Then, the relationship in Eq. (5.166) has the form

$$\Delta p = \Delta p_i = \text{const} \quad \text{and} \quad Q = \sum_{i=1}^{k}\frac{W_i H_i^3}{12\eta_{si}^*}\frac{\Delta p_i}{L_i} \tag{5.258}$$

where Δp is the total pressure drop; Δp_i is the pressure drop in the ith elementary channel; Q is the total flow rate; Q_i is the flow rate in the ith elementary channel;

W_i, H_i, L_i are the dimensions of the elementary channel; $\overset{*}{\eta}_{si}$ is the viscosity in the elementary channel.

Assuming that the viscosity is the same in the entire flow area, i.e. $\overset{*}{\eta}_{si} = \overset{*}{\eta}_s =$ const, and that each elementary channel has the same height and length, i.e. $H_i = H =$ const and $L_i = L =$ const, we get

$$Q = \frac{H^3 \Delta p}{12\overset{*}{\eta}_s L} \sum_{i=1}^{k} W_1 \tag{5.259}$$

and, after transformation

$$\Delta p = \frac{12\overset{*}{\eta}_s L}{W^* H^3} Q \tag{5.260}$$

where W^* is the equivalent width of the flat channel, which is expressed by

$$W^* = \sum_{i=1}^{k} W_1 \tag{5.261}$$

The equivalent width of the selected ribbed profile is given in Figure 5.36.

This method can be used for many other channels, including those in which the shape of elementary channels is not the same. An example of such a channel is shown in Figure 5.37.

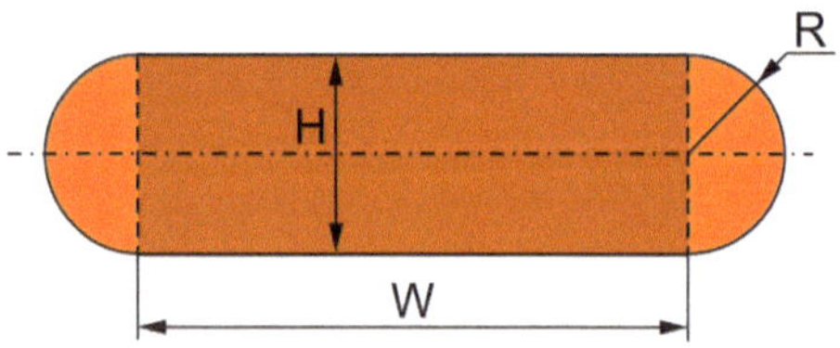

Figure 5.37
Division of cross-section of the channel composed of a rectangle and two semi-circles

This channel can be divided into two elementary channels, a flat channel and a circular channel. Then, based on the dependence in Eq. (5.166), we have

$$Q = Q_s + Q_c \tag{5.262}$$

and next, based on Eqs. (4.17) and (4.78) and applying the representative viscosity, we get

$$Q = \left(\frac{WH^3}{12\overset{*}{\eta}_s} + \frac{\pi R^4}{8\overset{*}{\eta}_c} \right) \frac{\Delta p}{L} \tag{5.263}$$

where Q_s, Q_c, $\overset{*}{\eta}_s$, $\overset{*}{\eta}_c$ are the flow rates and the representative viscosities in the flat and circular channels, respectively.

After transformation of Eq. (5.263), we obtain the expression determining the pressure drop:

$$\Delta p = \frac{QL}{\left(\frac{WH^3}{12\overset{*}{\eta}_s} + \frac{\pi R^4}{8\overset{*}{\eta}_c}\right)} \tag{5.264}$$

Since the pressure drop in both parts of the channel is the same, i.e. $\Delta p = \Delta p_s = \Delta p_c = \text{const}$, we can then write

$$\frac{\Delta p}{L} = \frac{12\overset{*}{\eta}_s Q_s}{WH^3} = \frac{8\overset{*}{\eta}_c Q_c}{\pi R^4} \tag{5.265}$$

and, on the basis of this relationship, the representative viscosities $\overset{*}{\eta}_s$ and $\overset{*}{\eta}_c$ can be determined in the iterative procedure.

The presented method of dividing the cross-sections of flow channels has practical applications, e.g. for calculations of typical annular dies, such as mandrel supported dies, screen pack dies, and spiral mandrel dies.

The basic method of dividing the cross-section is often extended to division in the longitudinal direction, and then the classical system of parallel and serial connections of elementary channels is obtained (Figure 5.30). Of course, calculations become more complex then and one must use iterative computer methods.

5.3.6.2 Computation by Shape Correction

The basis of the shape correction method is flow modeling in the channels with complex geometry using numerical methods, e.g. finite difference methods (FDM) or finite element methods (FEM).

The shape correction method consists in applying for pressure drop calculations in the channels of various cross-sections, e.g. rectangular, semicircle, ellipse etc. (Figure 5.38), the flat flow solutions corrected by the **correction factor**, which depends on the shape and dimensions of these channels. Such factors are determined, as functions of the characteristic dimensions of channels, by performing accurate numerical calculations of the pressure gradient in various channels. These calculations are referred to the pressure calculations in the flat channel with dimensions equal to the characteristic dimensions.

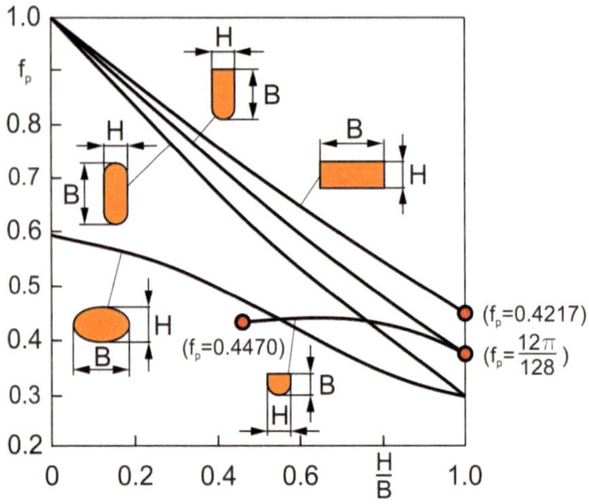

Figure 5.38 Correction factor f_p for different channels as a function of characteristic dimensions, *H* and *B*, adapted from [30]

Thus, the correction factor can be described by the formula

$$f_p = \frac{(\mathrm{d}p/\mathrm{d}x)_{flat}}{(\mathrm{d}p/\mathrm{d}x)_{true}} \tag{5.266}$$

where f_p is the shape correction factor; $(\mathrm{d}p/\mathrm{d}x)_{flat}$ is the pressure gradient calculated according to the flat flow solution (in the channel of height *H* and width *B*); $(\mathrm{d}p/\mathrm{d}x)_{true}$ is the true pressure gradient calculated by using the FDM or FEM method (or measured) in the channel considered (characterized by two dimensions, height *H* and width *B*).

Calculations according to this method consist in determining the pressure gradient for a flat channel with dimensions of width *B* and height *H*, equal to the characteristic dimensions of the considered flow channel, and then improving the obtained result with the shape correction factor f_p.

This procedure corresponds to the transformation of Eq. (5.266) to give

$$(\mathrm{d}p/\mathrm{d}x)_{true} = \frac{(\mathrm{d}p/\mathrm{d}x)_{flat}}{f_p} \tag{5.267}$$

where $(\mathrm{d}p/\mathrm{d}x)_{true}$ is determined by Eq.(4.18) as

$$(\mathrm{d}p/\mathrm{d}x)_{flat} = \frac{12\eta}{BH^3}Q \tag{5.268}$$

where η is the polymer viscosity; Q is the volume flow rate; B, H are the characteristic dimensions.

After substituting Eq. (5.268) into Eq. (5.267), the formula is obtained that allows us to calculate the pressure drop in the channels of various cross-sections:

$$\Delta p = \frac{12\eta QL}{BH^3}\frac{1}{f_p} \tag{5.269}$$

where Δp is the pressure drop in the channel of the length L and the characteristic dimensions, width B and height H; $f_p = f_p(H/B)$ is the shape factor.

The shape factor for various channels, depending on the ratio of characteristic dimensions, i. e. H/B, is shown in Figure 5.38. It is worth adding that such dependencies can be easily approximated with appropriate regression equations and can be used in this form in calculations.

Although determining the shape coefficients is time consuming, using them in calculations is quick and easy, and most importantly it gives good results.

The simplest and least accurate method used to calculate the flows in irregularly shaped channels is the general cross-section method using the concept of hydraulic radius.

This method consists in performing calculations according to the flow model through the circular channel whose diameter is determined based on the **hydraulic radius** of the channel with an irregular cross-section. This radius is defined by

$$R_h = \frac{2A}{U} \tag{5.270}$$

where R_h is the hydraulic radius; A is the channel cross-section; U is the circumference of the channel cross section.

After substituting Eq. (5.270) into the Newtonian solution of the flow through the circular channel (Eq. (4.78)), we have

$$\frac{\Delta p}{L} = \frac{2\eta(\dot{\gamma}_h)U^2 Q}{A^3} \tag{5.271}$$

where $\eta(\dot{\gamma}_h)$ is the viscosity determined for the hydraulic shear rate $\dot{\gamma}_h$. Shear rate can be determined using Eq. (4.81)

$$\dot{\gamma}_h = \frac{2QU}{A^2} \tag{5.272}$$

5.4 Global Modeling

Extrusion is a continuous process of co-operation of the extruder/die system. Physical phenomena occurring in the extruder determine the flow in the extrusion die and vice versa, the flow in the die determines the phenomena in the extruder. Any change in the processing conditions in the extruder causes a change in the processing conditions in the die and vice versa. Thus, modeling of the extrusion process cannot be limited to modeling the flow in the extruder and must include the flow in the extrusion die.

The term global modeling means modeling of the interacting phenomena occurring in the extruder and the die, that is, modeling of the extruder/die system (or screw/die system).

The classical global modeling of extrusion process is based on the separate models for each section of the screw, i.e. solid transport section, melting and pre-melting sections, and melt flow section, and the global model consists of these elementary models (Figure 5.1). In this case, the computations are performed step-by-step in the elementary segments of the screw, and the output process parameters from the current segment are the input parameters for the next one.

Global modeling requires the use of a computation algorithm appropriate for a given type of extrusion. For classical (flood fed) extrusion, the forward scheme of computation is suitable. In this case, the flow rate of the material is not known, and it is the result of the screw/die co-operation and must be determined in multiple iterative computations. For extrusion with starvation, the backward scheme of computation is applied, i.e. the inverse computation algorithm since the pressure profile is not continuous here, and there is no continuous the flow rate/pressure relation. In this case, the flow rate is established and equal to the flow rate of the material metered by the dosing device. The issue of global modeling was recently discussed in detail in the literature [32].

References

1. Schenkel, G., *Schneckenpressen für Kunststoffe* (1959) Carl Hanser Verlag, Munich
2. McKelvey, J., *Polymer Processing* (1962) John Wiley & Sons, New York
3. Schenkel, G., *Plastics Extrusion* (1966) Illife Books Ltd., London
4. Pearson, J.R.A., *Mechanical Principles of Polymer Melt Processing* (1966) Pergamon Press, Oxford
5. Fenner, R.T., *Extruder Screw Design* (1966) Illife-Butterworth, London
6. Tadmor, Z., Klein, I., *Engineering Principles of Plasticating Extrusion* (1970) Van Nostrand Reinhold Company, New York
7. Plajer, O., *Praktische Rheologie für Kunststoffschmelzen* (1970) Zechner und Hüthig, Zürich
8. Lenk, R.S., *Rheologie der Kunststoffe* (1971) Carl Hanser Verlag, Munich
9. Torner, R.W., *Grundprocesse der Verarbeitung von Polymeren* (1974) VEB Deutscher Verlag, Leipzig

10. Han, Ch. D., *Rheology in Polymer Processing* (1976) Academic Press, New York
11. Pearson, J.R. A., Richardson, S.M., *Computational Analysis of Polymer Processing* (1983) Elsevier Publisher, London
12. Stevens, M.J., *Extruder Principles and Operation* (1985) Elsevier Publisher, London
13. Hensen, F., Knappe, W., Potente, H., *Handbuch der Kunststoff-Extrusiontechnik. Grundlagen* (1989) Carl Hanser Verlag, Munich
14. White, J.L., *Principles of Polymer Engineering Rheology* (1990) John Wiley & Sons, New York
15. Knappe, W., Lampl, A., Heuel, O., *Kunststoff-Verarbeitung and Werkzeugbau* (1992) Carl Hanser Verlag, Munich
16. O'Brien, K.T., *Computer Modeling for Extrusion and Other Continuos Polymer Processes* (1992) Hanser Publishers, New York
17. White, J.L., *Rubber Processing* (1995) Hanser Publishers, Munich
18. Baird, D.G., Collias, D.I., *Polymer Processing. Principles and Design* (1995) Butterworth-Heinemann, Oxford
19. Belofsky, H., *Plastics Product Design and Process Engineering* (1995) Hanser Publishers, Munich
20. Griskey, R.G., *Polymer Process Engineering* (1995) Chapman and Hall, New York
21. White, J.L., Potente, H., *Screw Extrusion* (2002) Carl Hanser Verlag, Munich
22. Tadmor, Z., Gogos, C.G., *Principles of Polymer Processing* (2006) John Wiley & Sons, New York
23. Osswald, T., Hernandez-Ortiz, J.P., *Polymer Processing: Modeling and Simulation* (2006) Hanser Publishers, Munich
24. Chung, Ch., I., *Extrusion of Polymers* (2010) Hanser Publishers, Munich
25. Vlachopoulos, J., *Polymer Rheology and Extrusion* (2011) Mc Master University, Hamilton
26. Rao, N.S., Schott, N.R., *Plastics Engineering Calculations* (2012) Hanser Publishers, Munich
27. Campbell, G.A., Spalding M.A., *Analyzing and Troubleshooting Single-Screw Extruders* (2013) Hanser Publishers, Munich
28. Rauwendaal, Ch., *Polymer Extrusion* (2014) Hanser Publishers, Munich
29. Agassant, J.F., Avenas, P., Carreau, P.J., Vergnes, B., Vincent, M., *Polymer Processing: Principles and Modeling* (2017) Hanser Publishers, Munich
30. Hopmann, Ch., Michaeli, W., *Extrusion Dies for Plastics and Rubber* (2017) Hanser Publishers, Munich
31. Vlachopoulos, J., Polychronopoulos, N.D., *Understanding Rheology and Technology of Polymer Extrusion* (2019) Polydynamics Inc., Dundas
32. Wilczyński, K., Nastaj, A., Lewandowski, A., Wilczyński, K.J., Buziak, K., *Fundamentals of Global Modeling for Polymer Extrusion, Polymers* (2019), 11(12), 2106; https://doi.org/10.3390/polym11122106

6 Computer Modeling for Polymer Processing

6.1 Overview of Computer Modeling Software

Software for computer modeling of polymer processing can be divided into two groups: **generally oriented systems** (i.e. general purpose) and **systems oriented on a specific processing technique** (e.g. extrusion or injection molding). The operation of these systems is based on solving equations of continuum mechanics (mass, momentum, and energy) for specific material models under specific flow conditions.

In the group of generally oriented systems, i.e. CFD systems (Computational Fluid Dynamics), several program packages can be distinguished, e.g.:

- ANSYS Polyflow (ANSYS, Inc., Canonsburg, PA, USA),
- ANSYS Fluent (ANSYS, Inc., Canonsburg, PA, USA),
- Autodesk CFD (Autodesk, Inc., San Rafael, CA, USA),
- OpenFOAM (The OpenFOAM Foundation, Ltd., London, UK),
- FLOW-3D (Flow Science, Inc., Pasadena, CA, USA),
- COMSOL Multiphysics (COMSO, Inc., Burlington, MA, USA).

The second group of modeling systems concerns clearly formulated issues and is limited to a specific type of processing techniques, e.g. extrusion and injection molding.

In the group of systems for simulation of injection molding (more precisely for simulation of flow in injection molds), the following program packages can be distinguished:

- Autodesk Moldflow (Autodesk, Inc., San Rafael, CA, USA),
- Moldex3D (CoreTech System Co., Ltd., Chupei City, Hsinchu County, Taiwan),
- CADMOULD 3D-F (Simcon GmbH, Wuerselen, Germany),
- SIGMASOFT Virtual Molding (SIGMA Engineering GmbH, Aachen, Germany),

- SIMULIA (Dassault Systemes, Vélizy-Villacoublay, France),
- REM 3D (TRANSFALOR S. A., Sophia Antipolis, France).

It is important to note that the only available software for simulation of flow in the injection molding machine is

- PSI (Paderborn University, Paderborn, Germany).

In the group of systems for simulation of extrusion, the following program packages for simulation of single screw extrusion can be distinguished:

- EXTRUD (Scientific Process & Research, Inc., Somerset, NJ, USA) - historically the first,
- NEXTRUCAD (Polydynamics, Inc., Dundas, Ontario, Canada),
- REX (Paderborn University, Paderborn, Germany),
- COMPUPLAST (Compuplast International Inc., Zlin, Czech Republic),
- WIN SSD (Stevens Institute of Technology, Hoboken, NJ, USA),
- CEMEXTRUD (CEMEF - Centre for Materials Processing, Ecole des Mines, Sophia Antipolis, France),
- GSEM (Warsaw University of Technology, Warsaw, Poland).

In the case of twin screw extrusion, the following program packages for simulation of co-rotating extrusion can be distinguished:

- AKRO-CO-TWIN SCREW (The University of Akron, Akron, OH, USA) - historically the first,
- SIGMA (Paderborn University, Paderborn, Germany),
- Ludovic (Sciences Computers Consultants, Inc., Saint-Etienne, France),
- WinTXS (PolyTech, New Haven, CT, USA).

as well as for simulation of counter-rotating extrusion:

- AKRO-COUNTER-TWIN SCREW (University of Akron, Akron, OH, USA) - historically the first,
- TSEM (Warsaw University of Technology, Warsaw, Poland).

Modeling with generally oriented systems is different from modeling with specific oriented systems. The basis of the first type of modeling is to define the considered problem each time, while in the second case the problem to be solved is already defined. Apart from this difference, modeling using various systems is similar. In any case, it consists in entering into the program data on the flow geometry, flow conditions, and material characteristics, and then performing computations.

The basic flow problems in polymer processing can be solved using CFD software, and the software for modeling of injection molding and extrusion. The selected flow problems presented in the further part of this chapter were solved using

ANSYS Polyflow [28] for CFD modeling, Autodesk Moldflow [29] for injection molding, and MULTI-SCREW [30] for extrusion, which includes GSEM and TSEM programs. Recently, several fundamental issues of modeling in the polymer processing were presented in a review paper [31].

6.2 CFD Modeling

Co-authored by Adrian Lewandowski

6.2.1 ANSYS Polyflow – Program Overview

ANSYS Polyflow is a finite-element computational fluid dynamics (CFD) program for simulating applications where viscous and viscoelastic flows play an important role. The flows can be isothermal or non-isothermal, two- or three-dimensional, steady-state or time-dependent. ANSYS Polyflow is used primarily to solve flow problems in polymer and rubber processing; however, it is also used to solve rheology problems of other materials such as glass or foodstuffs.

ANSYS Polyflow can be used to solve various flow problems which are encountered in polymer processing, e. g.:

- three-dimensional flows in extrusion dies and extruder plasticating units,
- extrudate swelling,
- extrudate swelling inverse problem,
- co-extrusion,
- blow molding,
- thermoforming,
- compression molding,
- fiber forming,
- film casting,
- mixing.

ANSYS Polyflow can be used to solve several types of flows, ranging from Newtonian inelastic to non-Newtonian viscoelastic. Several shear-rate-dependent and temperature-dependent viscosity models are supplied. It is, however, essential to be aware that none of these models leads to realistic predictions in all types of deformation of any particular non-Newtonian fluid.

ANSYS Polyflow offers a wide variety of constitutive models, which can be divided into three groups:

a) the generalized Newtonian models (Section 1.2.4.2): Newtonian, power law, Bird-Carreau, Cross, Bingham, Herschel-Bulkley, log-log, Carreau-Yasuda,

b) the differential viscoelastic models (Section 1.3.5): Maxwell, Oldroyd-B, White-Metzner, Phan Thien-Tanner (PTT), Giesekus, FENE-P, POM-POM, Leonov,

c) the integral viscoelastic models (Section 1.3.5): Doi-Edwards, Lodge-Maxwell, Papanastasiou-Scriven-Macosko (PSM), K-BKZ (Kaye, Bernstein, Kearsley, Zapas).

If the flow is non-isothermal, the viscosity dependence on temperature must be taken into account (Section 1.2.3.1) using e.g. the Arrhenius law or Williams-Landel-Ferry law (WLF).

The ANSYS Polyflow package includes the following modules:

a) DesignModeler, SpaceClaim, and Meshing, the preprocessors for geometry modeling and mesh generation,

b) Polydata, the preprocessor for problem definition,

c) Polymat, the preprocessor for material data specification,

d) Polyflow, the solver,

e) Polystat, the statistical postprocessor for quantitative comparison of flows,

f) CFD-Post, the graphical postprocessor for examining results.

6.2.2 Modeling Procedure

Modeling with ANSYS Polyflow consists of:

- modeling of flow geometry and mesh generation (DesignModeler, SpaceClaim, Meshing),
- defining the problem and specifying material data (Polydata, Polymat),
- solving the problem (Polyflow) and examining the results (CFD-Post).

The essence of CFD modeling is the ability to model a variety of issues. However, it is necessary to properly define the problem currently under consideration, i.e. to create its appropriate model. Defining the problem is performed using the Polydata module and consists in general definition of the problem and then determination of the boundary (and/or initial) conditions for the equations of momentum and energy. The rules for general definition are common to different types of flow, while the rules for determining boundary/initial conditions change depending on the problem under consideration.

Generally defining the problem consists in defining the main parameters of the task, which determine:

- flow geometry, e.g. 2D or 2½D axisymmetric flow, 2D or 2½D flat flow, 3D flow,
- dependence on time, i.e. steady-state or unsteady-state flow,
- dependence on temperature, i.e. isothermal or non-isothermal flow,
- rheological model, i.e. the generalized Newtonian model, the differential viscoelastic model, or the integral viscoelastic model.

Defining the boundary/initial conditions consists in defining the flow conditions, for the equation of momentum and energy, on the flow boundaries.

Solving a problem in ANSYS Polyflow requires that you prescribe information about the flow along the boundary of the computational domain for each sub-task. Solving a time-dependent flow also requires that you prescribe initial conditions within the domain.

The basic boundary conditions available in ANSYS Polydata for the equation of momentum are as follows:

a) **zero wall velocity** – allows you to set the normal and tangential velocity components on the boundary section to zero,

b) **interface condition** – establishes the continuity between two sides of an intersection between different sub-tasks, and it guarantees continuity of the velocity field and the contact forces,

c) **normal and tangential velocities condition (v_n & v_s)** – allows you to specify the normal and tangential velocity components on the boundary section, the default condition corresponds to the zero wall velocity condition ($v_n = 0$, $v_s = 0$),

d) **normal and tangential force condition (f_n & f_s)** – allows you to specify the normal and tangential force components on the boundary section, these are the surface force densities expressed, e.g., in Pa,

e) **normal velocity and tangential force condition (v_n & f_s)** – allows you to specify the normal velocity and tangential force on the boundary section,

f) **normal force and tangential velocity condition (f_n & v_s)** – allows you to specify the normal force and tangential velocity on the boundary section,

g) **slip condition** – when the slip condition is applied to non-moving parts, a zero normal velocity component is imposed simultaneously with one of three relationships between the shear force and the tangential relative velocity,

h) **axis of symmetry condition** – establishes the symmetry of flow relative to the axis of symmetry (axisymmetric flow), and it is equivalent to imposing zero values for the normal velocity and tangential force on the boundary section,

i) **plane of symmetry condition** – establishes the symmetry of flow relative to the plane of symmetry, and it is equivalent to imposing zero values for the normal velocity and tangential force on the boundary section,

j) **inflow condition** - allows you to specify a flow rate across a boundary section surrounded by other boundary sections with well-defined adhesion conditions (slip, no-slip, symmetry, and so on),

k) **outflow condition** - for generalized Newtonian flow it is similar to the normal force and tangential velocity condition with ($f_n = 0$, $v_s = 0$),

l) **free surface condition** - when the free surface condition is applied along boundary sections, surface forces are imposed while the fluid cannot cross the free surface.

m) **Cartesian velocity condition (v_x & v_y & v_z)** - allows you to prescribe a rigid-body velocity field on the boundary section.

The basic boundary conditions available in ANSYS Polydata for the equation of energy are as follows:

a) **temperature imposed condition** - specifies the temperature on the boundary section, as a constant value for temperature or a value dependent on the coordinates,

b) **flux density imposed condition** - allows you to specify the heat flux on the boundary section by providing its three components: conduction, convection, and radiation,

c) **insulated boundary condition** - establishes the zero heat flux conduction component on the boundary section,

d) **outflow condition** - is equivalent to the insulated boundary option, but it does not allow defining of the convective component of the heat flux.

6.2.3 Examples of Modeling

ANSYS Polyflow can be used to solve various flow problems in the polymer processing. In the further part of this section, the examples of modeling of important class of flows are presented, i. e. the pressure flow, the extrudate swell, and extrudate swell inverse problem, as well as the single screw extrusion flow and the twin screw extrusion flow, both co-rotating and counter-rotating. Some details of this modeling are presented in the literature [32–35].

6.2.3.1 Pressure Flow

Pressure flows are an important class of flows in rheology and polymer processing (Section 4.2.1). These are defined as the flows that take place in a closed channel due to the pressure difference occurring along this channel, i. e. these are pressure driven flows. These flows occur, among others, in capillary rheometers, extrusion dies, and injection molds, as well as in plasticating units of extruders and injection molding machines.

In this section, the convergent pressure flows (cylindrical and rectangular) are considered, both Newtonian and non-Newtonian (shear thinning and shear thickening).

The flow problem has been defined as a steady-state, isothermal, 3D generalized Newtonian flow, and the power law model of Ostwald-de Waele was assumed (Eq. (1.158)):

$$\eta(\dot{\gamma}) = m\dot{\gamma}^{n-1} \tag{6.1}$$

where m, n are the model parameters; m is the consistency index, $(N{\cdot}s^n)/m^2$; n is the power law exponent (dimensionless).

The material parameters are as follows:

- Newtonian viscosity, μ — $6 \cdot 10^3$ Pa·s
- consistency index, m — $6 \cdot 10^3$ Pa·s^n
- power law exponent, n — 0.39 / 1.3
- density, ρ — 740 kg/m^3

and the following flow conditions are assumed:

- flow rate, G — 20 kg/h
- flow geometry:
  a) length of sections — 30/55/75 mm
  b) diameter of sections (inflow, outflow) — 45/10 mm
  c) width/height of sections (inflow, outflow) — 45/20/30/5 mm

Cylindrical Flow

The geometry of cylindrical flow is depicted in Figure 6.1. The flow is bounded by three boundaries, and the following boundary conditions were applied at the flow boundaries:

- at the inlet to the channel: boundary BC1 – **inflow,** the flow rate $(Q_{in}) = (Q)$ is imposed,
- at the outlet of the channel: boundary BC2 – **outflow,** vanishing normal forces and tangential velocities are imposed $(f_n \,\&\, v_s) = (0, 0)$, which means that the flow rate $(Q_{out}) = (Q)$ is imposed,
- on the channel wall: boundary BC3 – **zero wall velocity,** vanishing normal and tangential velocities are imposed $(v_n \,\&\, v_s) = (0, 0)$, which means that the velocity vanishes at the channel wall.

Simulations were performed to study the flow parameters distributions for Newtonian flow ($n = 1$) and non-Newtonian flow, both shear thinning ($n < 1$) and shear thickening ($n > 1$). The simulation results are shown in the cross-sections marked in Figure 6.2.

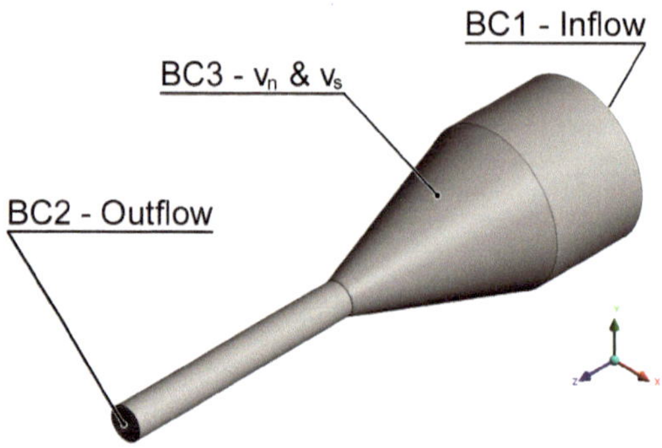

Figure 6.1 The flow geometry and boundary conditions for convergent cylindrical pressure flow: BC1 - inflow, BC2 - outflow, BC3 - zero wall velocity

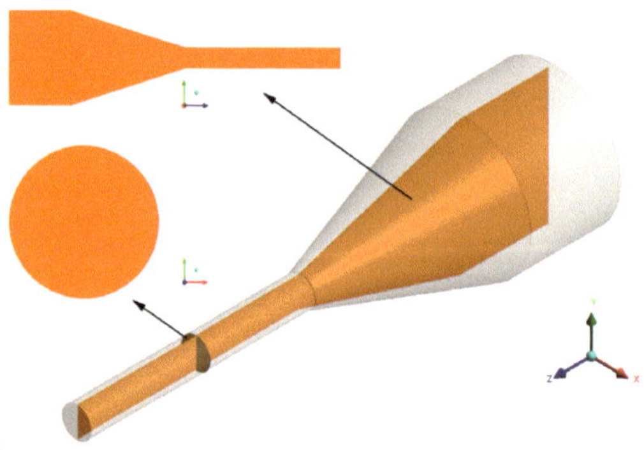

Figure 6.2 Cross-sections of results presentation for convergent cylindrical flow

Distributions of the velocity, shear rate, and viscosity for the Newtonian and non-Newtonian (shear thinning and shear thickening) flow are depicted in Figure 6.3 to Figure 6.5. These are compared with each other in Figure 6.6 to Figure 6.8.

Distribution of the shear rate, which is a derivative of the velocity, is linear in the case of Newtonian flow and non-linear in the case of non-Newtonian flow (Figure 6.7). Since the shear stress distribution is also linear, the result is that the viscosity is constant in the case of Newtonian flow and varies in the case of non-Newtonian flow (Figure 6.8). For the pseudoplastic flow ($n < 1$) we have a maximum in the centerline, and in the case of dilatant flow ($n > 1$) we have a minimum in the centerline. This discussion based on the FEM computations can be compared with analytical computations presented in Section 4.2.1.2.

Pressure distributions are depicted in Figure 6.9. It is clearly seen that the highest pressure gradient is observed for the dilatant flow ($n > 1$), and the smallest one is for the pseudoplastic flow ($n < 1$). Moreover, when the cylinder diameter decreases, i. e. the flow resistance increases, the pressure gradient increases.

Distributions of the velocity, shear rate, and viscosity along the cylinder for the Newtonian and non-Newtonian (shear thinning and shear thickening) flow are depicted In Figure 6.10 to Figure 6.12. It is interesting to note that for the Newtonian flow the viscosity is constant throughout the entire volume of flow. For the non-Newtonian flow, both pseudoplastic ($n < 1$) and dilatant ($n > 1$), the viscosity varies, and it is different at each point of the flow,

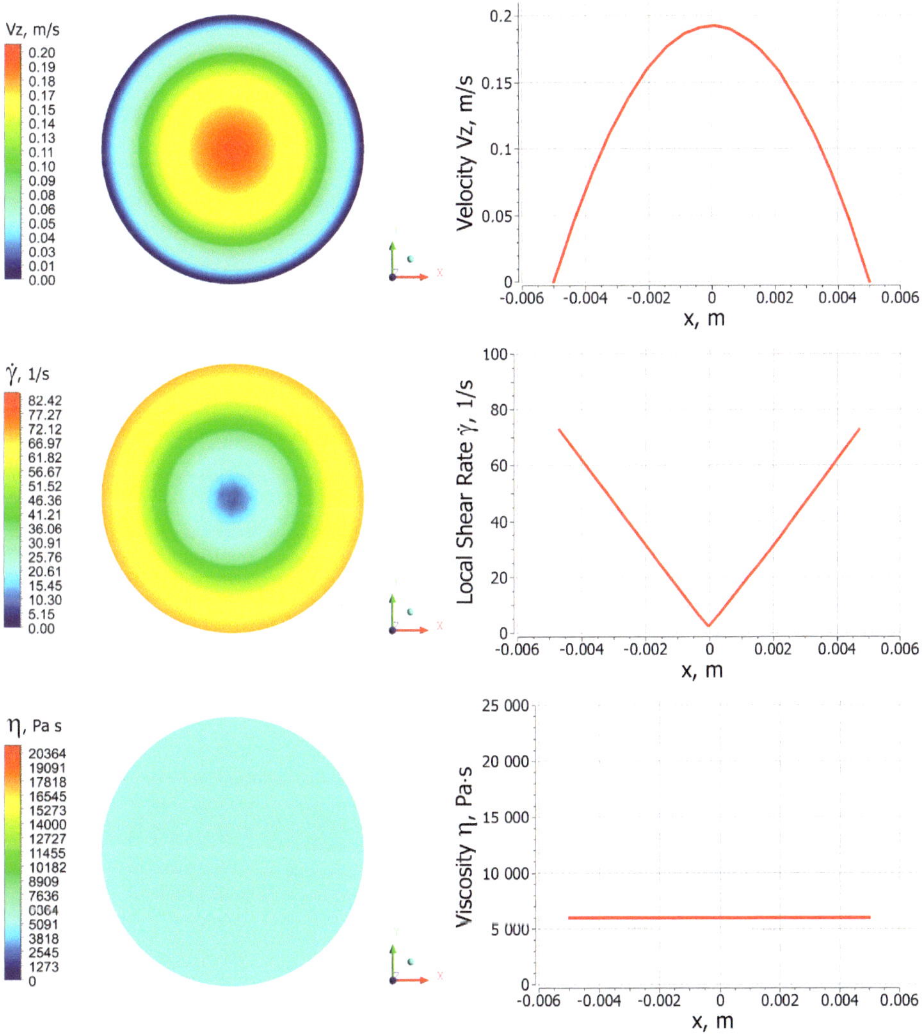

Figure 6.3 Distributions of the velocity, shear rate, and viscosity for the Newtonian flow ($n = 1$)

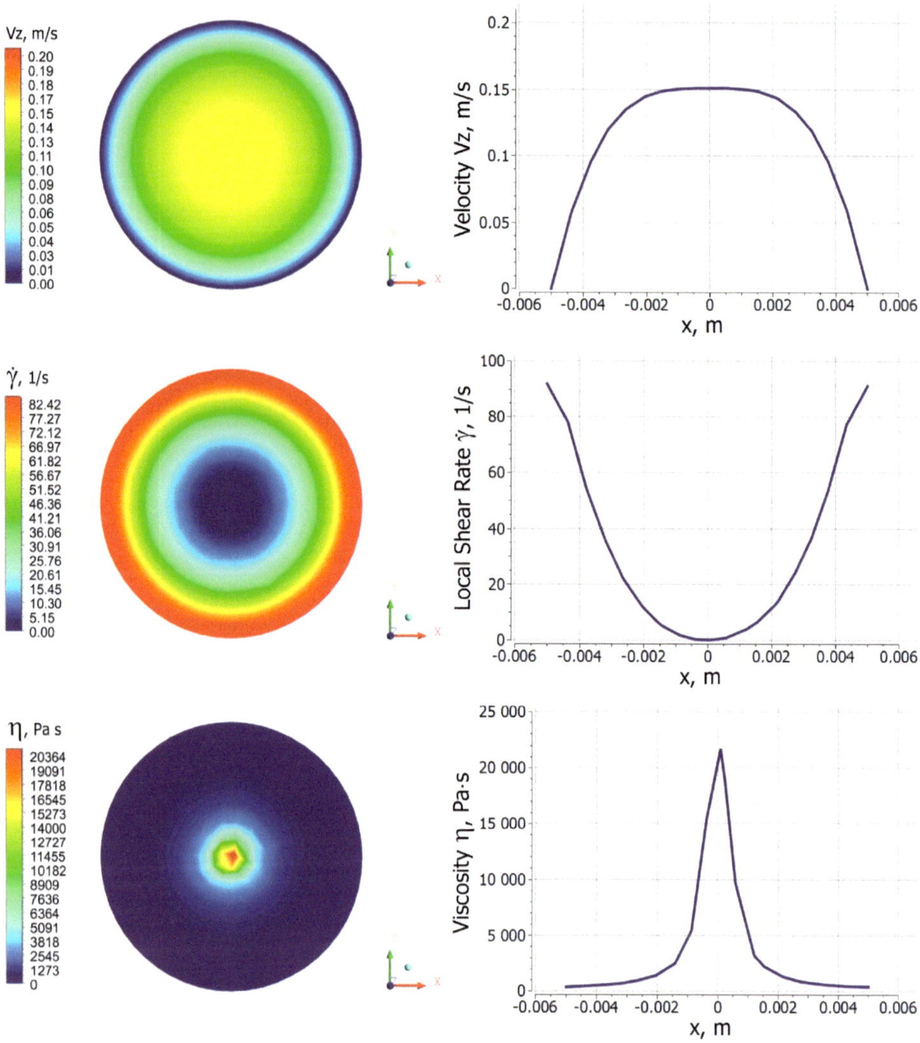

Figure 6.4 Distributions of the velocity, shear rate, and viscosity for the non-Newtonian pseudoplastic flow ($n = 0.39$)

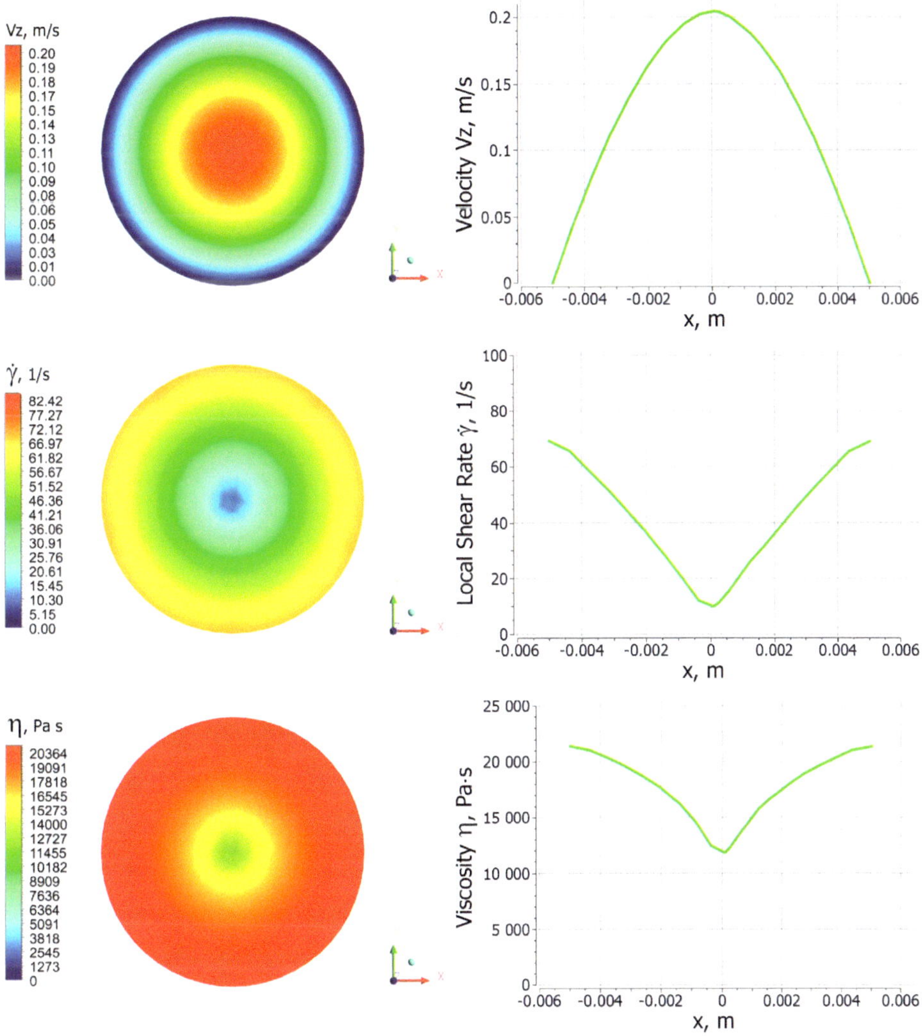

Figure 6.5 Distributions of the velocity, shear rate, and viscosity for the non-Newtonian dilatant flow ($n = 1.3$)

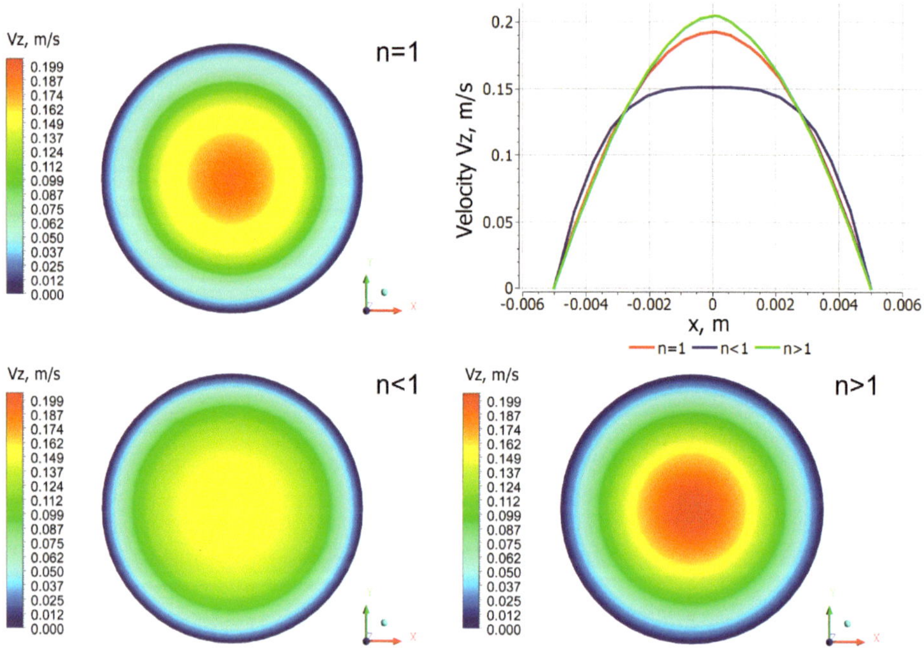

Figure 6.6 Velocity distributions for the Newtonian flow ($n = 1$), non-Newtonian pseudoplastic flow ($n = 0.39$), and non-Newtonian dilatant flow ($n = 1.3$)

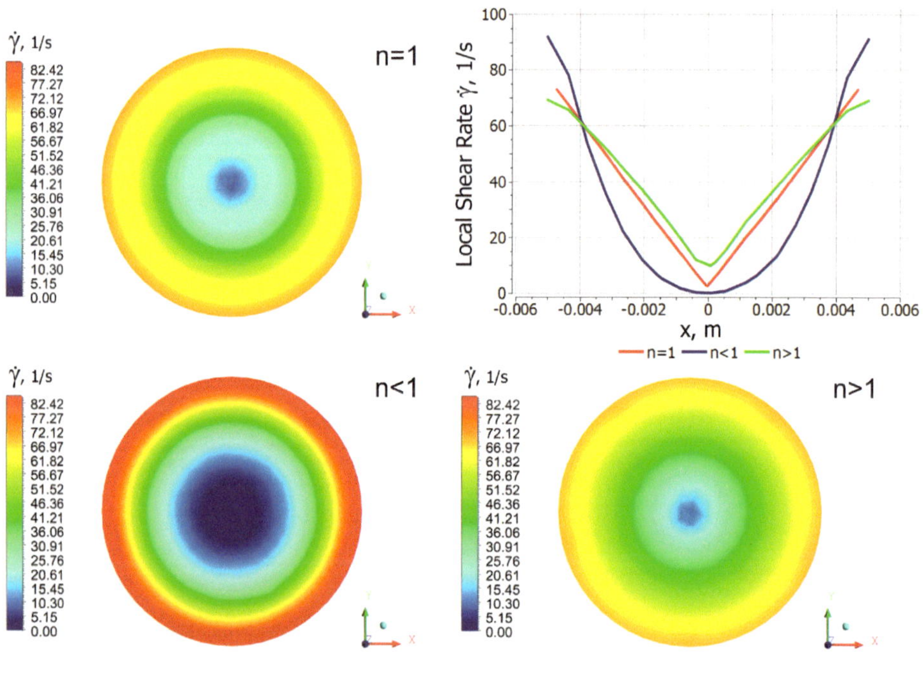

Figure 6.7 Shear rate distributions for the Newtonian flow ($n = 1$), non-Newtonian pseudoplastic flow ($n = 0.39$), and non-Newtonian dilatant flow ($n = 1.3$)

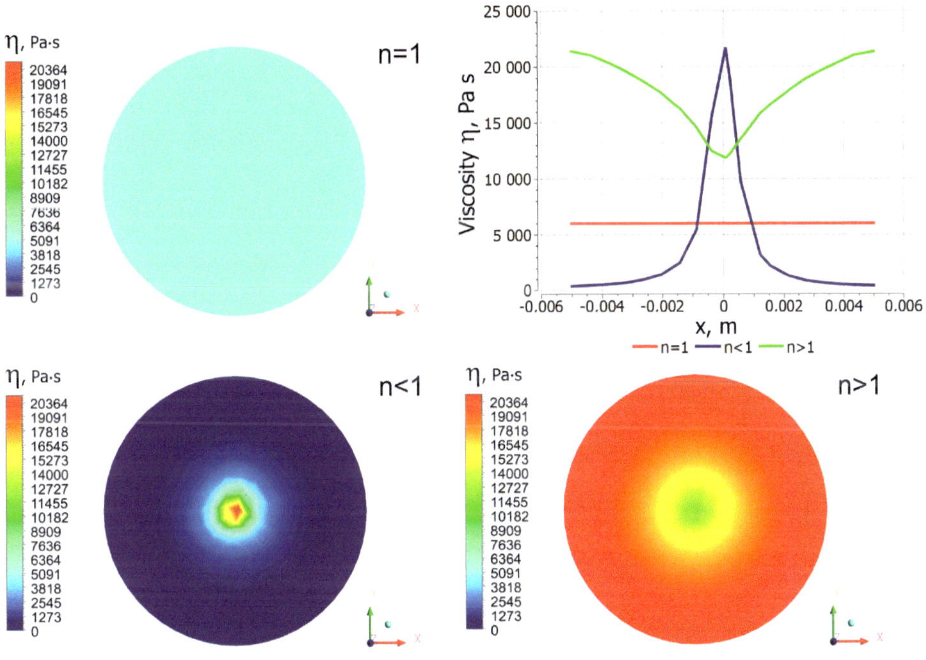

Figure 6.8 Viscosity distributions for the Newtonian flow ($n = 1$), non-Newtonian pseudoplastic flow ($n = 0.39$), and non-Newtonian dilatant flow ($n = 1.3$)

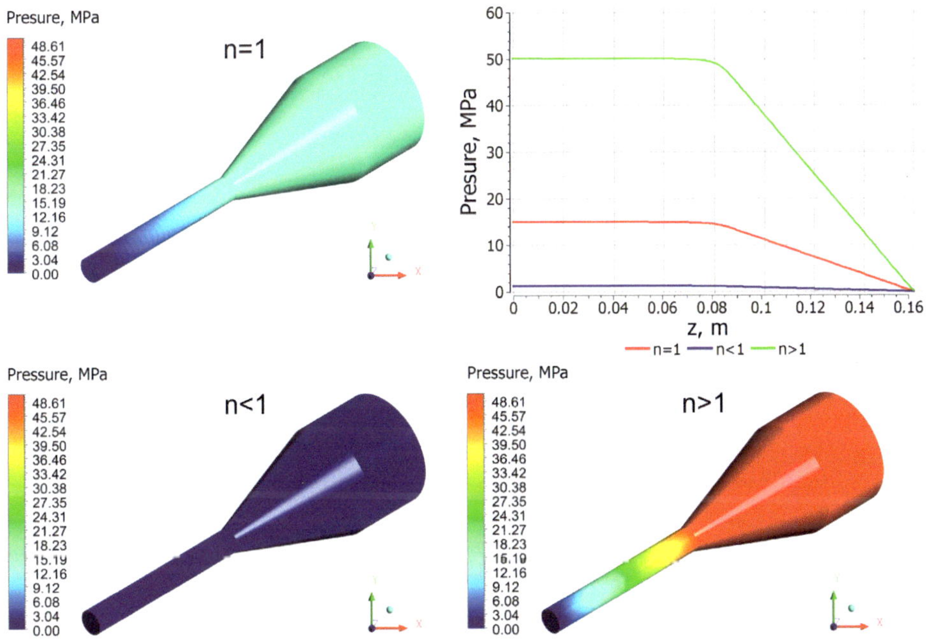

Figure 6.9 Pressure distributions for the Newtonian flow ($n = 1$), non-Newtonian pseudoplastic flow ($n = 0.39$), and non-Newtonian dilatant flow ($n = 1.3$)

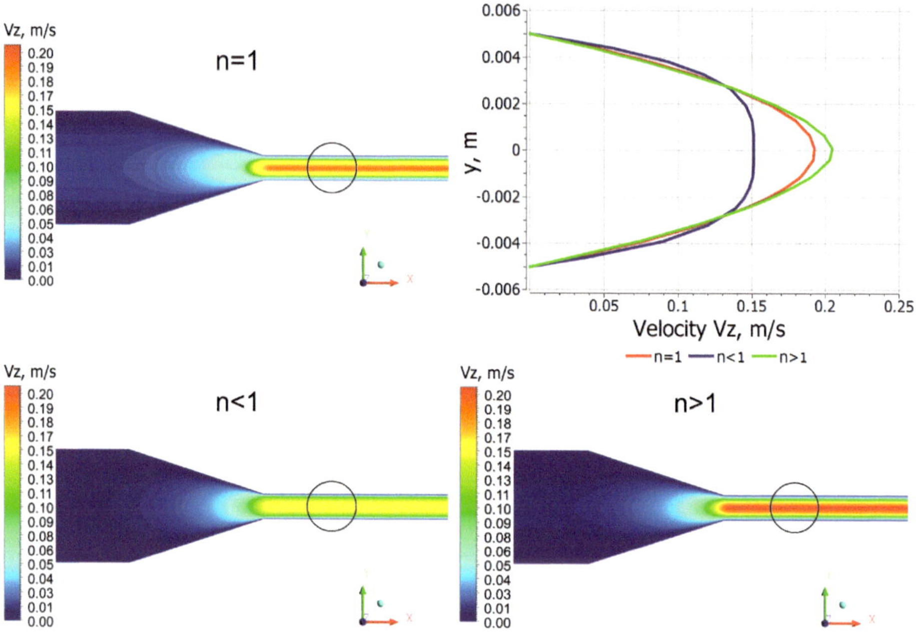

Figure 6.10 Velocity distributions along the cylinder for the Newtonian flow ($n = 1$), non-Newtonian pseudoplastic flow ($n = 0.39$). and non-Newtonian dilatant flow ($n = 1.3$)

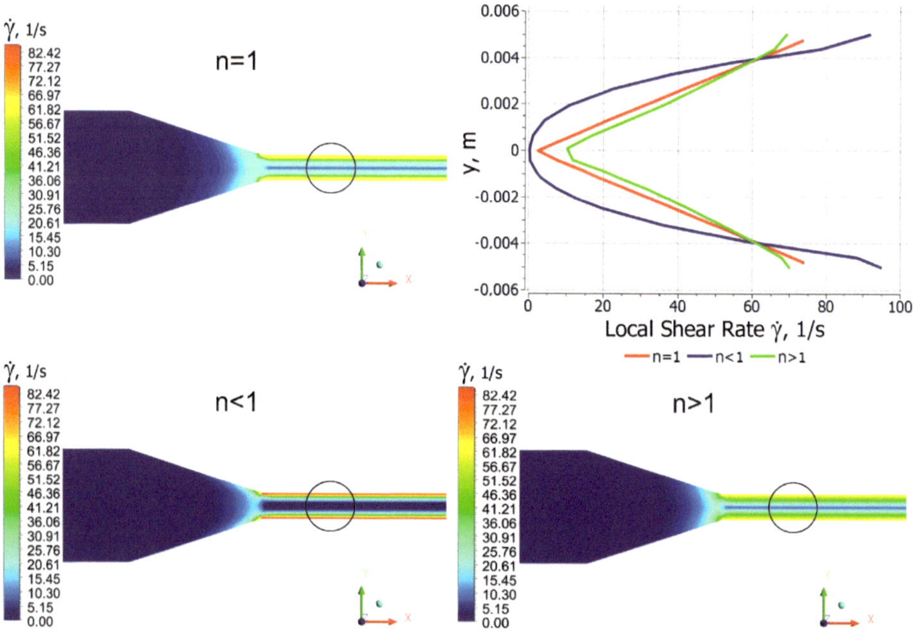

Figure 6.11 Shear rate distributions along the cylinder for the Newtonian flow ($n = 1$), non-Newtonian pseudoplastic flow ($n = 0.39$), and dilatant flow ($n = 1.3$)

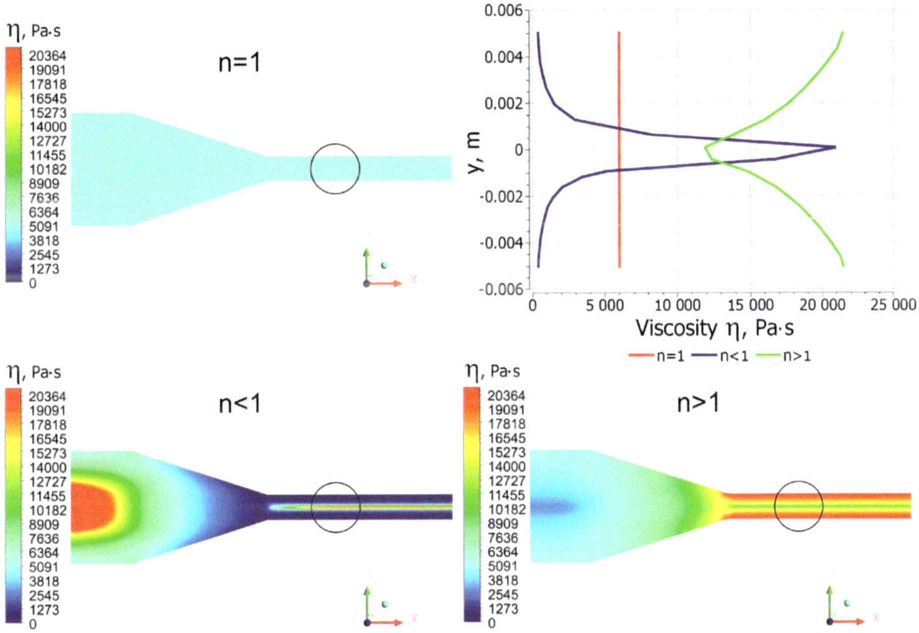

Figure 6.12 Viscosity distributions along the cylinder for the Newtonian flow ($n = 1$), non-Newtonian pseudoplastic flow ($n = 0.39$), and non-Newtonian dilatant flow ($n = 1.3$)

Rectangular Flow

The geometry of rectangular flow is depicted in Figure 6.13. The flow is bounded by three boundaries, and the following boundary conditions were applied at the flow boundaries:

- at the inlet to the channel: boundary BC1 - **inflow,** the flow rate $(Q_{in}) = (Q)$ is imposed,
- at the outlet of the channel: boundary BC2 - **outflow,** vanishing normal forces and tangential velocities are imposed $(f_n \ \& \ v_s) = (0, 0)$, which means that the flow rate $(Q_{out}) = (Q)$ is imposed,
- on the channel wall: boundary BC3 - **zero wall velocity,** vanishing normal and tangential velocities are imposed $(v_n \ \& \ v_s) = (0, 0)$, which means that the velocity vanishes at the channel wall.

Simulations were performed to study the flow parameters distributions for Newtonian flow ($n = 1$) and non-Newtonian flow, both shear thinning ($n < 1$) and shear thickening ($n > 1$). The simulation results are shown in the cross-sections marked in Figure 6.14.

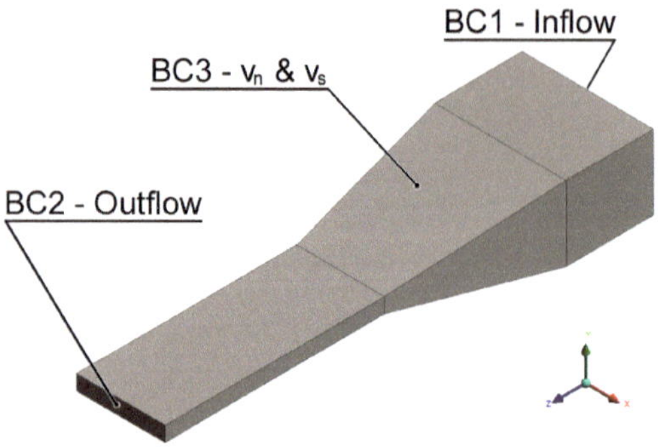

Figure 6.13 The flow geometry and boundary conditions for convergent rectangular pressure flow: BC1 - inflow, BC2 - outflow, BC3 - zero wall velocity

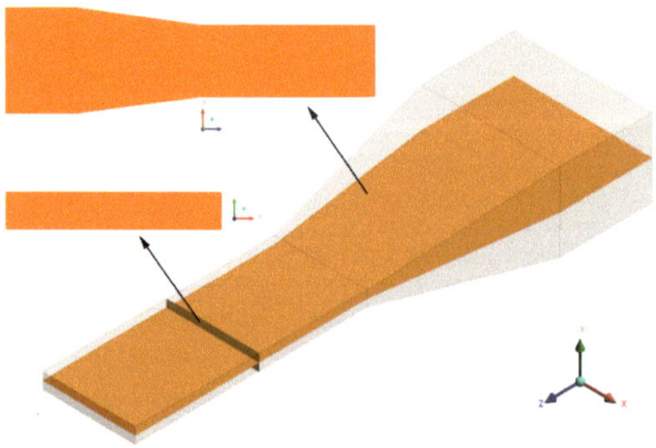

Figure 6.14 Cross-sections of results presentation for convergent rectangular flow

Distributions of the velocity, shear rate, and viscosity for the Newtonian and non-Newtonian (shear thinning and shear thickening) flow are depicted in Figure 6.15 to Figure 6.17. These are compared with each other in Figure 6.18 to Figure 6.20.

The analysis of simulation results in the case of rectangular flow is similar to that of cylindrical flow. However, it is important to note that distributions of flow parameters are here substantially different in the cross-section along the height of the channel and the cross-section along the width of the channel. Distribution of the shear rate, which is a derivative of the velocity, is linear in the case of Newtonian flow and non-linear in the case of non-Newtonian flow (Figure 6.19). Since the shear stress distribution is also linear, the result is that the viscosity is constant in the case of Newtonian flow and varies in the case of non-Newtonian flow (Figure 6.20). For the pseudoplastic flow ($n < 1$) we have a maximum in the centerline, and in the case of dilatant flow ($n > 1$) we have a minimum in the centerline. This discussion based on the FEM computations can be compared with analytical computations presented in Section 4.2.1.1.

Pressure distributions are depicted in Figure 6.21. It is clearly seen that the highest pressure gradient is observed for the dilatant flow ($n > 1$), and the smallest one is for the pseudoplastic flow ($n < 1$). Moreover, when the channel dimensions decrease, i.e. the flow resistance increases, the pressure gradient increases.

Distributions of the velocity, shear rate, and viscosity along the channel for the Newtonian and non-Newtonian (shear thinning and shear thickening) flow are depicted in Figure 6.22 to Figure 6.24. It is interesting to note that for the Newtonian flow the viscosity is constant throughout the entire volume of flow. For the non-Newtonian flow, both pseudoplastic ($n < 1$) and dilatant ($n > 1$), the viscosity varies, and it is different at each point of the flow.

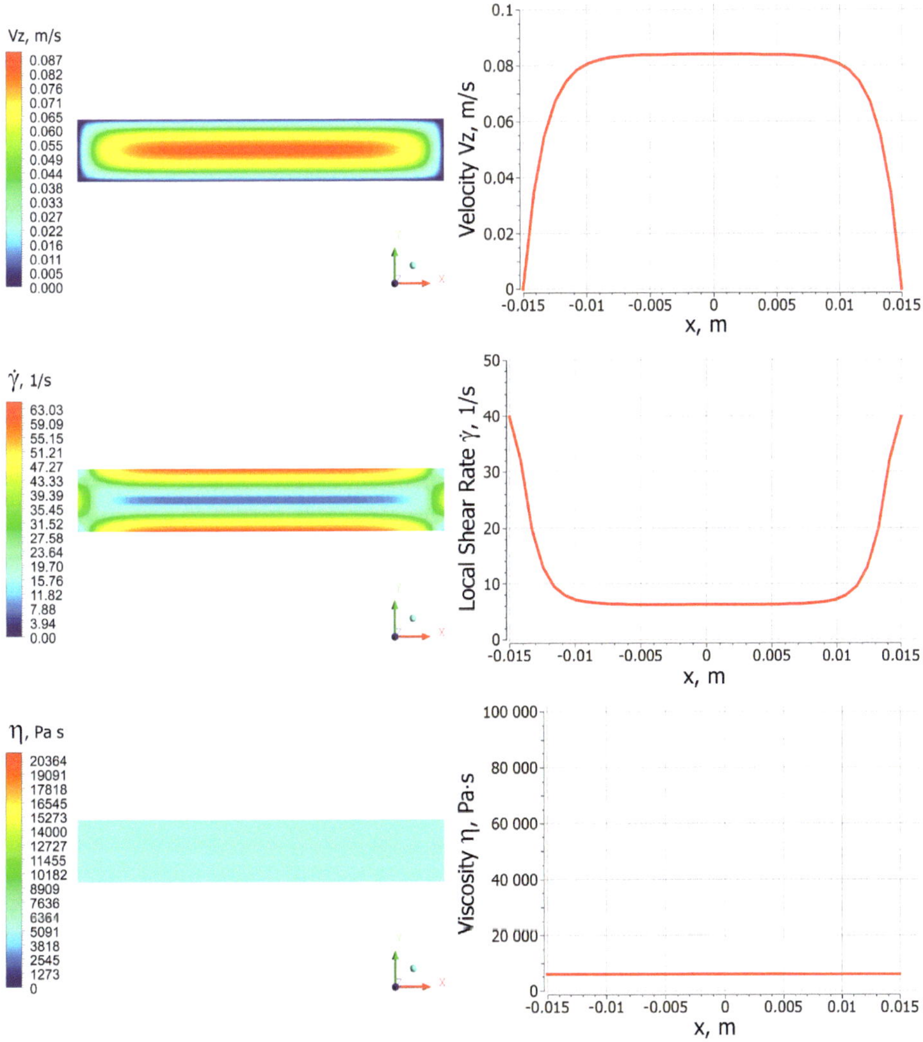

Figure 6.15 Distributions of the velocity, shear rate, and viscosity for the Newtonian flow ($n = 1$)

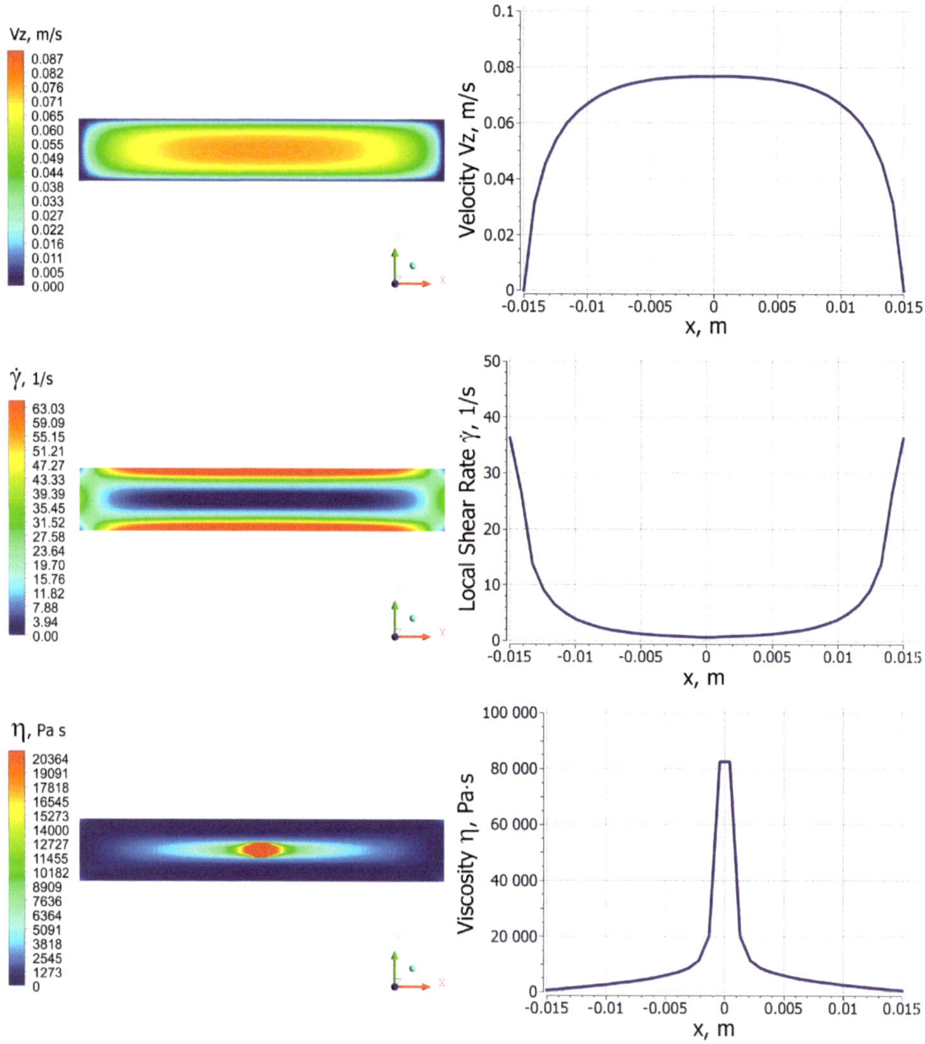

Figure 6.16 Distributions of the velocity, shear rate, and viscosity for the non-Newtonian pseudoplastic flow ($n = 0.39$)

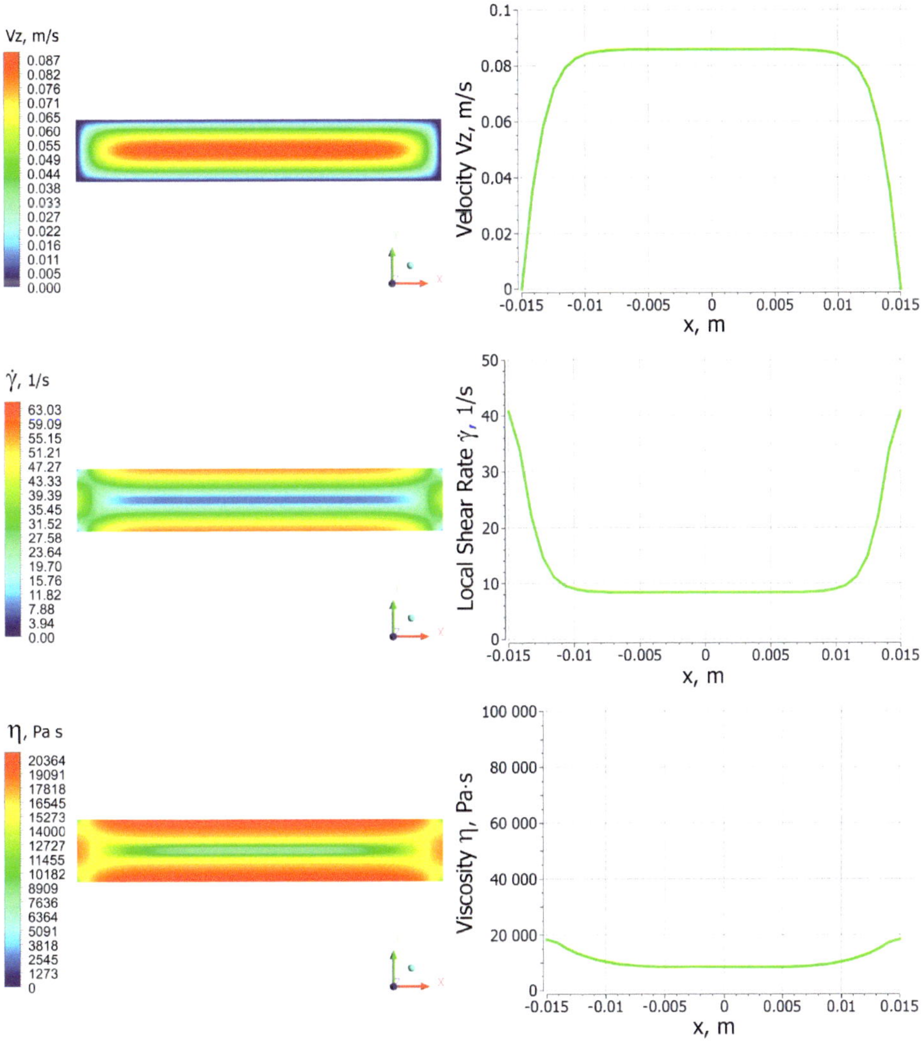

Figure 6.17 Distributions of the velocity, shear rate, and viscosity for the non-Newtonian dilatant flow ($n = 1.3$)

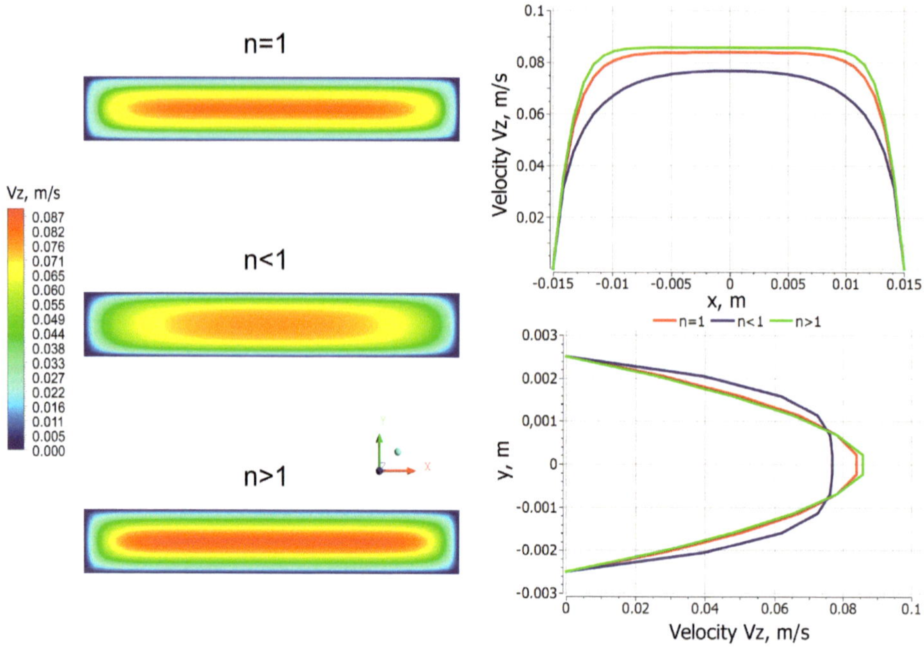

Figure 6.18 Velocity distributions for the Newtonian flow ($n = 1$), non-Newtonian pseudoplastic flow ($n = 0.39$), and non-Newtonian dilatant flow ($n = 1.3$)

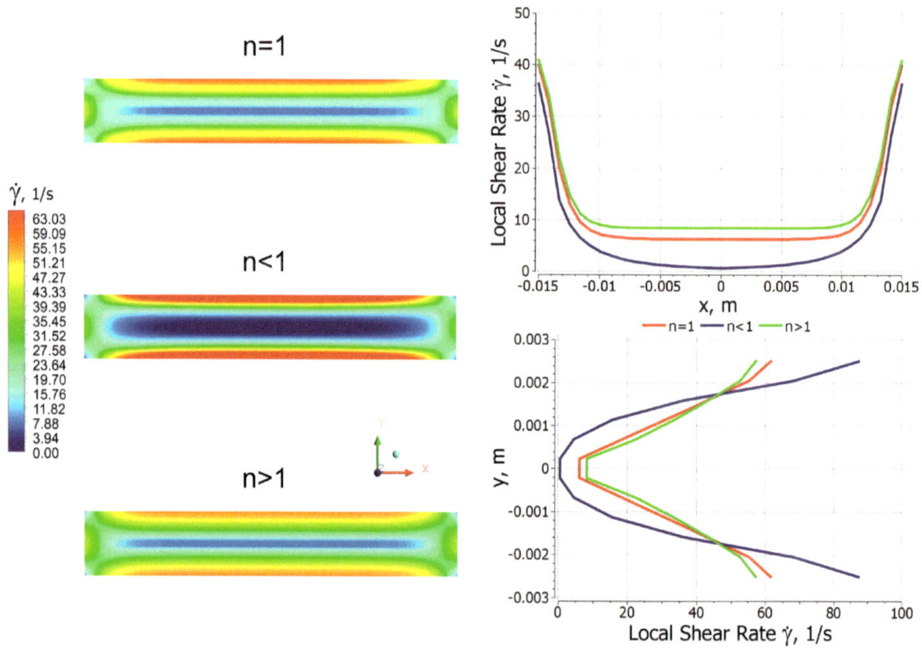

Figure 6.19 Shear rate distributions for the Newtonian flow ($n = 1$), non-Newtonian pseudoplastic flow ($n = 0.39$), and non-Newtonian dilatant flow ($n = 1.3$)

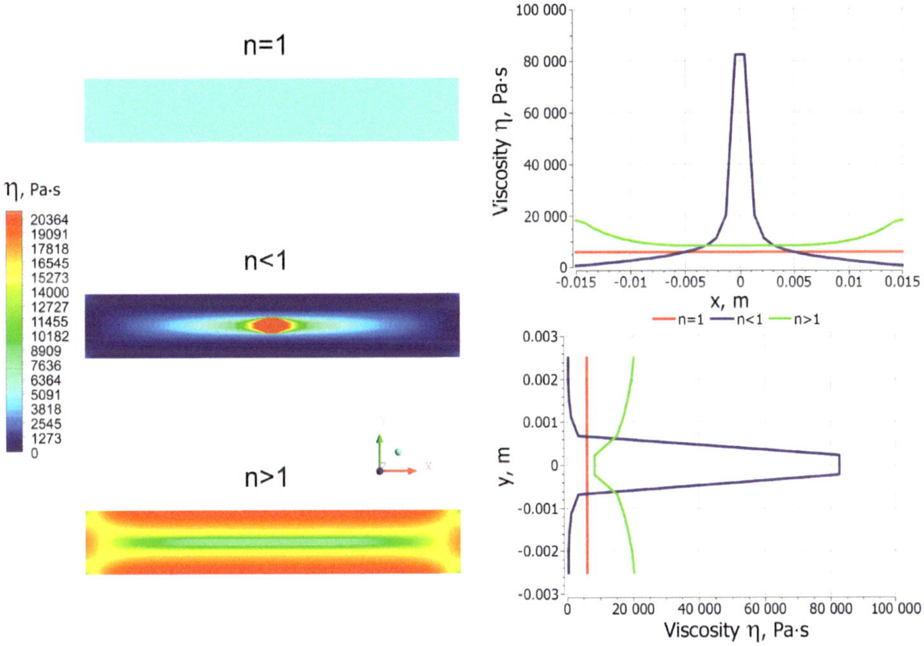

Figure 6.20 Viscosity distributions for the Newtonian flow ($n = 1$), non-Newtonian pseudoplastic flow ($n = 0.39$), and non-Newtonian dilatant flow ($n = 1.3$)

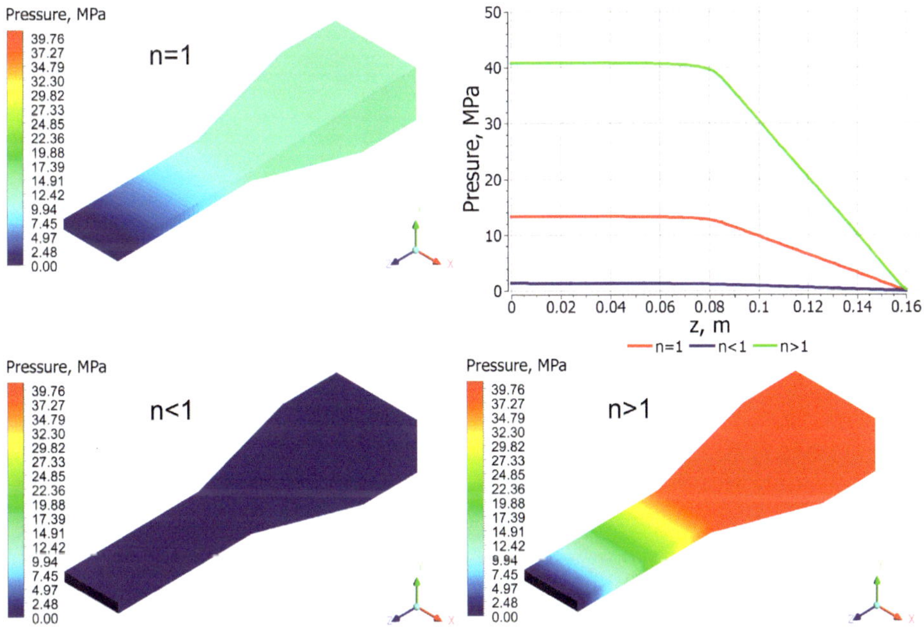

Figure 6.21 Pressure distributions for the Newtonian flow ($n = 1$), non-Newtonian pseudoplastic flow ($n = 0.39$), and non-Newtonian dilatant flow ($n = 1.3$)

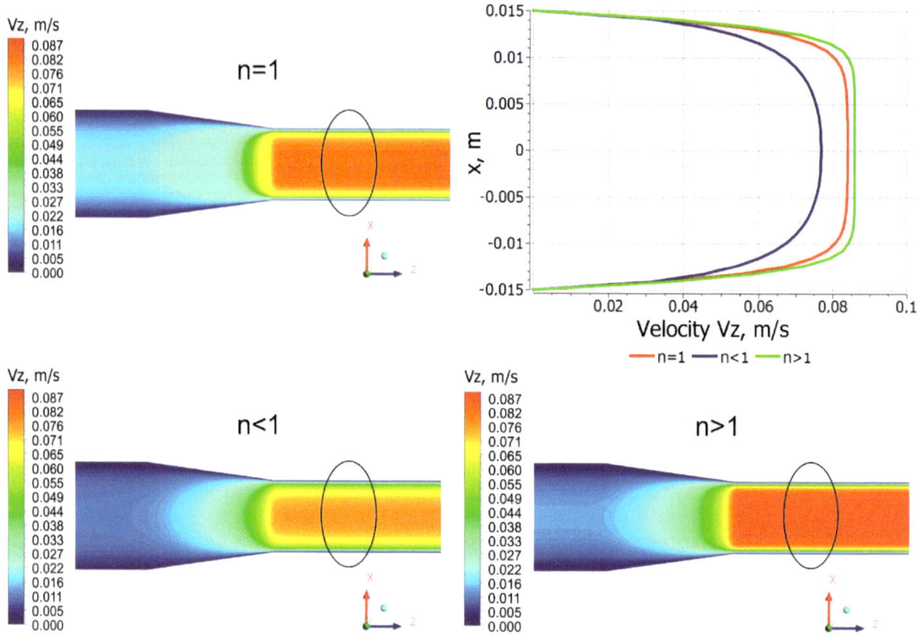

Figure 6.22 Velocity distributions along the cylinder for the Newtonian flow ($n = 1$), non-Newtonian pseudoplastic flow ($n = 0.39$), and dilatant flow ($n = 1.3$)

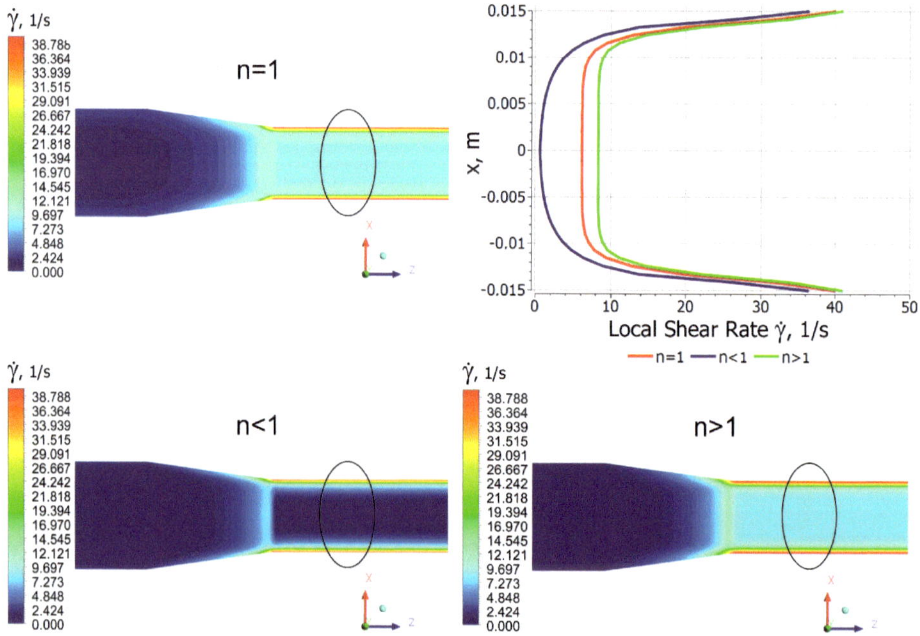

Figure 6.23 Shear rate distributions along the rectangular channel for the Newtonian flow ($n = 1$), non-Newtonian pseudoplastic flow ($n = 0.39$), and dilatant flow ($n = 1.3$)

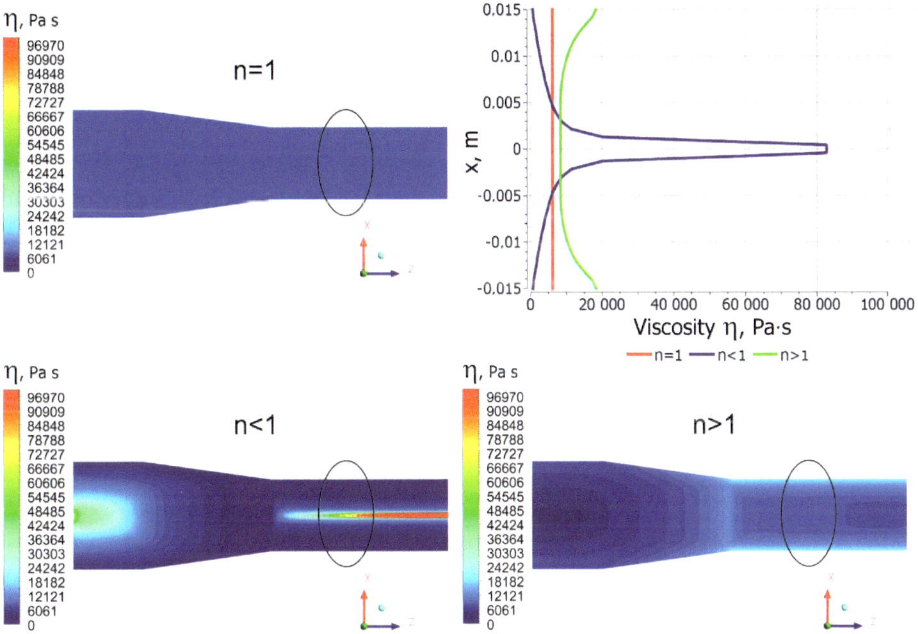

Figure 6.24 Viscosity distributions along the rectangular channel for the Newtonian flow ($n = 1$), non-Newtonian pseudoplastic flow ($n = 0.39$), and non-Newtonian dilatant flow ($n = 1.3$)

6.2.3.2 Extrudate Swell

Extrudate swell is a phenomenon of expansion of the stream of liquid flowing out of the flow channel, e.g. a capillary or an extrusion die. This is characteristic for viscoelastic materials, but also to a certain extent occurs in the case of Newtonian liquids (Section 1.3.2). Figure 6.25 shows the cross-section of the extrudate against the background of the die cross-section.

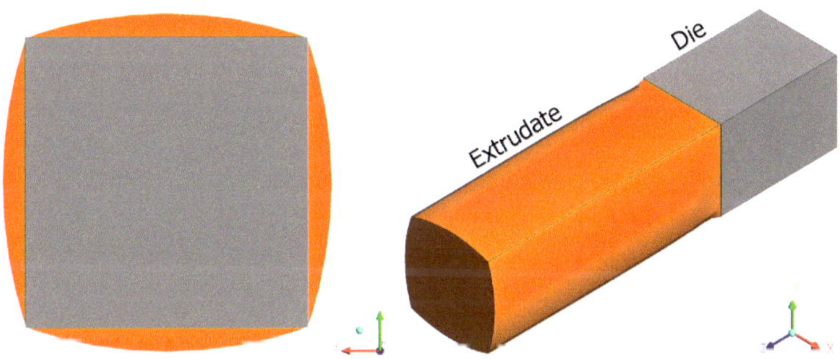

Figure 6.25 Extrudate swell phenomenon (die swell)

Extrudate swelling is a free surface problem. In an extrusion process, where the shape of the extrudate is not known in advance, a free surface is used to represent the outer surface of the extrudate. A free surface problem involves a boundary whose position is computed as a part of the solution since it is not known in advance.

The basis for modeling such issues using ANSYS Polyflow is to distinguish two sub-domains of the flow, fixed and movable. In the movable sub-domain, one of the boundaries, i. e. the free surface, is not defined and forms a part of the task solving.

As an example of the extrudate swell phenomenon, the rectangular pressure flow is considered, non-Newtonian and shear thinning (pseudoplastic). The flow geometry is depicted in Figure 6.26.

The flow problem has been defined as a steady-state, isothermal, 3D generalized Newtonian flow, and the power law model of Ostwald-de Waele was assumed (Eq. 6.1)).

The material parameters are as follows:

- consistency index, m $30 \cdot 10^3$ Pa·s^n
- power law exponent, n 0.75

and the following flow conditions are assumed:

- volume flow rate, Q 10 cm^3/s
- flow geometry:
  a) length of sections (die / extrudate) 30/60 mm
  b) die dimensions (width / height) 20/20 mm

The flow is bounded by four boundaries, and the following boundary conditions were applied at the flow boundaries (Figure 6.26):

- at the inlet to the channel: boundary BC1 – **inflow,** the flow rate $(Q_{in}) = (Q)$ is imposed,
- at the outlet of the channel: boundary BC2 – **normal and tangential forces,** vanishing normal and tangential forces are imposed $(f_n \;\&\; f_s) = (0, 0)$, which means that a uniform velocity profile is reached at the outlet, and the melt is not subjected to any externally applied stress at the outlet,
- on the channel wall: boundary BC3 – **zero wall velocity,** vanishing normal and tangential velocities are imposed $(v_n \;\&\; v_s) = (0, 0)$, which means that the velocity vanishes at the channel wall,
- on the free surface: boundary BC4 – **free surface.**

This flow involves a free surface whose shape is unknown a priori, but will be calculated together with the flow equations. A portion of the mesh is affected by the relocation of this boundary, so a special remeshing technique is applied on this part of the mesh. The free surface is entirely contained within the extrudate subdomain, and therefore only this subdomain is affected by the relocation of the free surface. The purpose of the remeshing technique is to relocate the internal nodes according to the displacement of the boundary nodes due to the motion of the free surface, since a part of the mesh is deformed.

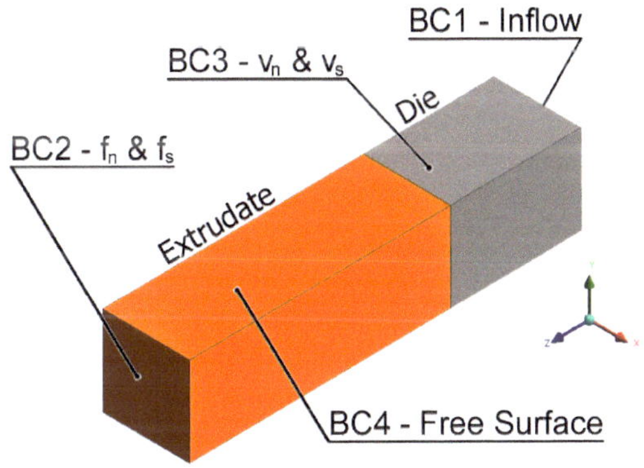

Figure 6.26 The flow geometry and boundary conditions for extrudate swell problem: BC1 – inflow, BC2 – normal and tangential forces, BC3 – zero wall velocity, BC4 – free surface

The results of simulation of the extrudate swell phenomenon are depicted in Figure 6.27 to Figure 6.30. Pressure distribution is shown in Figure 6.27. Distributions of the velocity and shear rate are depicted in Figure 6.28 and Figure 6.29. And finally, distributions of the velocity and shear rate along the die and extrudate are shown in Figure 6.30.

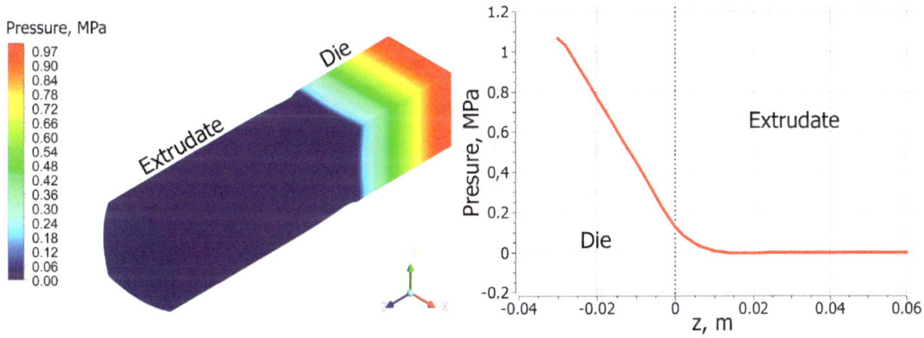

Figure 6.27 Pressure distribution: extrudate swell

It is worth noting that when the polymer leaves the die, the pressure drops to zero, while the distribution of the velocity and shear rate substantially reconfigures. The velocity distribution in the die shows the maximum in the central part of the die and decreases to zero at the die wall. After the polymer leaves the die, an increase in the velocity of boundary layer and a decrease in the velocity in the central part of the extrudate is observed. As a result, as the distance from the die outlet increases, the distribution of the velocity and shear velocity in the extrudate becomes balanced.

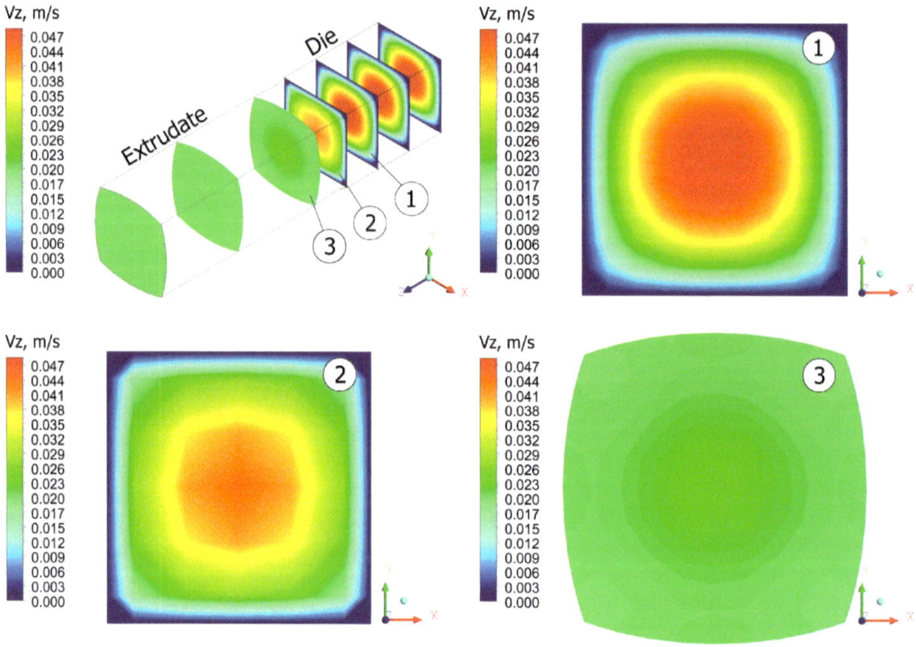

Figure 6.28 Velocity distribution: extrudate swell

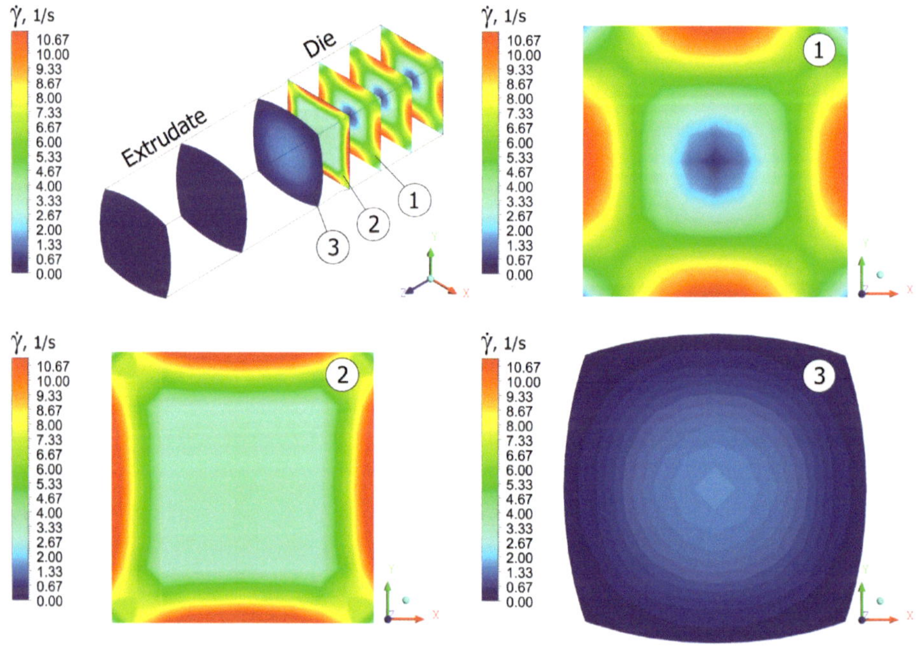

Figure 6.29 Shear rate distribution: extrudate swell

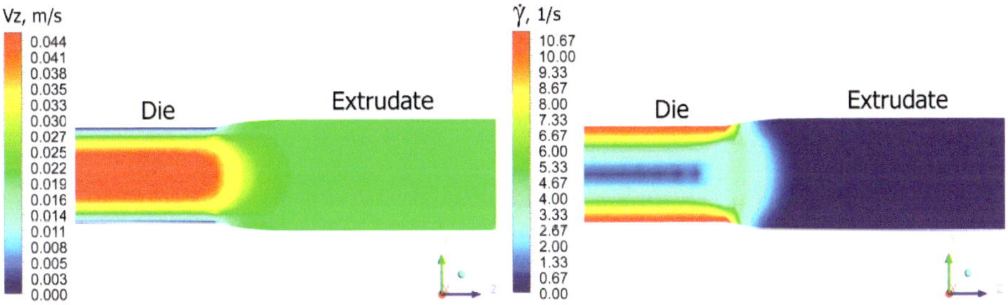

Figure 6.30 Distributions of velocity and shear rate along the die and extrudate

6.2.3.3 Extrudate Swell Inverse Problem

In many cases, we want to predict the die shape required in order to obtain a desired extrudate shape, rather than predicting the extrudate shape for a particular die. This is referred to as an **inverse extrusion problem,** which is schematically presented in Figure 6.31, where the cross-section of the extrudate is shown against the background of the die cross-section.

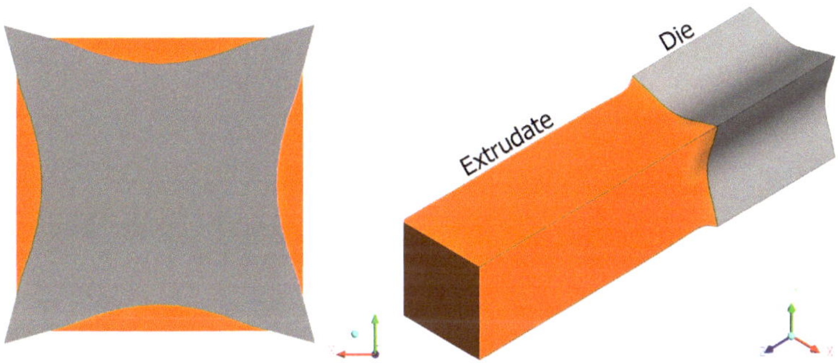

Figure 6.31 Extrudate swell phenomenon (die swell) – inverse problem

The procedure for setting up an inverse extrusion problem is similar to the procedure for a direct extrusion problem, but the remeshing will be applied to the die in addition to the extrudate, because the die shape is unknown a priori and will be computed based on the given extrudate shape. The extrudate must be remeshed as well, because only the exit shape is known.

As an example of extrudate swell inverse problem, the rectangular pressure flow is considered, non-Newtonian and shear thinning (pseudoplastic). The flow geometry is depicted in Figure 6.32.

The flow problem has been defined as a steady-state, isothermal, 3D generalized Newtonian flow, and the power law model of Ostwald-de Waele was assumed (Eq. (6.1)). The material parameters and flow conditions were the same as for the extrudate swell problem.

The flow is bounded by four boundaries, and the following boundary conditions were applied at the flow boundaries (Figure 6.32):

- at the inlet to the channel: boundary BC1 – **inflow,** the flow rate $(Q_{in}) = (Q)$ is imposed,
- at the outlet of the channel: boundary BC2 – **normal and tangential forces,** vanishing normal and tangential forces are imposed $(f_n \,\&\, f_s) = (0, 0)$, which means that a uniform velocity profile is reached at the outlet, and the melt is not subjected to any externally applied stress at the outlet,
- on the channel wall: boundary BC3 – **slip,** vanishing normal velocity is imposed, while the tangential force is a function of the slip velocity $(v_n \,\&\, f_s) = (0, f(v_s))$,
- on the free surface: boundary BC4 – **free surface,**

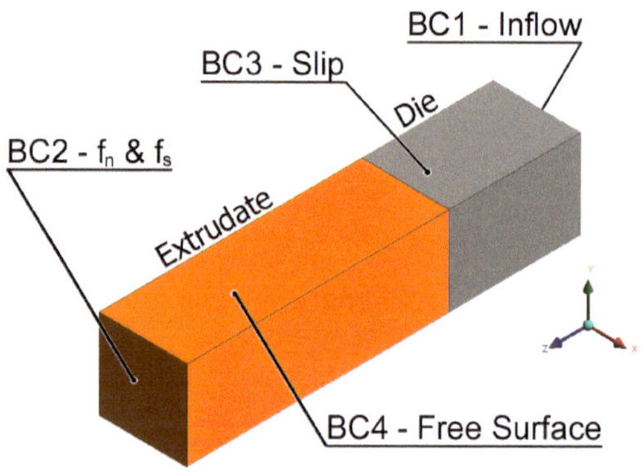

Figure 6.32 The flow geometry and boundary conditions for extrudate swell – inverse problem: BC1 – inflow, BC2 – normal and tangential forces, BC3 – slip, BC4 – free surface

When modeling the inverse extrusion problem, the evolution scheme is used to aid with computational convergence by starting with a low value of the slip coefficient and slowly increasing it to reach a no-slip condition. With a low value of the slip coefficient there is no swelling of the extrudate, simplifying the calculation. As the slip coefficient increases, the extrudate begins to swell because the fluid in contact with the wall slows down, which increases the velocity of the fluid in the die center.

The results of simulation of the inverse extrusion problem are depicted in Figure 6.33 to Figure 6.36. Pressure distribution is shown in Figure 6.33. Distributions of the velocity and shear rate are depicted in Figure 6.34 and Figure 6.35. And finally, distributions of the velocity and shear rate along the die and extrudate are shown in Figure 6.36.

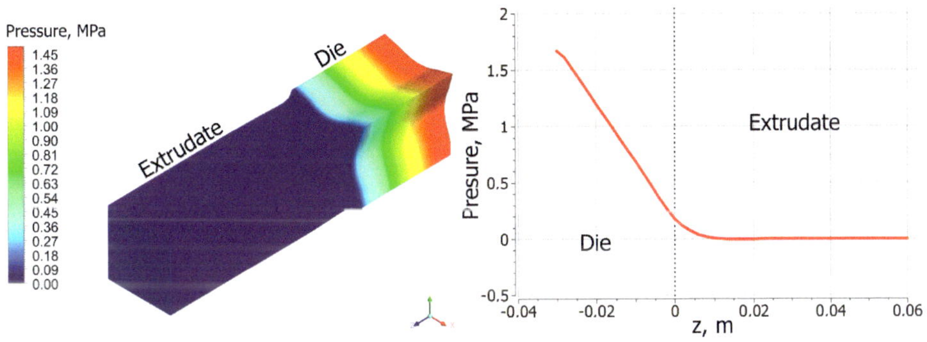

Figure 6.33 Pressure distribution: extrudate swell – inverse problem

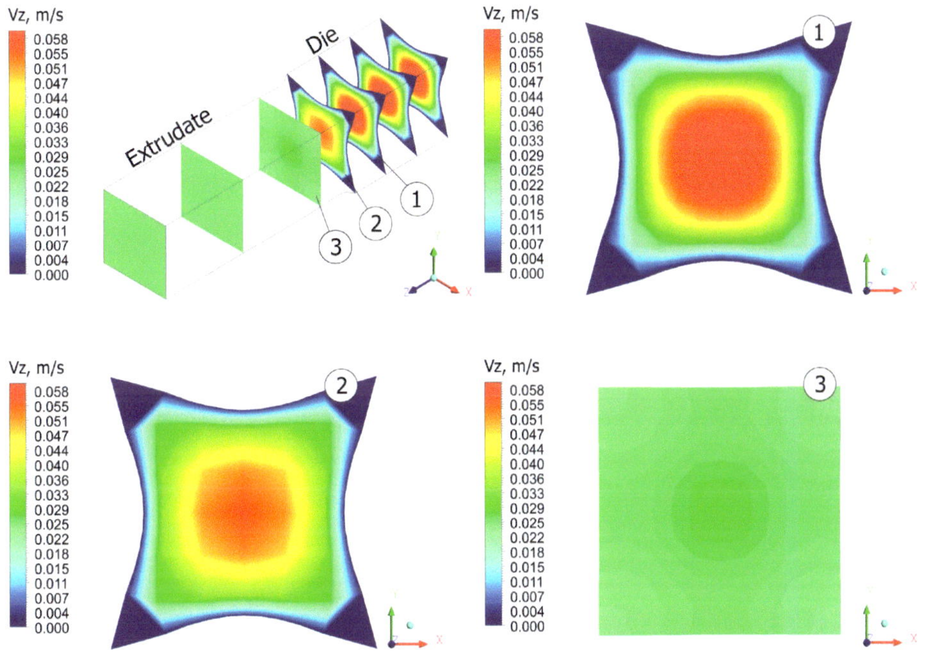

Figure 6.34 Velocity distribution: extrudate swell – inverse problem

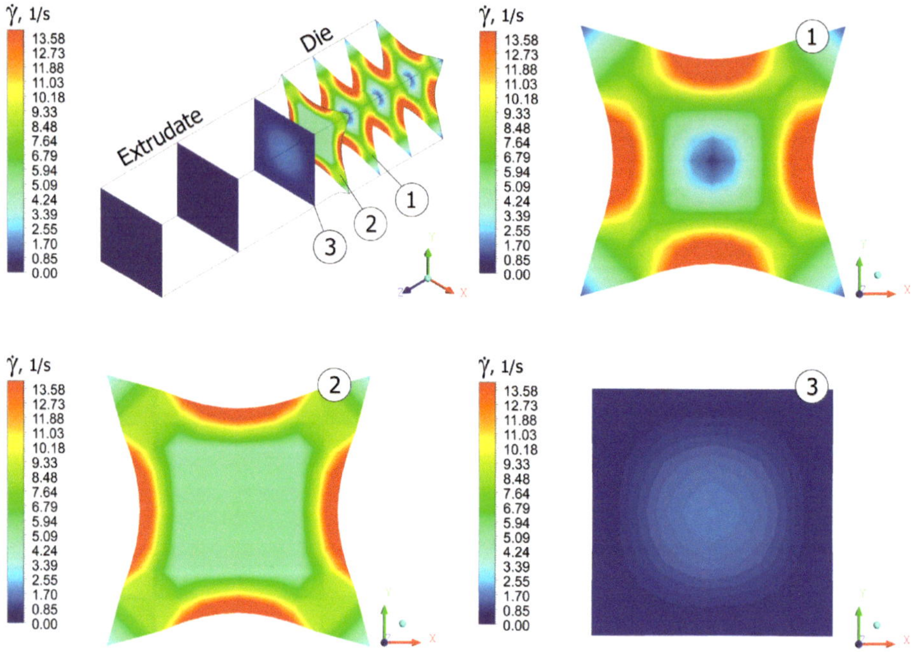

Figure 6.35 Shear rate distribution: extrudate swell – inverse problem

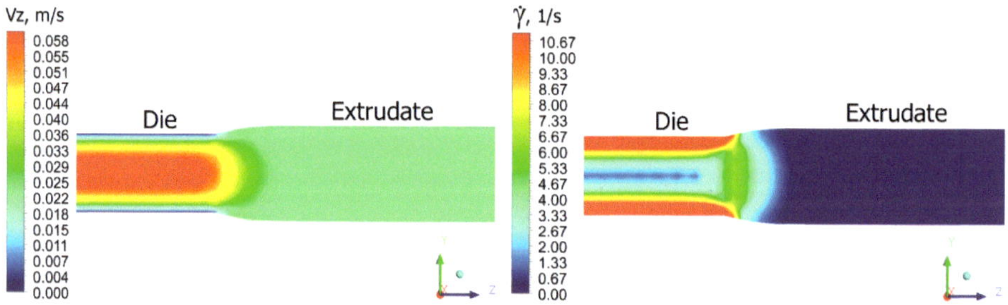

Figure 6.36 Distributions of velocity and shear rate along the die and extrudate

In the case of inverse extrusion, as in the case of direct extrusion, when the polymer leaves the die, the pressure drops to zero, while the distribution of the velocity and shear rate substantially reconfigures, especially at the die corners. The velocity distribution in the die shows the maximum in the central part of the die and decreases to zero at the die wall. After the polymer leaves the die, an increase in the velocity of boundary layer and a decrease in the velocity in the central part of the extrudate are observed. As a result, as the distance from the die outlet increases, the distribution of the velocity and shear velocity in the extrudate becomes balanced.

6.2.3.4 Single Screw Extrusion

The polymer flow in a single screw extruder results from the relative movement of the screw and barrel as well as from the pressure gradient developed there (Section 3.1.2, Figure 3.5). This is the drag-pressure flow.

In single screw extrusion, the flow is three-dimensional and unsteady since the screw is rotating. In order to simplify the set-up of the modeling of such a flow problem, a technique referred to as the mesh superposition technique (MST) has been implemented into the Polyflow software. This is a very useful technique to simulate flows with moving and intermeshing parts. A finite element mesh is generated for each part of the flow simulation: one for the flow domain (Subdomain 1) which is the inner part of the barrel without the screw, and one for the screw (Subdomain 2). At each time step, the MST algorithm updates and replaces the screw at their nominal positions, since the screw and the flow meshes are overlapping. The problem set-up consists in the definition of boundary conditions to specify the process. The screw is rotating at a given screw speed N, and the velocity is assumed to vanish at the barrel wall. A flow rate is imposed at the entry of the computational domain and it is also an outflow condition at the exit of the same domain.

The flow problem has been defined as an isothermal, 3D generalized Newtonian flow, and the power law model of Ostwald-de Waele was assumed (Eq. (6.1)).

The material parameters are as follows:

- consistency index, m — $6 \cdot 10^3$ Pa·s^n
- power law exponent, n — 0.39
- density, ρ — 740 kg/m^3

and the following flow conditions are assumed:

- screw speed, N — 60 rpm
- flow rate, G — 0/14.2/28.4 kg/h
- flow geometry (right-handed screw):
  a) screw diameter, D — 45 mm
  b) screw pitch, t — 45 mm
  c) screw channel depth, H — 3 mm
  d) screw flight width, e — 5 mm
  e) screw/barrel clearance, h_f — 0.2 mm
  f) screw section length, L — 210 mm

The computation scheme is depicted in Figure 6.37. Two subdomains are distinguished, for the flowing material (Subdomain 1) and for the screw (Subdomain 2). The following flow boundary conditions were applied to define the process:

- at the inlet to the flow domain: boundary BC1 – **inflow,** the flow rate $(Q_{in}) = (Q)$ is imposed,
- at the outlet of the flow domain: boundary BC2 – **outflow,** vanishing normal forces and tangential velocities are imposed $(f_n \,\&\, v_s) = (0, 0)$, which means that the flow rate $(Q_{out}) = (Q)$ is imposed,
- on the screw domain: boundary BC3 – **Cartesian velocity,** Cartesian velocities are imposed $(v_x \,\&\, v_y \,\&\, v_z) = (-N)$, which means that the screw rotates at the screw speed (N) counterclockwise,
- on the barrel wall: boundary BC4 – **zero wall velocity**, vanishing normal and tangential velocities are imposed $(v_n \,\&\, v_s) = (0, 0)$, which means that the velocity vanishes at the barrel wall.

The two first boundary conditions, BC1 – inflow $(Q_{in}) = (Q)$ and BC2 – normal forces and tangential velocities $(f_n \,\&\, v_s) = (0, 0)$, imply that pressure may be generated along the screw. Since we do not know the pressure at the end of the screw element, we calculate the pressure gradient, which is relative to the zero pressure at the element exit. Negative pressures resulting from computations do not mean negative pressures in the extrusion process.

Simulations were performed to study the flow parameters distributions for different pressure gradients: positive gradient ($\partial p/\partial z > 0$, i.e. $\Delta p < 0$), zero gradient ($\partial p/\partial z = 0$, i.e. $\Delta p = 0$), and negative gradient ($\partial p/\partial z < 0$, i.e. $\Delta p > 0$), where Δp is the pressure difference between the pressure at channel inlet and the pressure at the channel outlet.

To perform these, we started the simulations with the initial boundary condition $BC1_{initial}$ – outflow $(Q_{out}) = (Q)$, with other conditions unchanged, which implies that pressure is not generated along the screw, and the flow rate is computed which is equal to the drag flow Q_d, that is, there is no pressure flow and we have ($\partial p/\partial z = 0$, i.e. $\Delta p = 0$).

Then, the simulations were performed with the basic boundary condition BC1 – inflow $(Q_{in}) = (Q)$, for different values $Q = 0$, $Q = Q_d$, $Q = 2Q_d$, which corresponds to the different pressure gradients: positive gradient ($\partial p/\partial z > 0$, i.e. $\Delta p < 0$), zero gradient ($\partial p/\partial z = 0$, i.e. $\Delta p = 0$), and negative gradient ($\partial p/\partial z < 0$, i.e. $\Delta p > 0$).

Distributions of the velocity, shear rate, and viscosity for the flow with negative pressure gradient are depicted in Figure 6.38. In this case, the pressure flow is added to the drag flow (Section 5.2.5). The velocity component v_x is perpendicular to the screw axis, and the velocity component v_z is parallel to this. According to the velocity distributions, the shear rate, which is a derivative of the velocity, is highest at the screw and barrel surfaces and has a minimum in the middle of the screw channel. Thus, the viscosity, which for polymers decreases with an increase of the shear rate, is lowest at the screw and barrel surfaces and has a maximum in the middle of the screw channel.

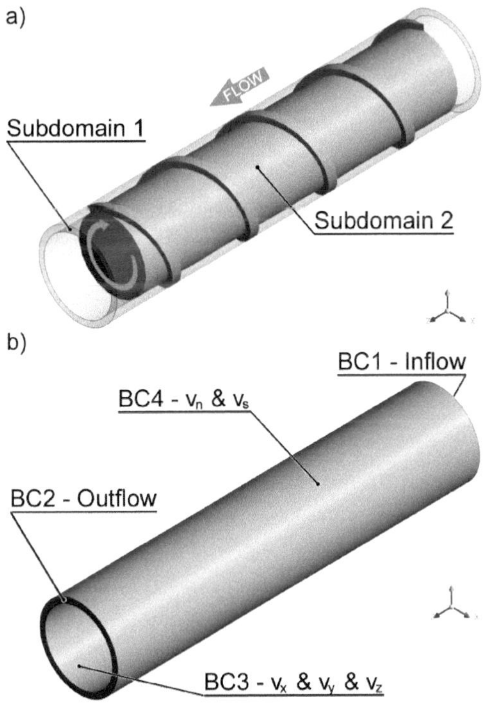

Figure 6.37
The flow geometry and boundary conditions for single screw extrusion: BC1 – inflow, BC2 – outflow, BC3 – Cartesian velocity, BC4 – zero wall velocity

Pressure distributions are depicted in Figure 6.39. For the positive pressure gradient ($\partial p/\partial z > 0$, i. e. $\Delta p < 0$), the pressure flow is subtracted from the drag flow, while for the negative gradient ($\partial p/\partial z < 0$, i. e. $\Delta p > 0$), the pressure flow is added to the drag flow. In the case of the zero pressure gradient ($\partial p/\partial z = 0$, i. e. $\Delta p = 0$), the pressure flow does not appear. It is worth noting the characteristic pressure pulsation for the screw, the so-called pressure saw, which is the result of changing screw geometry in a given area of an analysis. At a certain point in the flow space, the screw flight and the screw channel alternate, which generates pressure pulsation. This phenomenon is an inseparable and always occurring feature of screw flows. It is also the cause of pulsation in the extruder output. The frequency of this pulsation is equal to the frequency of screw rotation (rotational speed).

Distributions of the velocity components for different pressure gradients are shown in Figure 6.40 to Figure 6.42. The velocity component v_x, which is perpendicular to the screw axis, is decreased by the pressure flow at the positive pressure gradient, and is increased by the pressure flow at the negative gradient. The velocity component v_y, also perpendicular to the screw axis, varies depending on the pressure gradient. The velocity component v_z, which is parallel to screw axis, has a parabolic distribution, and increases with an increase of the pressure difference Δp. This distribution can be compared to Figure 5.21.

Distributions of the local shear rate and viscosity for different pressure gradients are presented in Figure 6.43 and Figure 6.44. For the positive pressure gradient, the shear rate increases from the screw to the barrel, while for the negative and zero gradients it has a minimum in the middle of the channel. These correspond to the velocity distributions v_x, v_y, and v_z (Figure 6.40 to Figure 6.42). The viscosity distribution results from the shear rate distribution. For the positive pressure gradient, it decreases from the screw to the barrel, while for the negative and zero gradients it has a maximum in the middle of the channel.

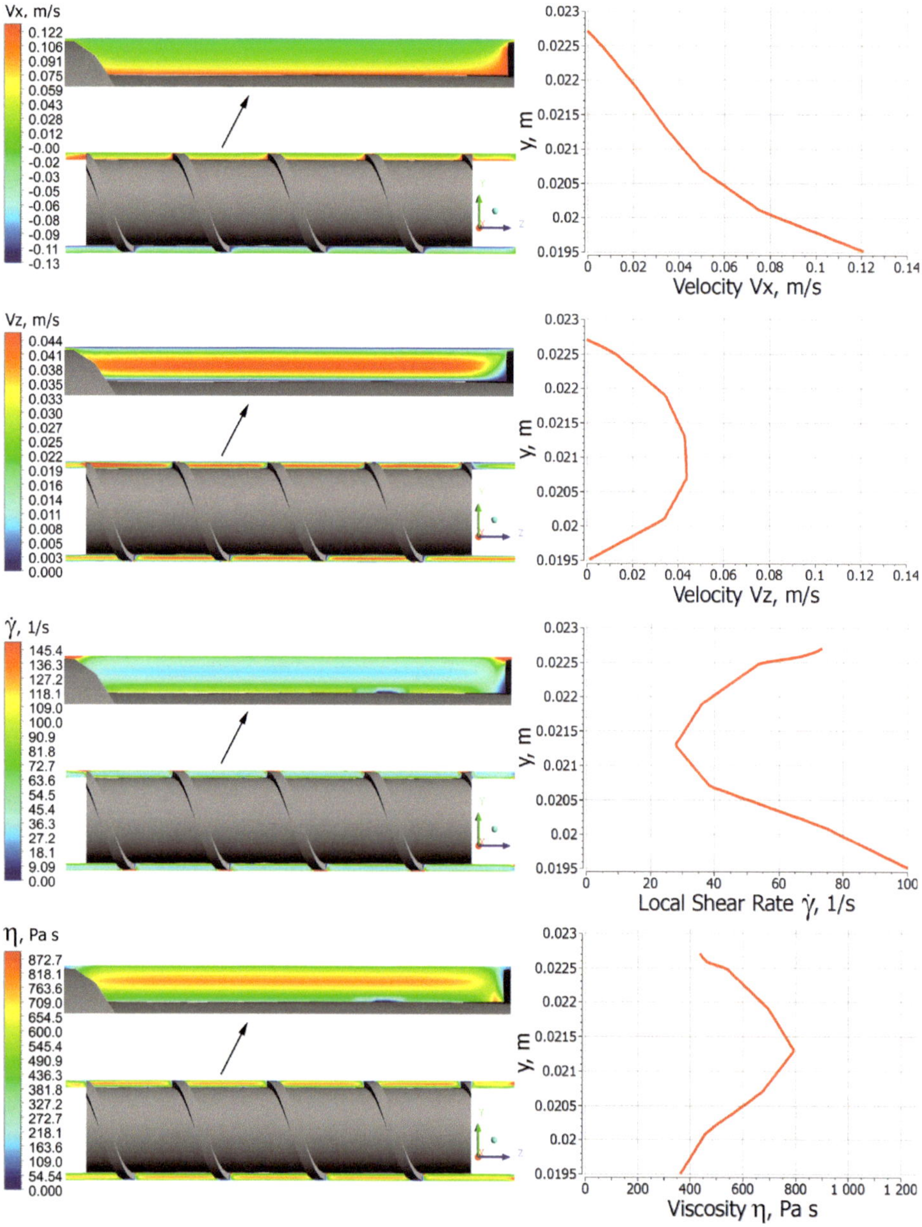

Figure 6.38 Distributions of the velocity, shear rate, and viscosity for the single screw extrusion, at negative pressure gradient ($\partial p/\partial z < 0$), i.e. ($\Delta p > 0$)

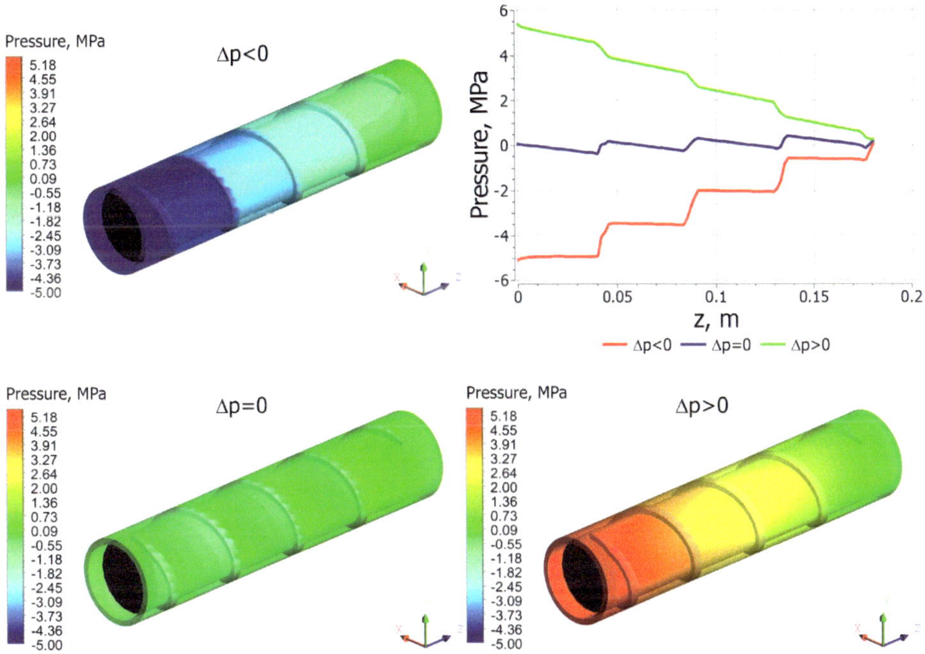

Figure 6.39 Pressure distributions for different pressure gradients ($\partial p/\partial z$): positive ($\Delta p < 0$), zero ($\Delta p = 0$), negative ($\Delta p > 0$), Δp - pressure difference

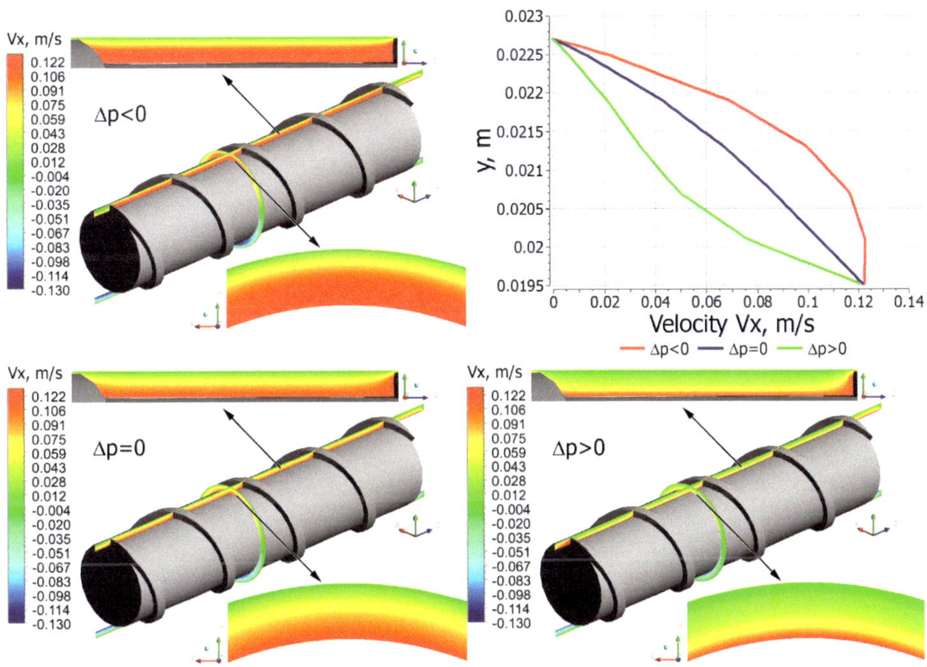

Figure 6.40 Distributions of the velocity v_x for the single screw extrusion

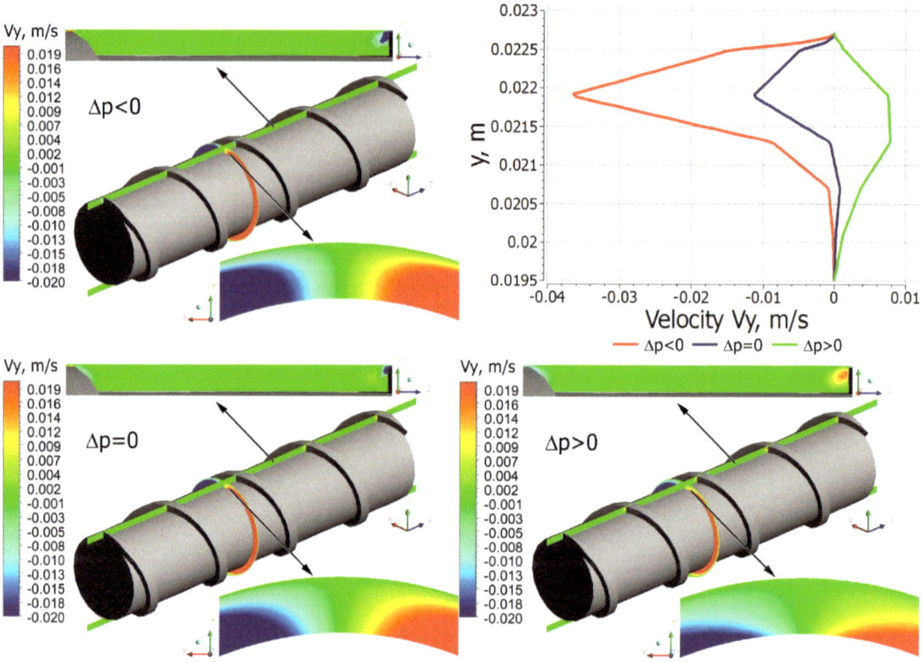

Figure 6.41 Distributions of the velocity v_y for the single screw extrusion (the chart represents the velocity distribution near the screw flight)

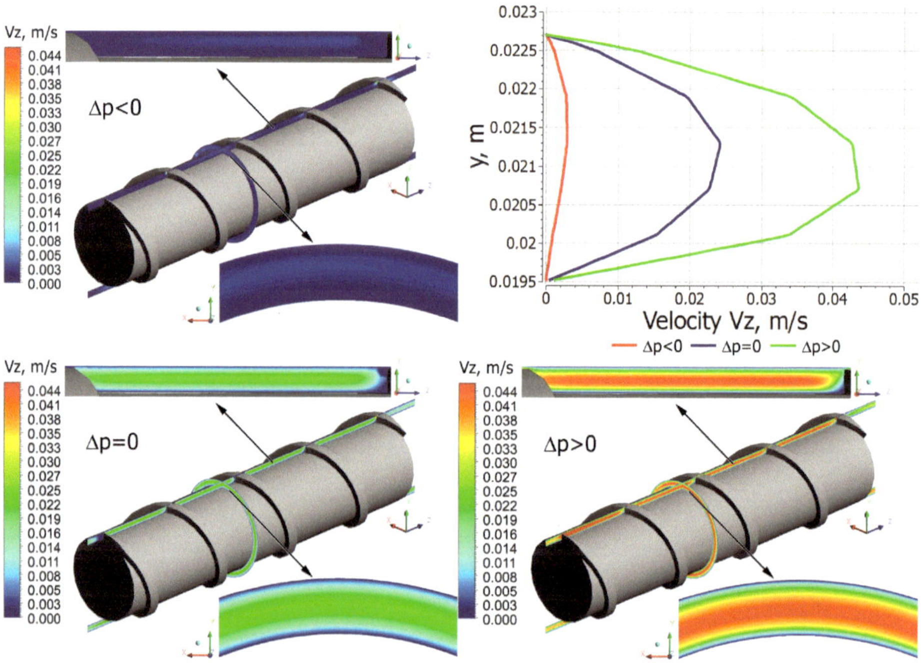

Figure 6.42 Distributions of the velocity v_z for the single screw extrusion

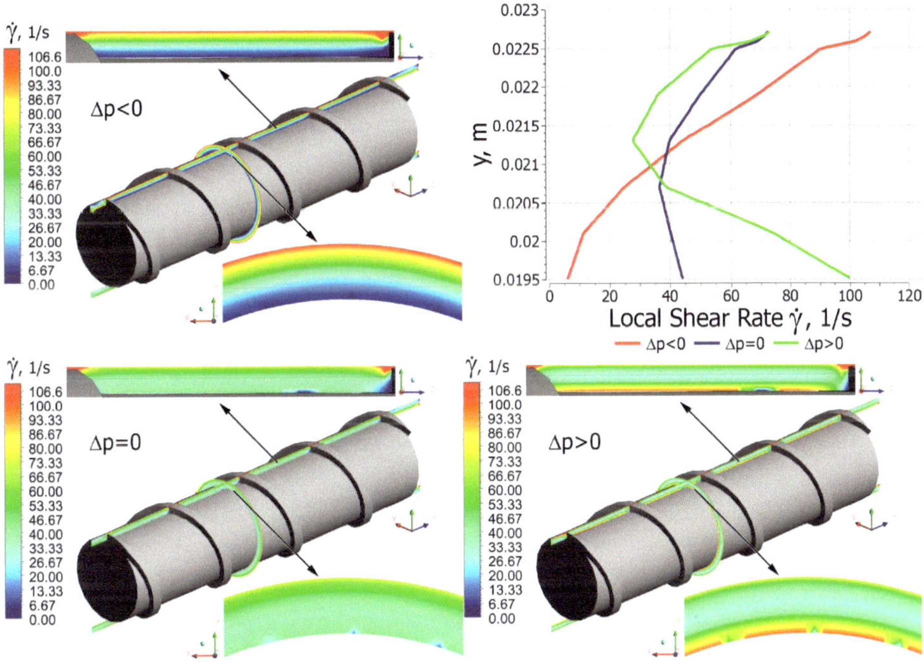

Figure 6.43 Distributions of the shear rate for the single screw extrusion

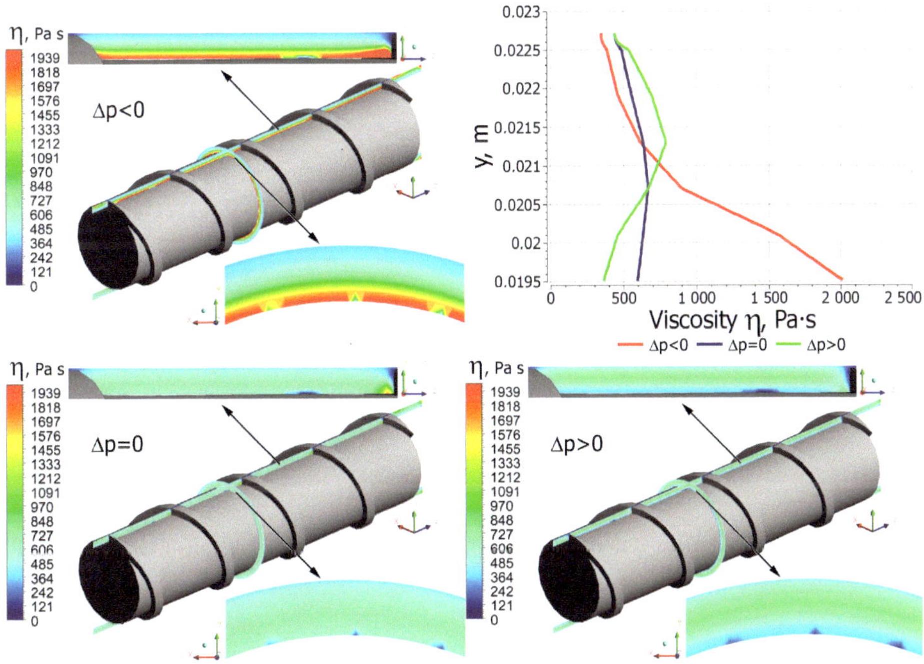

Figure 6.44 Distributions of the viscosity for the single screw extrusion

The basic flow model in a single screw extruder assumes that there is no gap between the tips of the screw and barrel. In fact, there is a gap, and the so-called **leakage flow** in the gap affects the flow rate. The two components of this flow can be distinguished: the drag component and the pressure one (Section 5.2.5.2). These may be expressed by Eq. (5.108).

The leakage flow is depicted in Figure 6.45 for different pressure gradients. For the positive pressure gradient, the pressure leakage flow decreases the flow rate, while for the negative gradient, it increases the flow rate. In the case of the zero pressure gradient, the pressure leakage flow does not appear. However, the drag component of the leakage flow still exists. It is also worth noting that the viscosity of polymer in the gap is much lower than the viscosity of polymer in the screw channel since the shear rate in the gap is much higher than in the screw channel.

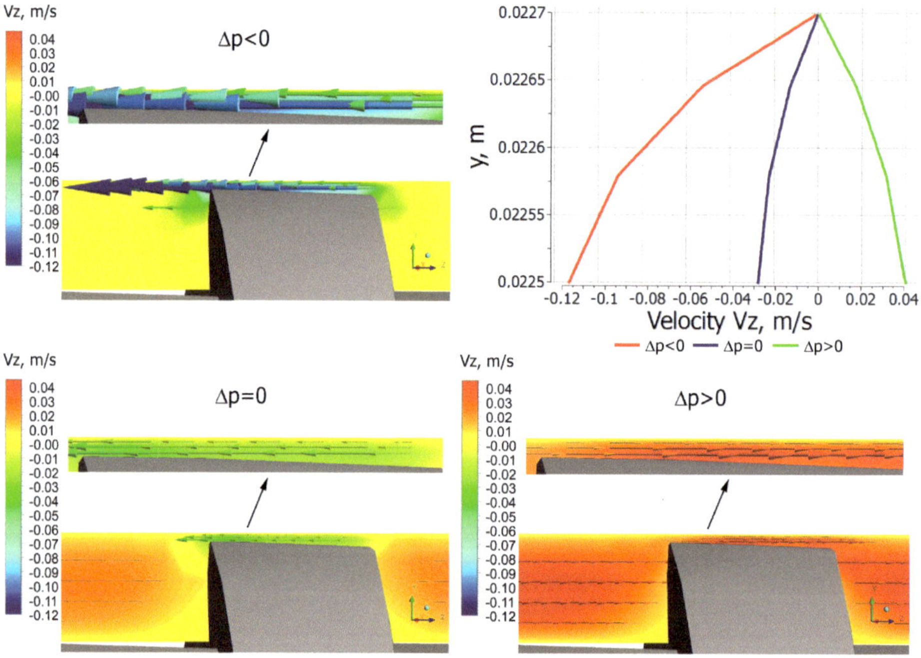

Figure 6.45 Distributions of the velocity v_z for the clearance flow in the single screw extrusion

6.2.3.5 Co-Rotating Twin Screw Extrusion

The flow of polymers in co-rotating twin screw extruders is the result of the relative movement of the screws and barrel and the pressure gradient in the extruder. It is the pressure-drag flow which takes place in open channels between screws (alternately one and the other) and the barrel (Section 3.1.3.1, Figure 3.7).

In co-rotating extruders, during each screw rotation, the polymer is transferred from one screw to the other, and moves in an open figure-eight pattern, along the so-called twisted eight. The flow of polymer is dragged, so that it resembles the flow in a single screw extruder in this respect. But, when moving between the screws, the polymer travels a longer path and undergoes higher shearing. There is a counter-rotating movement in the inter-screw gap, as a result of which high shear stresses are produced in this area.

In a co-rotating twin screw extrusion, the flow is three-dimensional and unsteady since the screws are rotating. In order to simplify the set-up of the modeling of such a flow problem, the mesh superposition technique (MST) has been implemented into the Polyflow software (Section 6.2.3.4). A finite element mesh is generated for each part of the flow simulation: one for the flow domain (Subdomain 1) which is the inner part of the barrel without the screws, and one for each screw (Subdomain 2, Subdomain 3). The problem set-up consists in the definition of boundary conditions. The screws are rotating at a given screw speed N, and the velocity is assumed to vanish at the barrel wall. A flow rate is imposed at the entry of the computational domain and it is also an outflow condition at the exit of the same domain.

The flow problem has been defined as an isothermal, 3D generalized Newtonian flow, and the power law model of Ostwald-de Waele was assumed (Eq. (6.1)).

The material parameters are as follows:

- consistency index, m — $6 \cdot 10^3$ Pa·s^n
- power law exponent, n — 0.39
- density, ρ — 740 kg/m^3

and the following flow conditions are assumed:

- screw speed, N — 60 rpm
- flow rate, G — 0/14/28 kg/h
- flow geometry (right-handed screws):
    a) screw diameter, D — 27 mm
    b) screw pitch, t — 25 mm
    c) screw channel depth, H — 4.5 mm
    d) screw flight width, e — 8 mm
    e) screw/barrel clearance, h_f — 0.5 mm
    f) screw section length, L — 120 mm

The computation scheme is depicted in Figure 6.46. Three subdomains are distinguished, for the flowing material (Subdomain 1) and for the screws (Subdomain 2, Subdomain 3). The following flow boundary conditions were applied to define the process:

- at the inlet to the flow domain: boundary BC1 – **inflow,** the flow rate (Q_{in}) = (Q) is imposed,
- at the outlet of the flow domain: boundary BC2 – **outflow,** vanishing normal forces and tangential velocities are imposed (f_n & v_s) = (0, 0), which means that the flow rate (Q_{out}) = (Q) is imposed,
- on the screw domain: boundary BC3 – **Cartesian velocity,** Cartesian velocities are imposed (v_x & v_y & v_z) = (–N), which means that the screw rotates at the screw speed (N) counterclockwise,
- on the screw domain: boundary BC4 – **Cartesian velocity,** Cartesian velocities are imposed (v_x & v_y & v_z) = (–N), which means that the screw rotates at the screw speed (N) counterclockwise,
- on the barrel wall: boundary BC5 – **zero wall velocity,** vanishing normal and tangential velocities are imposed (v_n & v_s) = (0, 0), which means that the velocity vanishes at the barrel wall.

The two first boundary conditions, BC1 – inflow (Q_{in}) = (Q), and BC2 – normal forces and tangential velocities (f_n & v_s) = (0, 0), imply that pressure may be generated along the screws. Since we do not know the pressure at the end of the screw element, we calculate the pressure gradient, which is relative to the zero pressure at the element exit. Negative pressures resulting from computations do not mean negative pressures in the extrusion process.

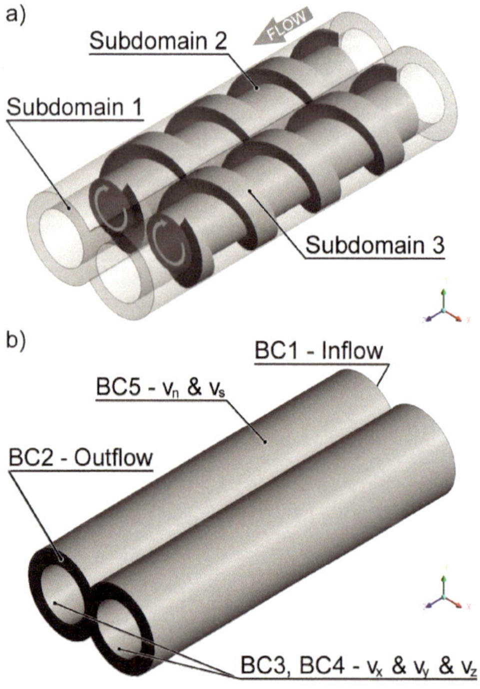

Figure 6.46
The flow geometry and boundary conditions for co-rotating twin screw extrusion: BC1 – inflow, BC2 – outflow, BC3/BC4 – Cartesian velocity, BC5 – zero wall velocity

As for a single screw extrusion, the simulations were performed to study the flow parameters distributions for different pressure gradients: positive gradient ($\partial p/\partial z > 0$, i. e. $\Delta p < 0$), zero gradient ($\partial p/\partial z = 0$, i. e. $\Delta p = 0$), and negative gradient ($\partial p/\partial z < 0$, i. e. $\Delta p > 0$), where Δp is the pressure difference between the pressure at channel inlet and the pressure at the channel outlet.

To perform these, we started the simulations with the initial boundary condition $BC1_{initial}$ - outflow (Q_{out}) = (Q), with other conditions unchanged, which implies that pressure is not generated along the screw, and the flow rate is computed which is equal to the drag flow Q_d, that is, there is no pressure flow and we have ($\partial p/\partial z = 0$, i. e. $\Delta p = 0$).

Then, the simulations were performed with the basic boundary condition BC1 - inflow (Q_{in}) = (Q), for different values $Q = 0$, $Q = Q_d$, $Q = 2Q_d$, which corresponds to the different pressure gradients: positive gradient, zero gradient, and negative gradient.

Pressure distributions are depicted in Figure 6.47. For the positive pressure gradient, the pressure flow is subtracted from the drag flow, while for the negative gradient, the pressure flow is added to the drag flow. In the case of the zero pressure gradient, the pressure flow does not appear. As in the case of single screw extrusion, the characteristic pressure pulsation is observed (Section 6.2.3.4). The frequency of this pulsation is equal to the frequency of screw rotation (rotational speed).

Distributions of the velocity components for different pressure gradients are shown in Figure 6.48 to Figure 6.50. The velocity component v_x, which is perpendicular to the screw axis, is decreased by the pressure flow at the positive pressure gradient, and is increased by the pressure flow at the negative gradient. It is worth noting that distributions are almost the same in both screws. The velocity component v_y, also perpendicular to the screw axis, varies depending on the pressure gradient. The velocity component v_z, which is parallel to screw axis, has a parabolic distribution, and increases with an increase of the pressure difference Δp. These distributions are similar but not the same in both screws.

Distributions of the local shear rate and viscosity for different pressure gradients are presented in Figure 6.51 and Figure 6.52. For the positive pressure gradient, the shear rate increases from the screw to the barrel, while for the negative and zero gradients it is highest at the screw and barrel surfaces with a minimum in the middle of the channel. These correspond to the velocity distributions v_x, v_y, and v_z (Figure 6.48 to Figure 6.50). The viscosity distribution results from the shear rate distribution. For the positive pressure gradient, it decreases from the screw to the barrel, while for the negative and zero gradients it is lowest at the screw and barrel surfaces with a maximum in the middle of the channel.

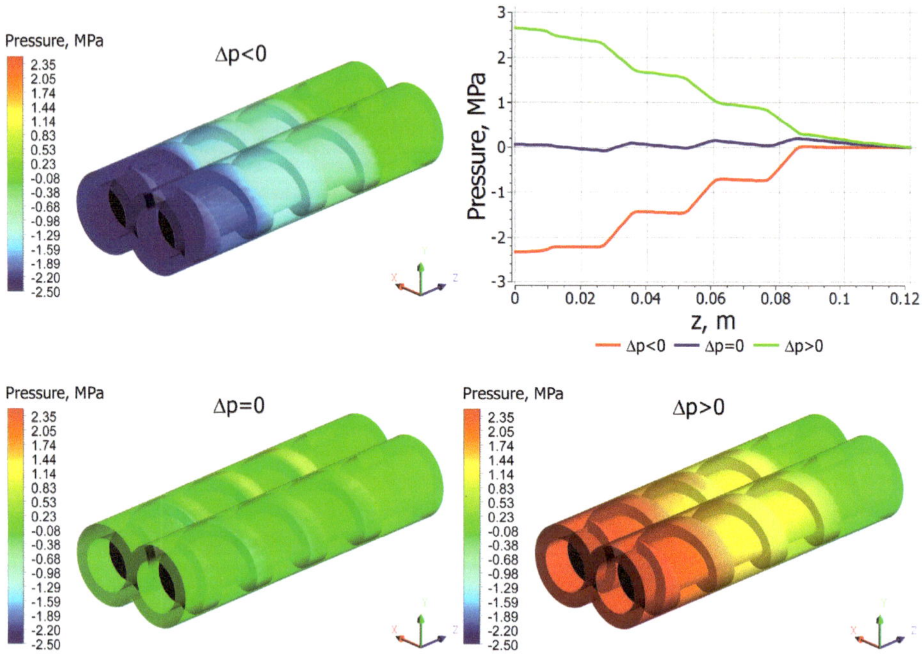

Figure 6.47 Pressure distributions for different pressure gradients ($\partial p/\partial z$): positive ($\Delta p < 0$), zero ($\Delta p = 0$), negative ($\Delta p > 0$), Δp - pressure difference

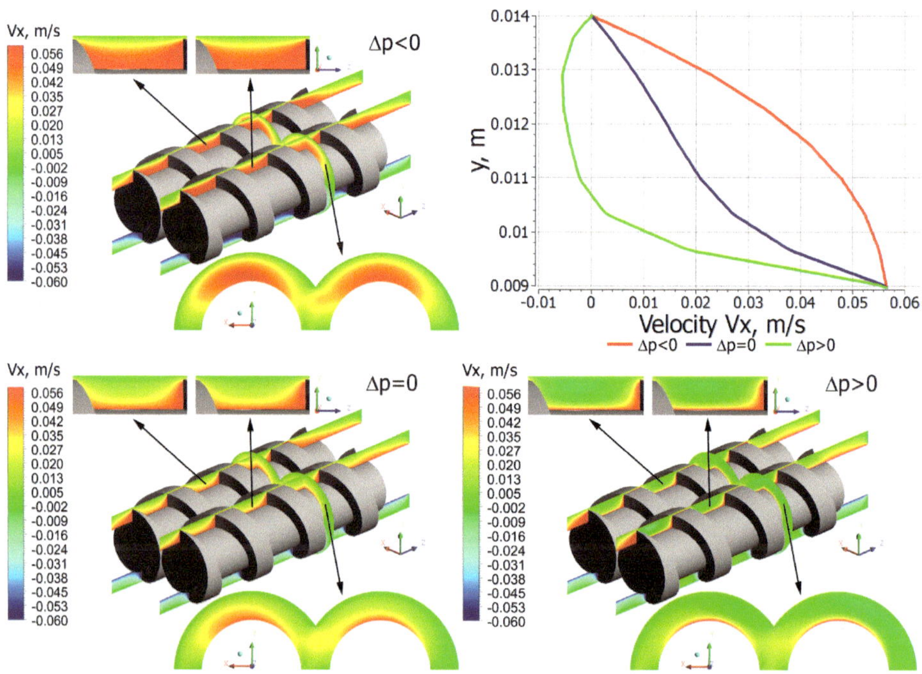

Figure 6.48 Distributions of the velocity v_x for the co-rotating twin screw extrusion

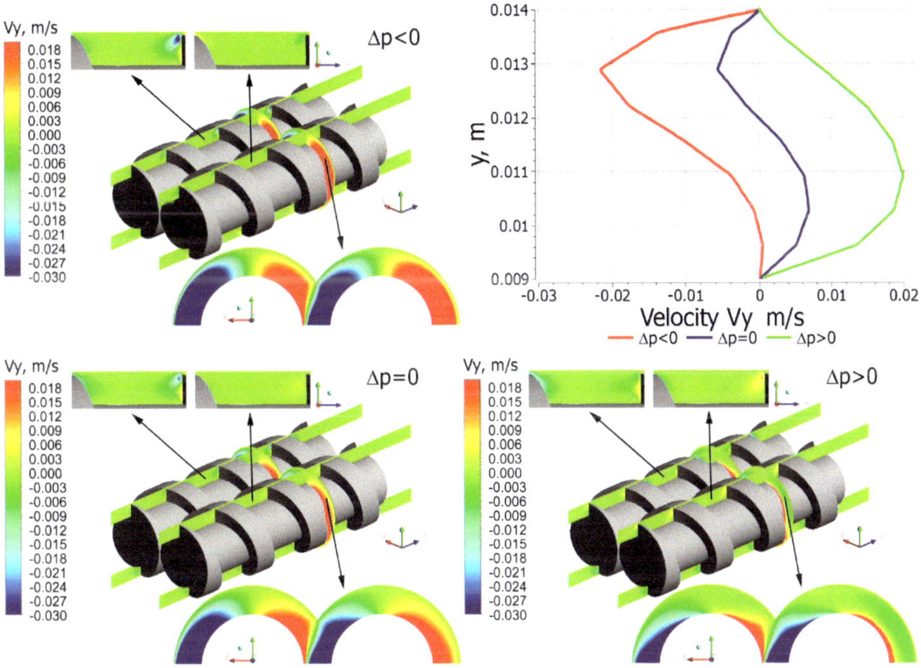

Figure 6.49 Distributions of the velocity v_y for the co-rotating twin screw extrusion (the chart represents the velocity distribution near the screw flight)

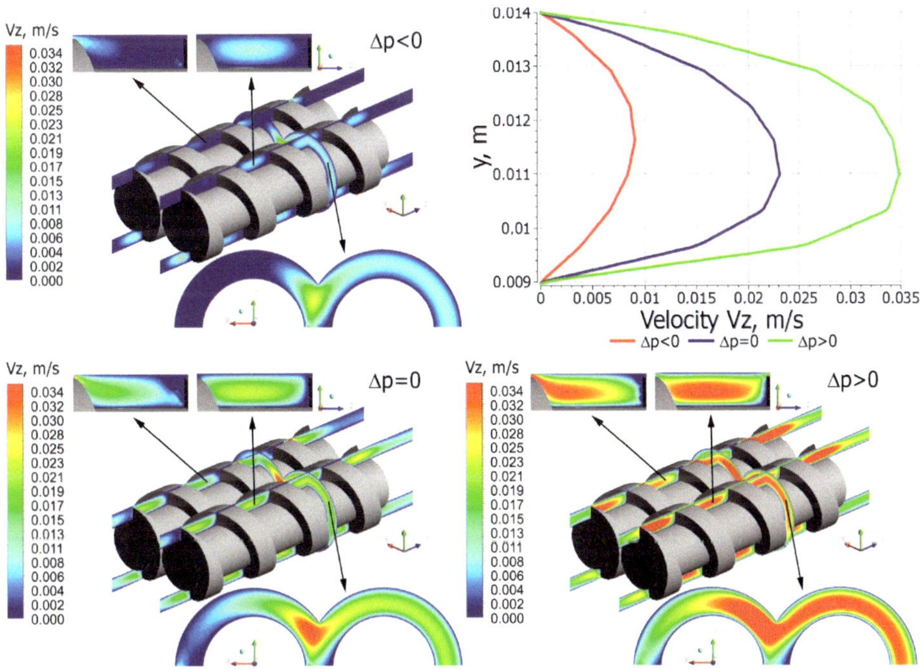

Figure 6.50 Distributions of the velocity v_z for the co-rotating twin screw extrusion

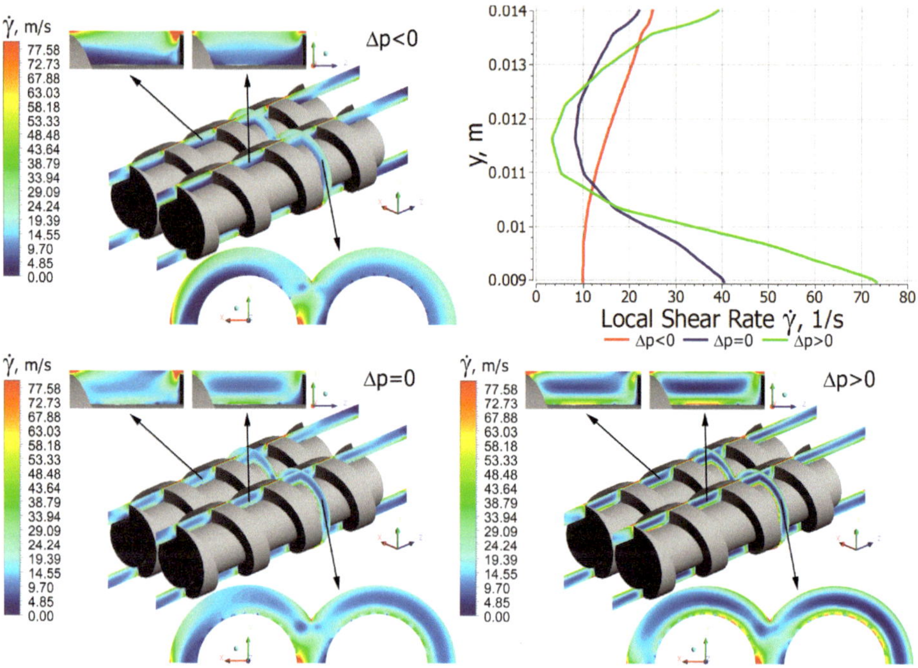

Figure 6.51 Distributions of the shear rate for the co-rotating twin screw extrusion

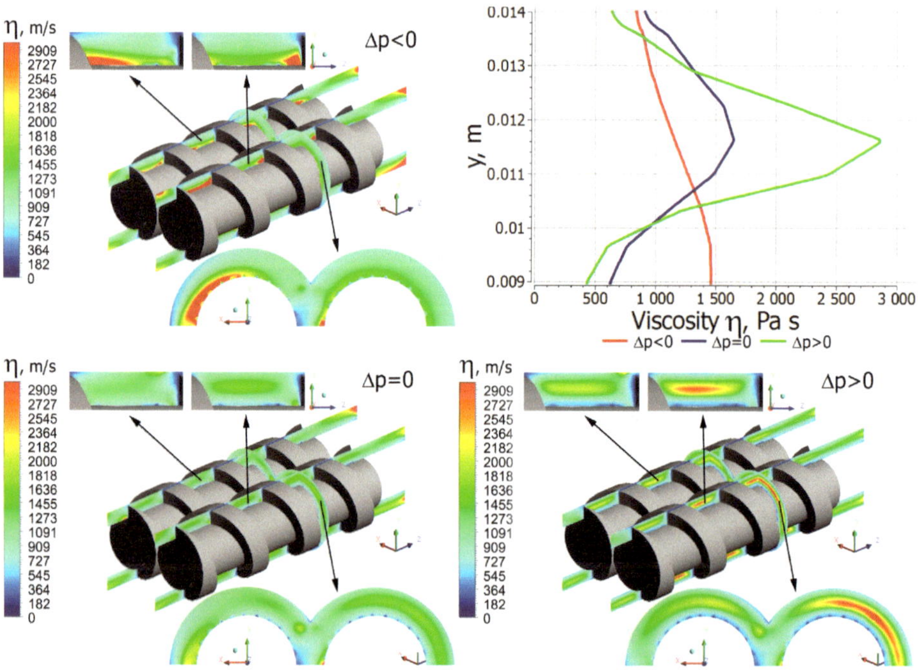

Figure 6.52 Distributions of the viscosity for the co-rotating twin screw extrusion

The flow of polymer in a co-rotating twin screw extruder is similar to some extent to the flow in a single screw extruder. This is the pressure-drag flow with the leakage flow in the gap between the tips of the screws and barrel, which affects the flow rate.

The leakage flow in a co-rotating twin screw extruder is depicted in Figure 6.53 for different pressure gradients. As for single screw extrusion, this is the pressure-drag flow. For the positive pressure gradient, the pressure leakage flow decreases the flow rate, while for the negative gradient, it increases the flow rate. In the case of the zero pressure gradient, the pressure leakage flow does not appear. However, the drag component of the leakage flow still exists. It is also worth noting that the viscosity of polymer in the gap is much lower than the viscosity of polymer in the channel, since the shear rate in the gap is much higher than in the screw channel.

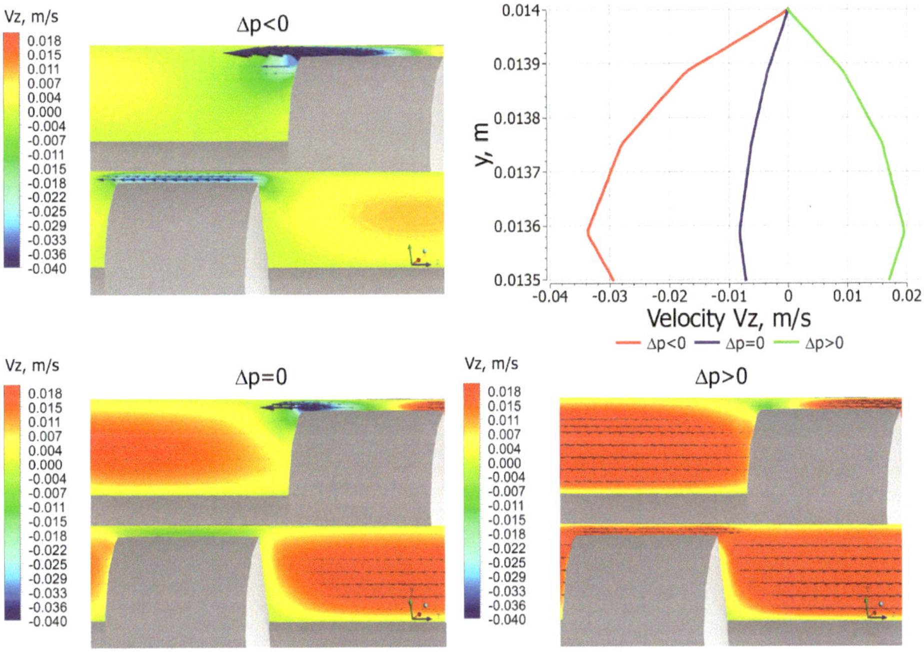

Figure 6.53 Leakage flows in the co-rotating twin screw extrusion

6.2.3.6 Counter-Rotating Twin Screw Extrusion

The flow of polymer in a counter-rotating twin screw extruder is totally different from that in a single screw extruder, but also different from that in a co-rotating twin screw extruder (Section 3.1.3.2). The essence of this flow is the so-called positive displacement mechanism that does not occur in other extrusion processes. The degree of positive displacement depends on the degree of screw meshing. It occurs most fully in the closely intermeshing counter-rotating extruders. There is no drag flow, so that the energy dissipation is insignificant.

The flow in a counter-rotating extruder is mainly performed in a closed space, in the C-shaped chamber, the so-called C-chamber, shown in Figure 3.9. Since the channels are closed, the material does not flow from one screw to the other except for some leakage flows, schematically shown in Figure 3.8, which reduce the degree of positive conveying. There is a co-rotating movement in the inter-screw gap, so that high shear stresses are not produced in this area, as in the case of co-rotating extrusion.

In a counter-rotating twin screw extrusion, as in the case of co-rotating one, the flow is three-dimensional and unsteady since the screws are rotating. The problem set-up is the same (Section 6.2.3.5), except that the screws are rotating in opposite directions.

The problem flow has been defined as an isothermal, 3D generalized Newtonian flow, and the power law model of Ostwald-de Waele was assumed (Eq. (6.1)).

The material parameters are as follows:

- consistency index, m — $6 \cdot 10^3$ Pa·s^n
- power law exponent, n — 0.39
- density, ρ — 740 kg/m^3

and the following flow conditions are assumed:

- screw speed, N — 60 rpm
- flow rate, G — 0/12.7/25.5 kg/h
- flow geometry (right-handed / left-handed screws):
  a) screw diameter, D — 27 mm
  b) screw pitch, t — 30 mm
  c) screw channel depth, H — 4.5 mm
  d) screw flight width, e — 6.5 mm
  e) screw/barrel clearance, h_f — 0.5 mm
  f) screw section length, L — 120 mm

The computation scheme is depicted in Figure 6.54. Three subdomains are distinguished, for the flowing material (Subdomain 1) and for the screws (Subdomain 2, Subdomain 3). The following flow boundary conditions were applied to define the process:

- at the inlet to the flow domain: boundary BC1 – **inflow,** the flow rate $(Q_{in}) = (Q)$ is imposed,
- at the outlet of the flow domain: boundary BC2 – **outflow,** vanishing normal forces and tangential velocities are imposed $(f_n \,\&\, v_s) = (0, 0)$, which means that the flow rate $(Q_{out}) = (Q)$ is imposed,

- on the screw domain: boundary BC3 - **Cartesian velocity,** Cartesian velocities are imposed (v_x & v_y & v_z) = (N), which means that the screw rotates at the screw speed (N) clockwise,
- on the screw domain: boundary BC4 - **Cartesian velocity,** Cartesian velocities are imposed (v_x & v_y & v_z) = ($-N$), which means that the screw rotates at the screw speed (N) counterclockwise,
- on the barrel wall: boundary BC5 - **zero wall velocity,** vanishing normal and tangential velocities are imposed (v_n & v_s) = (0, 0), which means that the velocity vanishes at the barrel wall.

The two first boundary conditions, BC1 - inflow (Q_{in}) = (Q), and BC2 - normal forces and tangential velocities (f_n & v_s) = (0, 0), imply that pressure may be generated along the screw. Since we do not know the pressure at the end of the screw element, we calculate the pressure gradient, which is relative to the zero pressure at the element exit. Negative pressures resulting from computations do not mean negative pressures in the extrusion process.

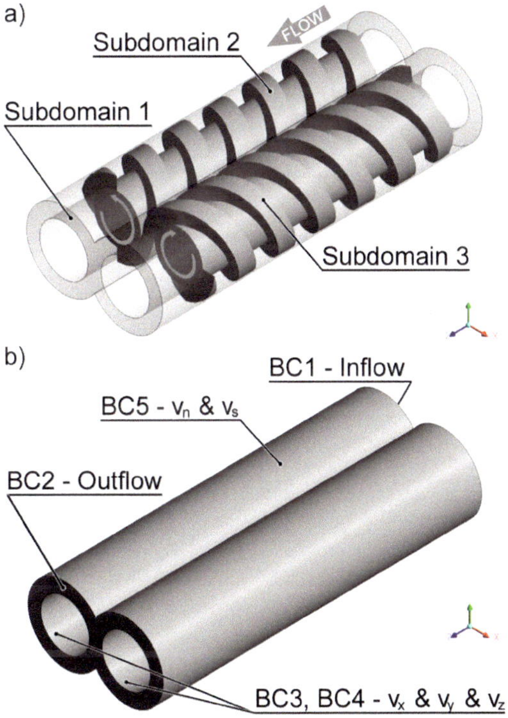

Figure 6.54
The flow geometry and boundary conditions for counter-rotating twin screw extrusion: BC1 - inflow, BC2 - outflow, BC3/BC4 - Cartesian velocity, BC5 - zero wall velocity

As for a co-rotating twin screw extrusion, the simulations were performed to study the flow parameters distributions for different pressure gradient: positive gradient ($\partial p/\partial z > 0$, i.e. $\Delta p < 0$), zero gradient ($\partial p/\partial z = 0$, i.e. $\Delta p = 0$), and negative gradient ($\partial p/\partial z < 0$, i.e. $\Delta p > 0$), where Δp is the pressure difference between the pressure at channel inlet and the pressure at the channel outlet.

To perform these, we started the simulations with the initial boundary condition $BC1_{initial}$ - outflow (Q_{out}) = (Q), with other conditions unchanged, which implies that pressure is not generated along the screw, and the flow rate is computed which is equal to the net flow Q_{net}, that is, there is no pressure flow and we have ($\partial p/\partial z = 0$, i.e. $\Delta p = 0$). The net flow is composed of the chamber flow and the drag components of leakage flows.

Then, the simulations were performed with the basic boundary condition BC1 - inflow (Q_{in}) = (Q), for different values $Q = 0$, $Q = Q_{net}$, $Q = 2Q_{net}$ which corresponds to the different pressure gradients: positive gradient, zero gradient, and negative gradient.

Pressure distributions are depicted in Figure 6.55. For the positive pressure gradient, the pressure leakage flows decrease the flow rate, while for the negative gradient these increase the flow rate. In the case of the zero pressure gradient, the pressure leakage flows do not appear. However, the drag components of the leakage flows still exist. As in the case of single screw extrusion and co-rotating twin screw extrusion, the characteristic pressure pulsation is observed (Sections 6.2.3.4 and 6.2.3.5). The frequency of this pulsation is equal to the frequency of screw rotation (rotational speed). However, in this case the pulsation is much lower.

Distributions of the velocity components for different pressure gradients are shown in Figure 6.56 to Figure 6.58. The velocity component v_x, which is perpendicular to the screw axis, is decreased at the positive pressure gradient, and is increased at the negative gradient. It is worth noting that distributions are totally different in both screws. The velocity component v_y, also perpendicular to the screw axis, varies depending on the pressure gradient. The velocity component v_z, which is parallel to screw axis, has a parabolic distribution, and increases with an increase of the pressure difference Δp. These distributions are similar in both screws but not the same.

Distributions of the local shear rate and viscosity for different pressure gradients are presented in Figure 6.59 and Figure 6.60. For the positive pressure gradient, the shear rate decreases from the screw to the barrel, while for the negative and zero gradients it is highest at the screw and barrel surfaces with a minimum in the middle of the channel. These correspond to the velocity distributions v_x, v_y, and v_z (Figure 6.56 to Figure 6.58). The viscosity distribution results from the shear rate distribution. For the positive pressure gradient, it increases from the screw to the barrel, while for the negative and zero gradients it is lowest at the screw and barrel surfaces with a maximum in the middle of the channel.

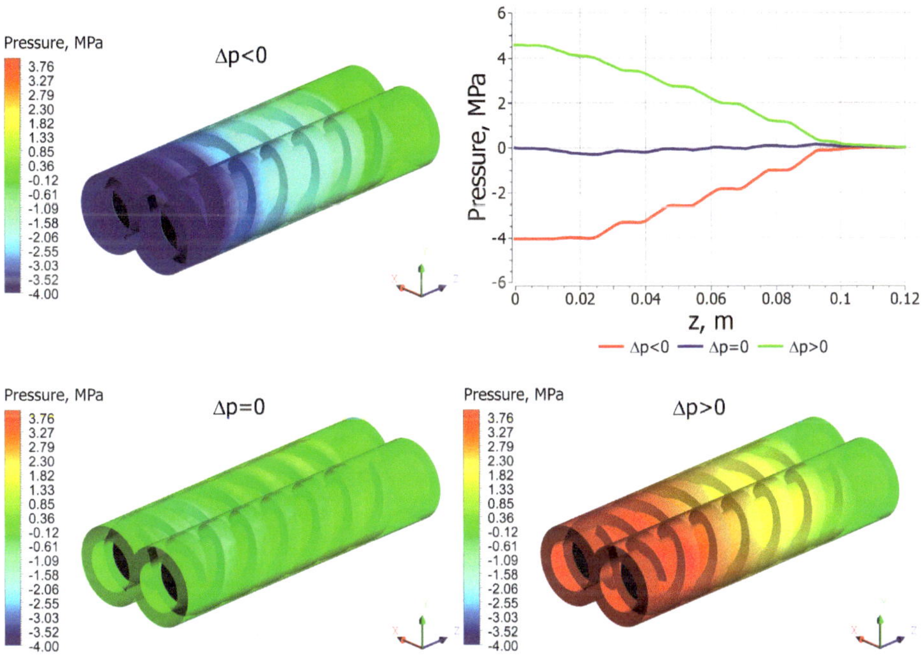

Figure 6.55 Pressure distributions for different pressure gradients ($\partial p/\partial z$): positive ($\Delta p < 0$), zero ($\Delta p = 0$), negative ($\Delta p > 0$), Δp - pressure difference

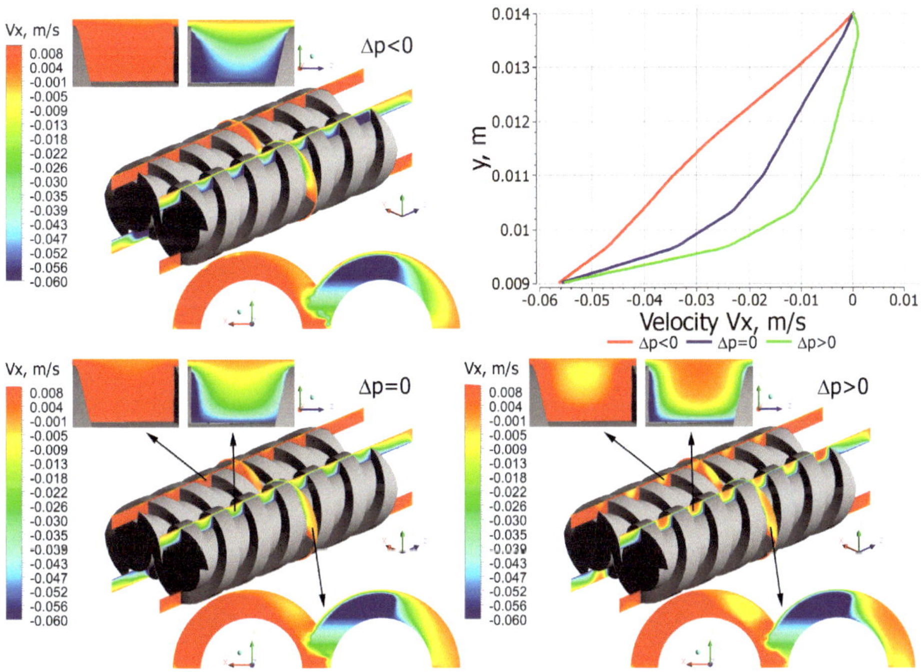

Figure 6.56 Distributions of the velocity v_x for the counter-rotating twin screw extrusion flow

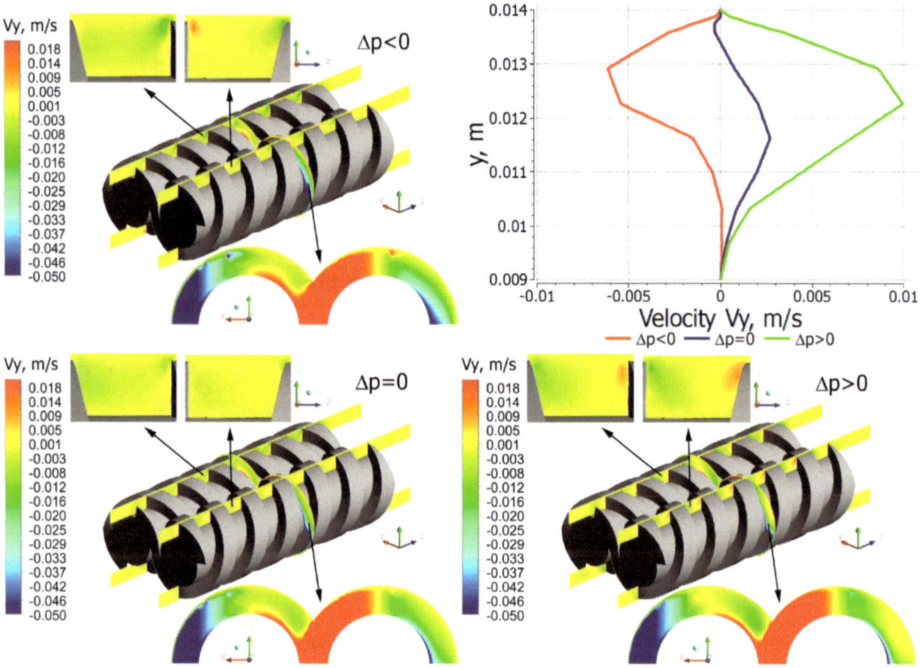

Figure 6.57 Distributions of the velocity v_y for the counter-rotating twin screw extrusion (the chart represents the velocity distribution near the screw flight)

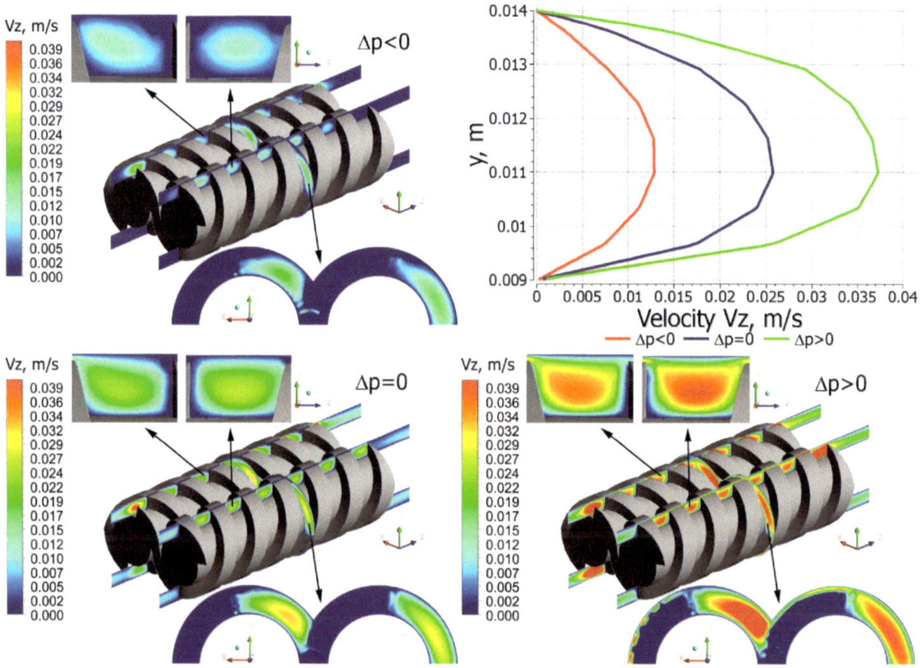

Figure 6.58 Distributions of the velocity v_z for the counter-rotating twin screw extrusion

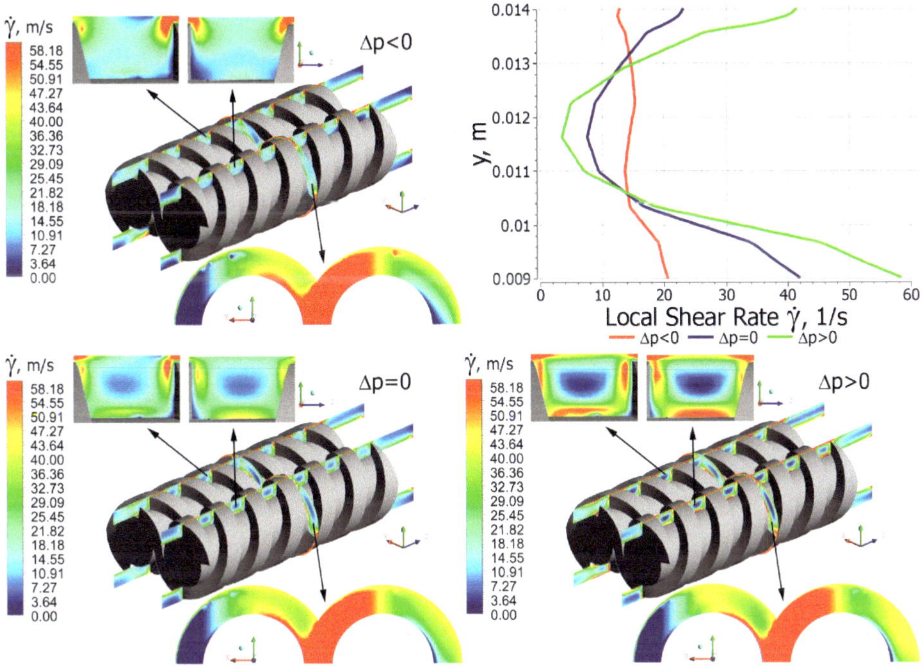

Figure 6.59 Distributions of the shear rate for the counter-rotating twin screw extrusion

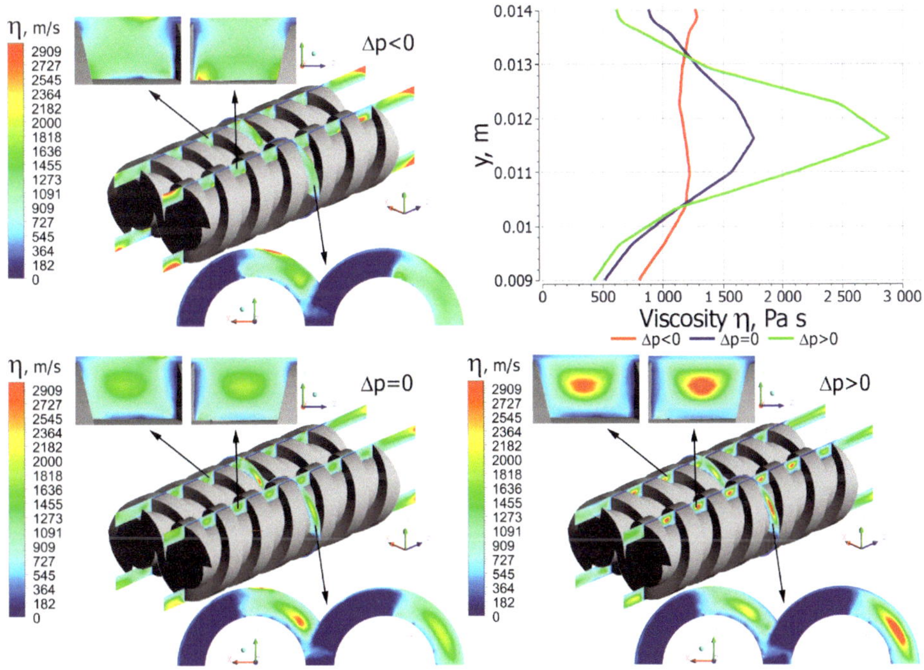

Figure 6.60 Distributions of the viscosity for the counter-rotating twin screw extrusion

The flow in a counter-rotating extruder is mainly performed in the C-chamber by positive displacement. However, there are also various leakage flows (Figure 3.8) which reduce the degree of positive conveying and affect the flow rate.

The following leakage flows can be distinguished:

- the calendering flow Q_c between the screw root and the screw flight,
- the flight flow Q_f over the screw flights, through the gap between the screw flight and the barrel,
- the back pressure inter-screw flow Q_t (tetrahedron flow) through the tetrahedral gap between the flight flanks of screws, in the radial direction,
- the side flow Q_s through the side gap between the flight flanks of screws, in the tangential direction.

The leakage flows are pressure-drag flows, except the tetrahedron flow which is a pressure flow, and the total leakage flow may be expressed as a sum of the drag flows and the pressure flows.

The leakage flows at the positive and negative pressure gradients are depicted in Figure 6.61. It is seen that at the positive pressure gradient the leakage flows are more intense. Moreover, the flight flow and the tetrahedron flow change direction when the sign of pressure gradient changes.

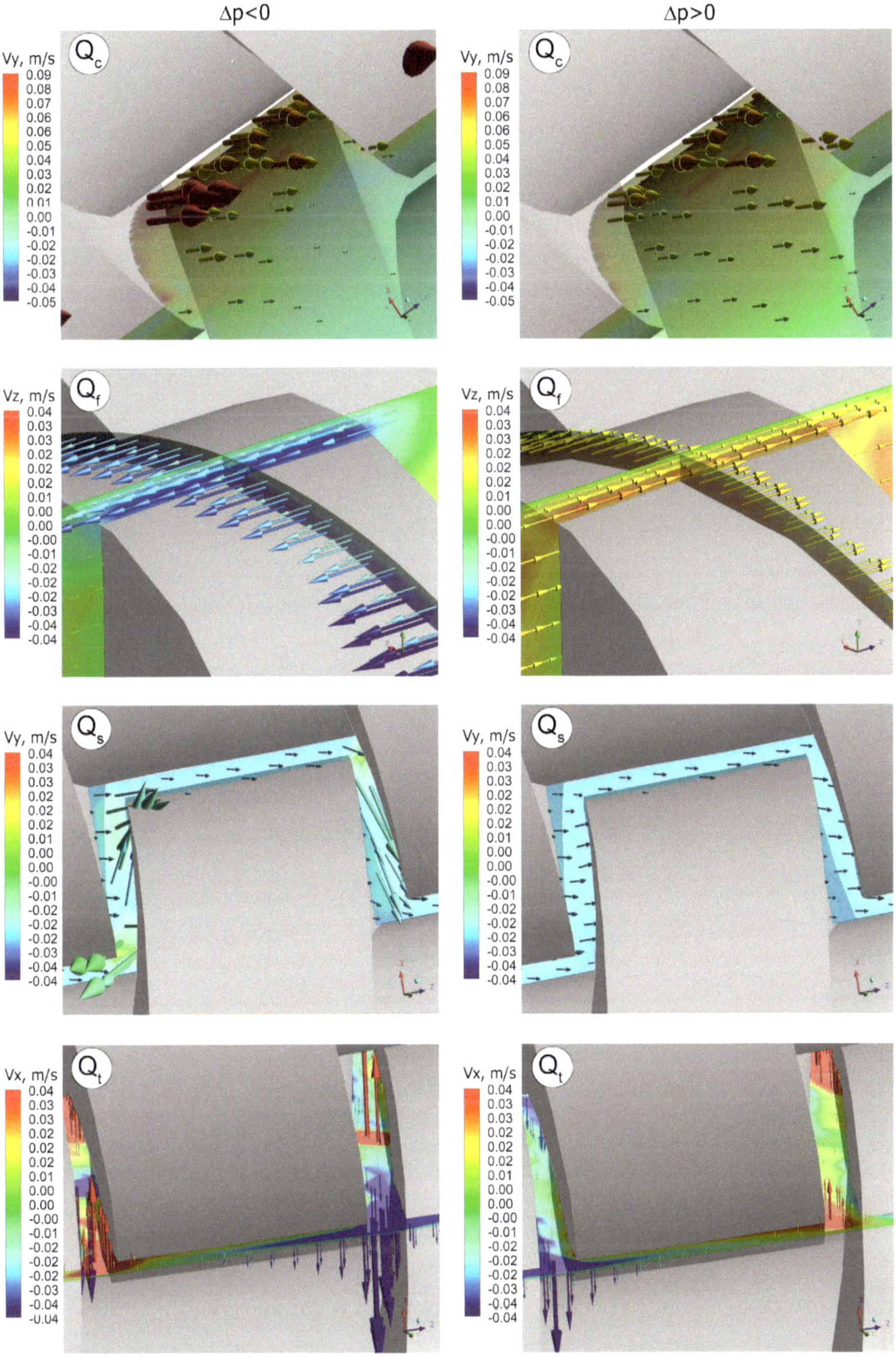

Figure 6.61 Leakage flows in the counter-rotating twin screw extrusion for different pressure gradients ($\partial p/\partial z$), positive ($\Delta p < 0$) and negative ($\Delta p > 0$): Q_c - calendering leakage flow, Q_f - flight leakage flow, Q_s - side leakage flow, Q_t - pressure leakage flow

6.3 Injection Molding

Co-authored by Krzysztof J. Wilczyński

6.3.1 Autodesk Moldflow – Program Overview

Injection molding is an important technology in the polymer processing industry, mainly due to the immense variety of manufactured products of various shapes, dimensions, and designs, as well as due to the large number of injection techniques used (Section 3.2).

Autodesk Moldflow is a finite-element program for simulating the polymer flow in injection molds which can be used to simulate various injection molding processes, e.g.:

- thermoplastics injection molding,
- overmolding,
- gas-assisted injection molding,
- co-injection molding,
- bi-injection molding,
- compression molding,
- reactive molding,
- microchip encapsulation,
- microcellular injection molding,
- resin transfer molding.

In general, Autodesk Moldflow makes it possible to simulate the subsequent phases of injection molding process:

- filling phase,
- holding (packing) phase,
- cooling phase,

and several process analyses can be performed, e.g.:

- fill analysis,
- pack analysis,
- cool analysis,
- gate location analysis,
- runner balance analysis,
- stress analysis,

- warp analysis,
- shrink analysis,
- venting analysis,
- fiber orientation analysis,
- crystallization analysis,
- molding windows analysis,
- process optimization analysis.

Autodesk Moldflow offers a wide variety of material models, e.g.:

a) the modified 2-domain Tait *pvT* model which is used to determine the density of the material as a function of the temperature and pressure,

b) the Cross-WLF viscosity model that describes the temperature, shear rate, and pressure dependency of the viscosity for thermoplastic materials,

c) the extension viscosity model that describes the dependence of the viscosity on the shear rate, temperature, pressure, and extension rate in 3D flow,

d) the Herschel-Bulkley-WLF viscosity model which can be used for thermoset materials that show a yield stress,

e) the matrix viscosity model which is used to determine the viscosity from measured data supplied at specific temperatures, shear rates, and pressures,

f) the Moldflow second order viscosity model that describes the temperature and shear rate dependence of the viscosity using a quadratic formulation.

6.3.2 Modeling Procedure

Modeling with Autodesk Moldflow consists of:

- selecting the molding process, e.g. thermoplastics injection molding,
- model preparation with the fundamental principle that the model must be equivalent to the flow in the real part, and the flow is dictated by the following elements, listed according to their significance, i.e. the part thickness, flow length, and volume,
- meshing the model using various analysis technologies, Midplane (a surface mesh with one-dimensional representation of the part through its midplane), Dual Domain (a surface mesh with one-dimensional representation on each surface of the part), or 3D (a volume mesh),
- selecting the analysis sequence,
- setting the injection location which represents the position where a polymer is injected, enabling the software to simulate the flow pattern inside the mold cavity,

- selecting the injection molding machine,
- setting the velocity/pressure switch-over point from the ram speed control to the packing pressure control which typically takes place before the cavity is filled,
- process settings, e. g. the Filling Control Profile setting which is used to enter filling profiles to vary the ram movement during the filling phase, the Pack Control Profile setting which is used to enter packing profiles to vary the pressure applied to the cavity during the packing phase, and the Mold Surface Temperature Profile setting which is used to enter the mold surface temperature profile of zones of the mold through the injection molding cycle,
- optimization analyses,
- checking the simulation results.

6.3.3 Examples of Modeling

Autodesk Moldflow can be used to simulate various injection molding processes to solve various flow problems and perform various process analyses. In the further part of this section, the examples of important analyses are presented, i. e. the gate location, the filling analysis, and the flow balancing. Some details of these analyses are presented in the literature [36–39].

6.3.3.1 Gate Location

The Gate Location analysis is used to recommend injection locations for the part. This analysis works for all analysis technologies and is used as a preliminary input for a full Fill & Pack analysis. A Runner Balance analysis can be used with a Fill analysis to ensure equal pressure is delivered at each cavity.

After running a Gate Location analysis, the recommended injection location is automatically displayed, as depicted in Figure 6.62 on an example of a tensile test specimen.

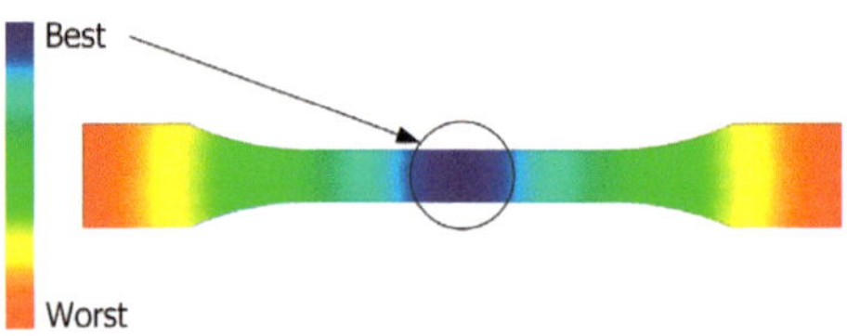

Figure 6.62
Gate Location analysis

6.3.3.2 Filling Analysis

The Fill analysis predicts the thermoplastic polymer flow inside the mold in the filling phase. This analysis is often run as the first part of a Fill & Pack analysis sequence. In this section, the Fill analysis is presented on an example of the double-cavity mold for tensile test specimens with two different gate locations with edge and lateral gating. It is performed for polypropylene.

This analysis calculates the flow front that grows through the part incrementally from the injection location, and continues until the velocity/pressure switch-over point has been reached. This is depicted in Figure 6.63.

Moreover, the filling analysis allows us to simulate the distribution of various flow parameters, e.g. velocity, pressure, temperature, etc., which are depicted in Figure 6.64 to Figure 6.66. It is seen that various gate locations result in various distributions of these parameters, and finally in various part deflections, which are shown in Figure 6.67.

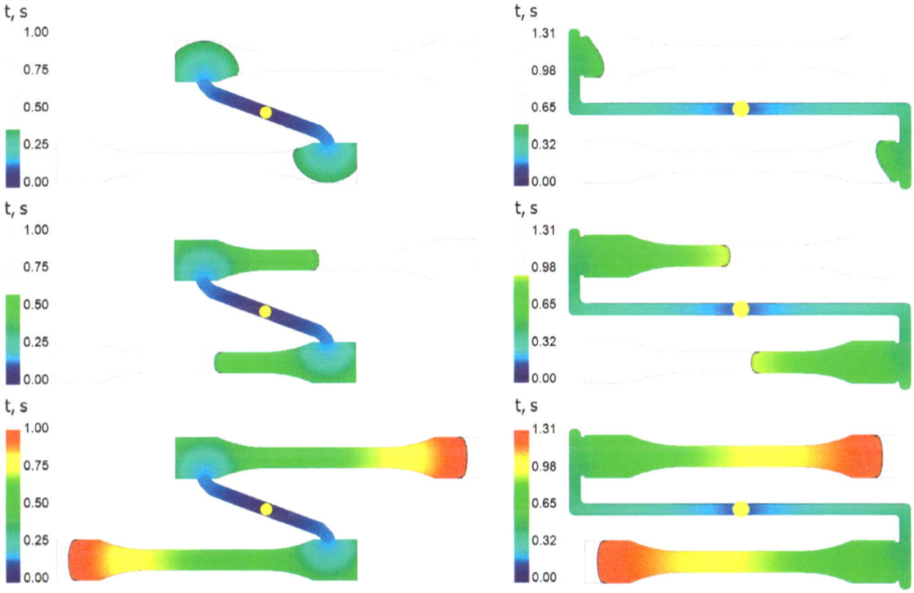

Figure 6.63 Fill analysis – edge vs lateral gating: growth of the flow front

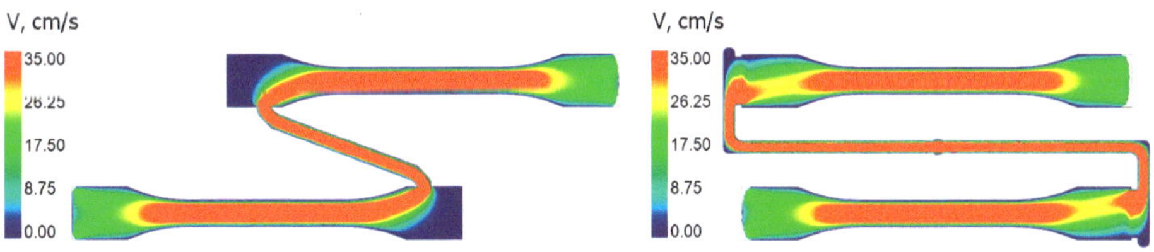

Figure 6.64 Fill analysis – edge vs lateral gating: velocity distribution

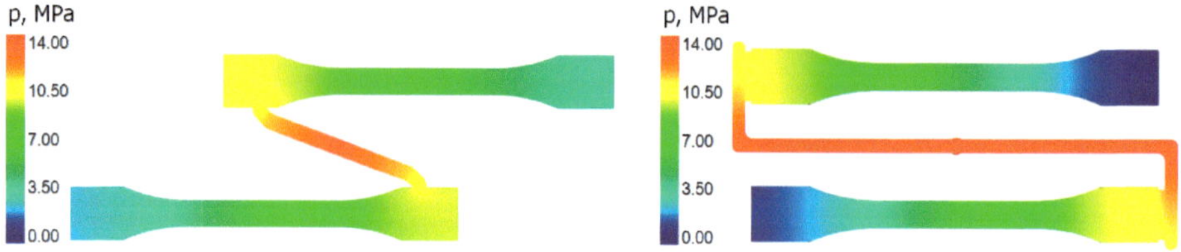

Figure 6.65 Fill analysis – edge vs lateral gating: pressure distribution

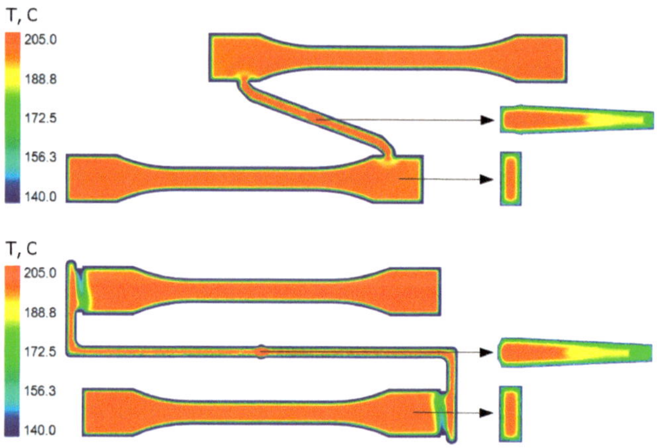

Figure 6.66 Fill analysis – edge vs lateral gating: temperature distribution

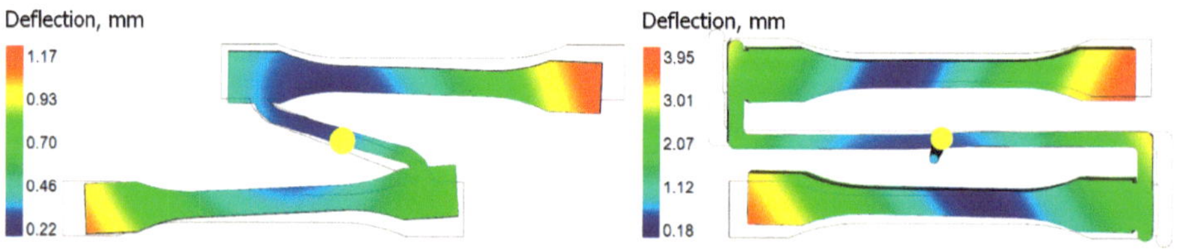

Figure 6.67 Fill analysis – edge vs lateral gating: deflection distribution

6.3.3.3 Flow Balancing

A part with a balanced flow has the plastic melt filling the extremities of the mold at the same time and pressure. A balanced flow leads to uniform orientation and shrinkage, less internal stress and warpage, and cost savings through reduced material usage. The gate location, the runner system design, and the part thickness affect the flow balance.

The Runner Balance analysis is used to determine the optimum volume for the sections of the runner system. This analysis ensures that the part fills evenly and there is an acceptable pressure magnitude in the cavity, minimizing the runner volume.

The Runner Balance analysis is for a model with an attached runner system. A runner system that is added after a Fill & Pack analysis helps you balance the cavities. The aim is to achieve runner dimensions with the following characteristics:

- the same pressure drop in all flow paths, so that all cavities fill at the same moment,
- controlled shear heating, which minimizes the stress level without using a high melt temperature,
- minimized volume of runner material relative to cavity volume, which ensures the highest possible pressure drops in the runner system.

In this section, the Runner Balance analysis is presented on an example of a six-cavity mold for plates.

In the case of the simulation presented in Figure 6.68, which was performed for polypropylene, the balanced flow has been obtained by diminishing the diameter of the middle runners from 4 mm to 1.73 mm. Changing the material from polypropylene to polybutylene required increasing the diameter of these runners from 1.73 mm to 1.96 mm, which is depicted in Figure 6.69.

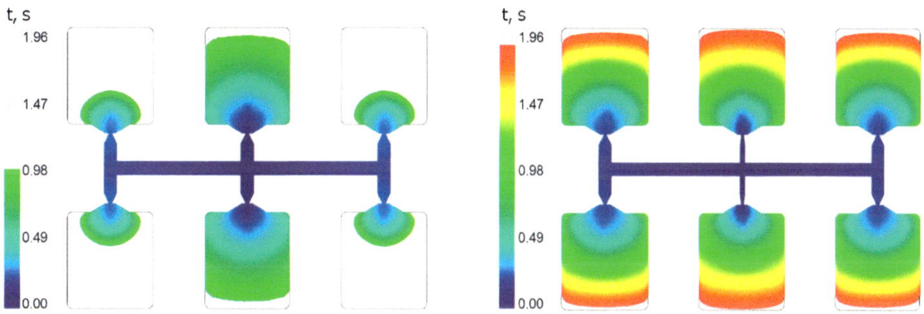

Figure 6.68 Runner Balance analysis for polypropylene (PP) – imbalanced vs balanced runner system: growth of the flow front

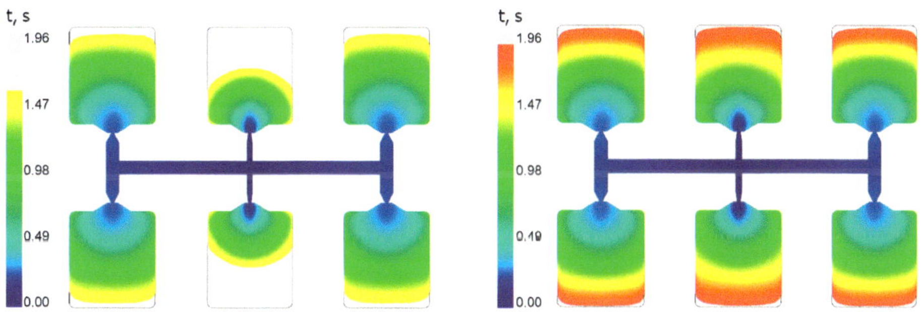

Figure 6.69 Runner Balance analysis for polybutylene (PBT) – imbalanced vs balanced runner system: growth of the flow front

Pressure distributions obtained by simulations performed for polypropylene are depicted in Figure 6.70. These distributions are more balanced for corrected runners. It finally results in a more balanced shrinkage distribution, which is shown in Figure 6.71.

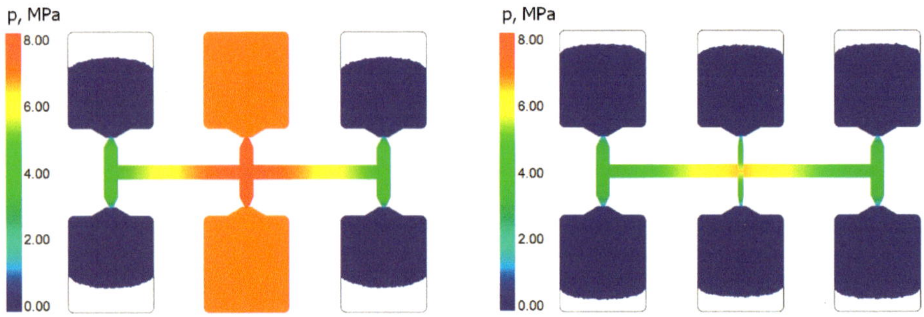

Figure 6.70 Runner Balance analysis - imbalanced vs balanced runner system: pressure distribution

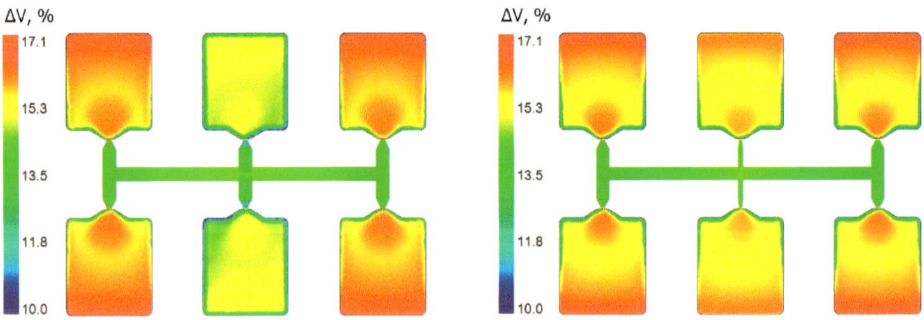

Figure 6.71 Runner Balance analysis - imbalanced vs balanced runner system: average volumetric shrinkage distribution

6.3.3.4 Melt Flipper

In the simulation presented in the previous section, the flow paths were not the same, so that the runner system was not geometrically balanced, and the consequence of this was uneven filling of the mold cavities.

It is interesting to note that filling imbalance can also be observed in molds equipped with geometrically balanced runner systems. This imbalance is observed in H-type runner layouts, with eight or more cavities. It is known that this imbalance results from the melt flow shear gradients developed in the runner system which lead to non-symmetrical temperature and viscosity distributions. The flow imbalance is created by the non-symmetrical shear distribution within the melt as it travels through the runner system. The non-symmetrical shear distribution results in a lower viscosity, higher temperature, high flow rate melt feeding the inner cavities of a multi-cavity mold. In the eight-cavity mold, an imbalance is created when the melt experiences a second flow division by flowing into a tertiary runner.

However, the commercial technique of melt rotation using the so-called Melt Flipper allows reduction of the imbalance [25]. This concept is depicted in Figure 6.72. The runner is modified by inserting the Melt Flipper at the intersection of the primary and secondary runner and provides a melt rotation as it flows into the secondary runner. Melt variations are rearranged such as they are symmetrical about the center. This creates symmetry at the following runner branch.

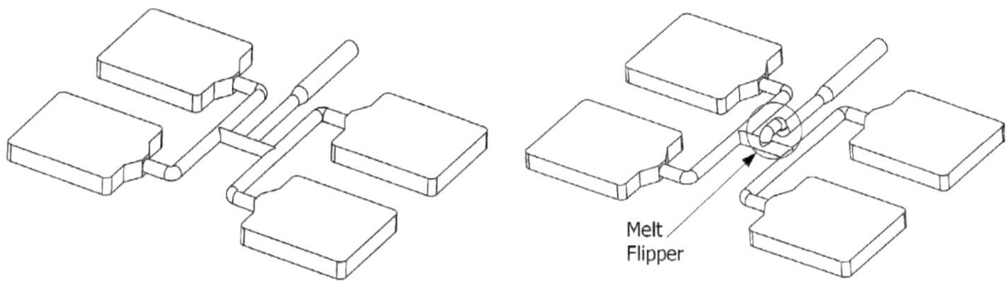

Figure 6.72 Schematic of the Melt Flipper concept: standard runner layout vs Melt Flipper runner layout

In this section, the Melt Flipper analysis is presented on an example of an eight-cavity mold for plates. It is performed for polypropylene.

Growth of the flow front for the standard runner layout and the Melt Flipper layout is depicted in Figure 6.73. The effect of the flow rate on the imbalance phenomenon is depicted in Figure 6.74. It is clearly seen for the standard runner layout that imbalance increases when the flow rate increases. However, it is important to note that using the Melt Flipper is not efficient over the entire flow rate range and an imbalance occurs at the highest rate. The outer cavities begin to fill faster.

Distributions of the velocity, shear rate, viscosity, and temperature are depicted in Figure 6.75 to Figure 6.79. These distributions are more balanced for corrected runners.

It is clearly seen from Figure 6.79 that inserting the Melt Flipper at the intersection of the primary and secondary runner rotates the melt and creates the flow symmetry.

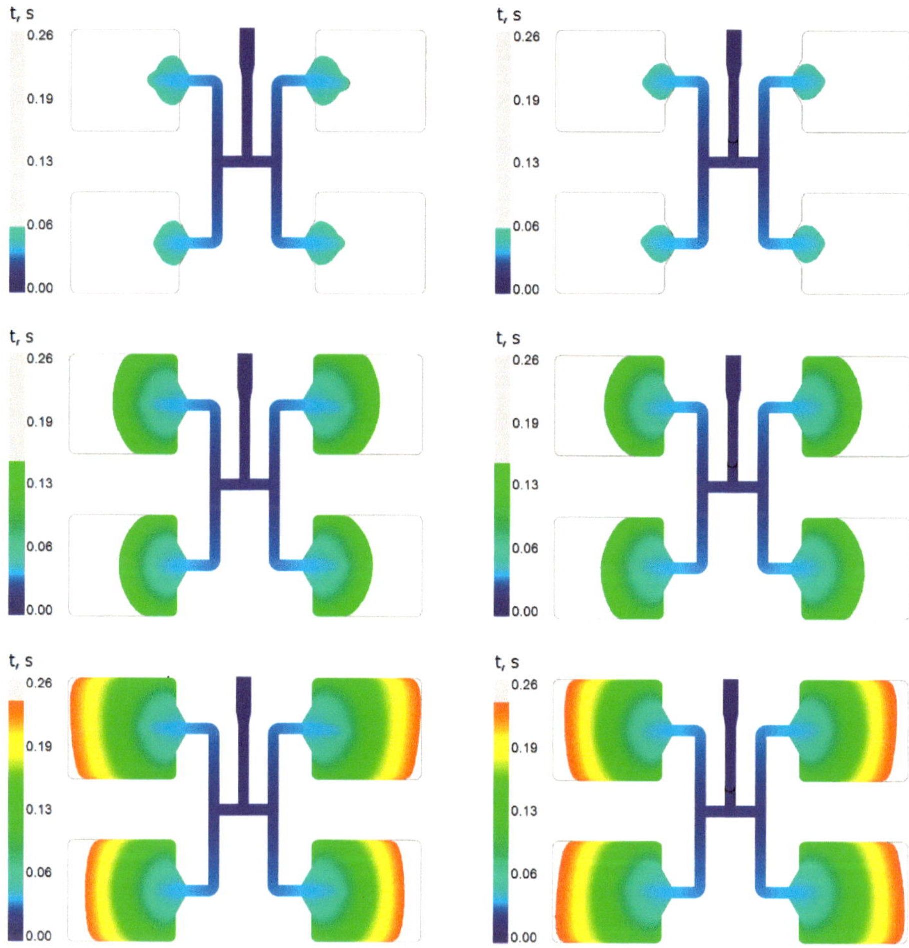

Figure 6.73 Flow Balance (Melt Flipper) analysis – imbalanced vs balanced runner system: growth of the flow front

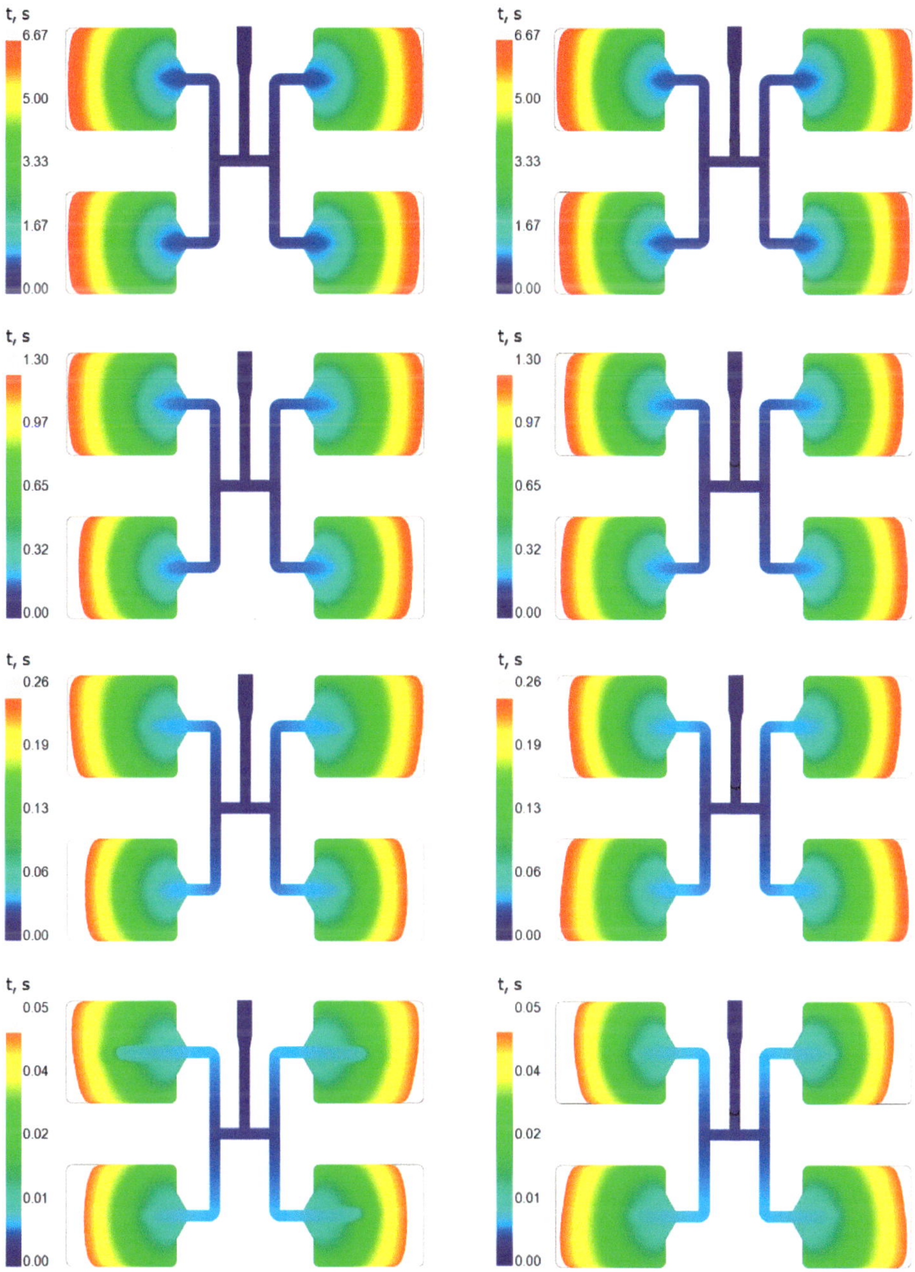

Figure 6.74 Flow Balance (Melt Flipper) analysis - imbalanced vs balanced runner system: effect of the flow rate on the growth of the flow front; $Q = 2\ cm^3/s$, $Q = 10\ cm^3/s$, $Q = 50\ cm^3/s$, $Q = 250\ cm^3/s$

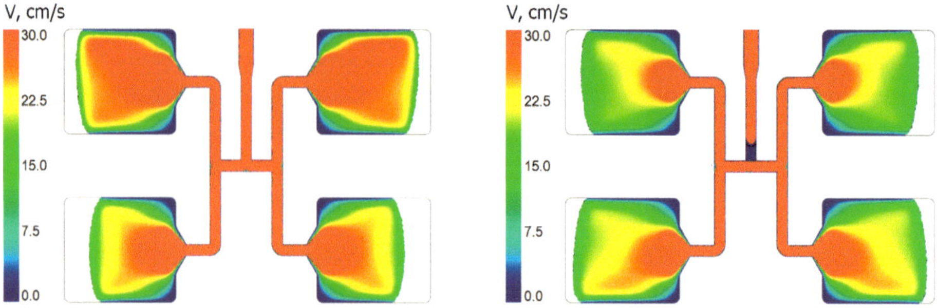

Figure 6.75 Flow Balance (Melt Flipper) analysis – imbalanced vs balanced runner system: velocity distribution, $Q = 50\ \text{cm}^3/\text{s}$

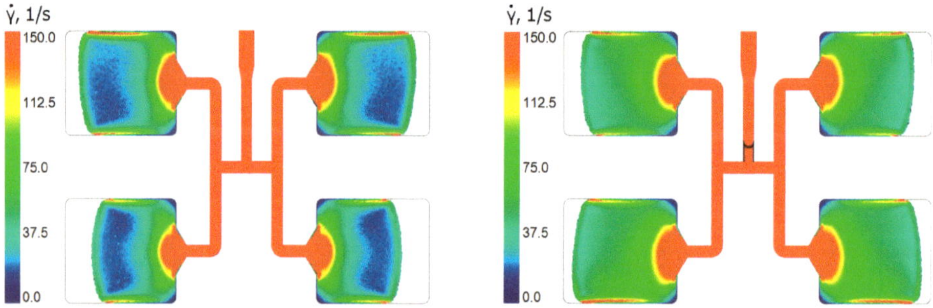

Figure 6.76 Flow Balance (Melt Flipper) analysis – imbalanced vs balanced runner system: shear rate distribution, $Q = 50\ \text{cm}^3/\text{s}$

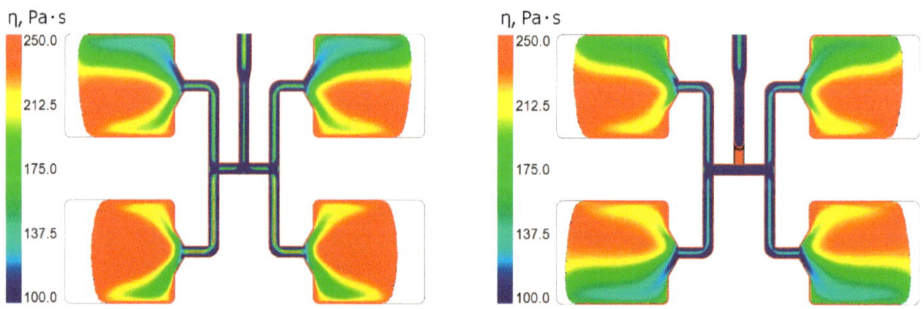

Figure 6.77 Flow Balance (Melt Flipper) analysis – imbalanced vs balanced runner system: viscosity distribution, $Q = 50\ \text{cm}^3/\text{s}$

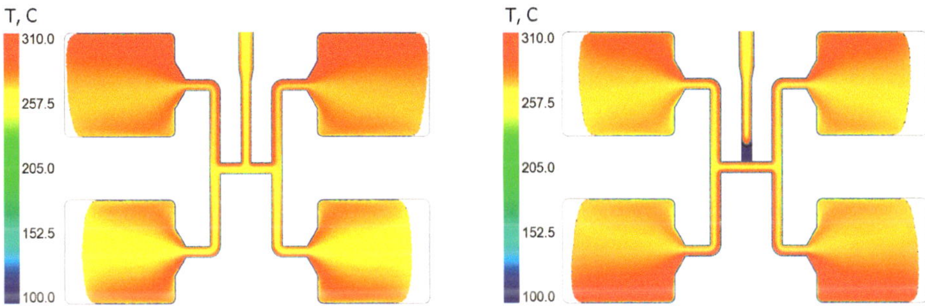

Figure 6.78 Flow Balance (Melt Flipper) analysis - imbalanced vs balanced runner system: temperature distribution, $Q = 50\ \text{cm}^3/\text{s}$

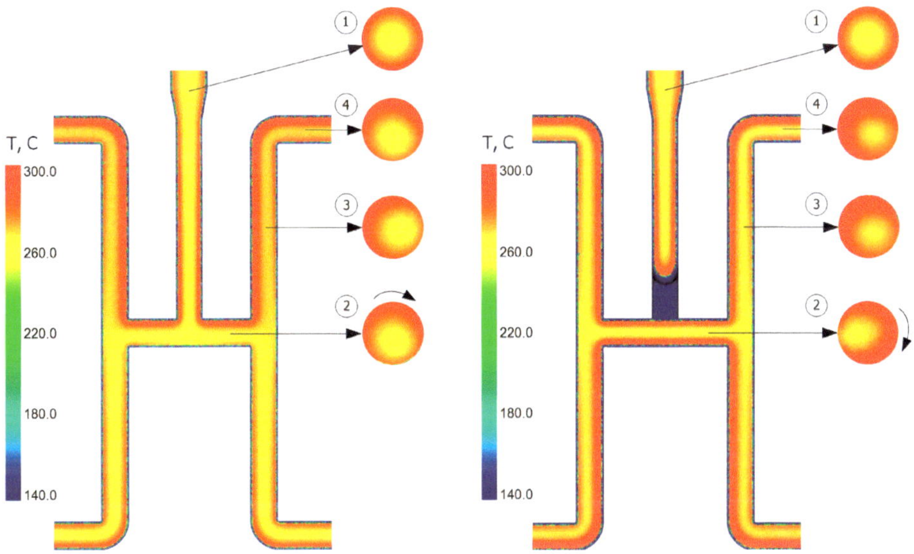

Figure 6.79 Flow Balance (Melt Flipper) analysis - imbalanced vs balanced runner system: temperature distribution, $Q = 50\ \text{cm}^3/\text{s}$

6.4 Extrusion

Co-authored by Andrzej Nastaj

6.4.1 MULTI-SCREW System - Program Overview

Extrusion is the most used technology by volume in the polymer processing industry since more than half of all plastics are extruded. It is also fundamental for compounding, i.e. mixing, pelletizing, filling, reinforcing, etc. (Section 3.1). Extrusion can be performed using single screw extruders or twin screw extruders, both co-rotating or counter-rotating. Single screw extruders are usually fed by gravity (flood fed extruders), while twin screw extruders are fed by metering (starve fed extruders).

The MULTI-SCREW system consists of three subsystems: GSEM program (Global Screw Extrusion Model) for simulation of single screw extrusion, TSEM program (Twin Screw Extrusion Model) for simulation of twin screw extrusion, and GASEO program (Genetic Algorithms Screw Extrusion Optimization) for extrusion process optimization.

GSEM *(Global Screw Extrusion Model)* is a program designed to simulate the single screw extrusion process, both flood fed and starve fed. The program allows simulation of extrusion using conventional and non-conventional screws equipped with mixing and shearing elements. The dies may have channels of various cross-sections. The program simulates an operation of the screw/die system making it possible to predict the extrusion throughput (in the case of flood feeding), the screw filling (in the case of metered feeding), the pressure and temperature profiles along the screw/die length, the melting of polymer, and the power consumption, etc. The simulation parameters are the material characteristics, the screw/die geometry, and the process operating conditions.

GSEM makes it possible to simulate an extrusion process using the following simulation options:

- **Extruder** option that simulates the extrusion process, i. e. co-operation of the screw/die system; in the case of the flood fed extrusion the extrusion throughput (the flow rate) is not known and is computed by simulations; however, for the starve fed extrusion the extrusion throughput is an input datum for simulations,
- **Screw** option that simulates the polymer flow in the plasticating unit of the extruder; in this case the extrusion throughput (the flow rate) is an input datum for simulations,
- **Die** option that simulates the polymer flow in the extruder die; in this case the extrusion throughput (the flow rate) is an input datum for simulations.

TSEM *(Twin Screw Extrusion Model)* is a program package designed to simulate the counter-rotating twin screw extrusion process. The program allows simulation of extrusion using conventional and non-conventional screw elements. The dies may have channels of various cross-sections. The program simulates an operation of the screw/die system making it possible to predict the screw filling, the pressure and temperature profiles along the screw/die length, the melting of polymer, and the power consumption, etc. The simulation parameters are the material characteristics, the screw/die geometry, and the process operating conditions.

GASEO *(Genetic Algorithms Screw Extrusion Optimization)* is a program package based on genetic algorithms designed to optimize the single screw extrusion process, both flood fed and starve fed. Optimization is based on the process simulation which is performed using the GSEM program.

6.4.2 Modeling Procedure

Modeling with the MULTI-SCREW system consists of:

- selecting the extrusion process, single screw extrusion or twin screw extrusion,
- selecting the feeding mode of the process, starve fed or flood fed,
- selecting the material characteristics; the rheological models of power law, Cross-WLF, and Klein are available,
- selecting the screw/die elements to configure the screw/die geometry,
- setting the process conditions, e.g. screw speed, barrel/die temperature profile, polymer flow rate,
- setting the simulation option, i.e. extruder, screw, or die,
- checking the simulation results.

6.4.3 Examples of Modeling

The MULTI-SCREW system can be used to simulate various extrusion processes to solve various flow problems and perform various process analyses. In the further part of this section, examples of important analyses are presented, both for single screw extrusion and twin screw extrusion. Some details of these analyses are presented in the literature [40–48].

6.4.3.1 Single Screw Extrusion

In this section, simulations for single screw extrusion, both flood fed and starve fed, are presented. The simulations were performed using a conventional three-sectional screw of diameter $D = 45$ mm, length/diameter ratio $L/D = 27$, and compression ratio, i.e. the ratio of the channel depth in the feeding zone to the channel depth in the metering zone, $CR = 8/3$ (Section 3.1.2), which is depicted in Figure 6.80.

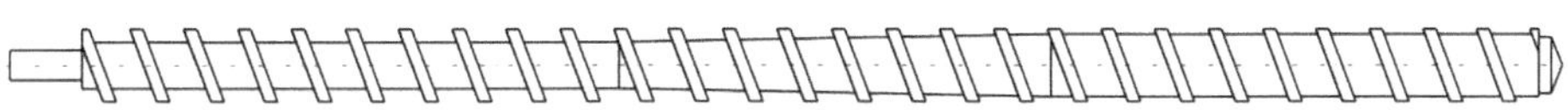

Figure 6.80 Conventional three-sectional screw: *FF 1-45* – single-flighted section of pitch $t = 45$ mm

Flood Fed Single Screw Extrusion

Simulations for flood fed single screw extrusion are presented in Figure 6.81 to Figure 6.84. The overall process characteristics (dimensionless), which include the pressure and temperature profiles, the screw filling profile, as well as the polymer melting profile represented by the ratio of the polymer solid bed width X to the screw channel width W, i. e. the solid bed profile $SBP = X/W$ (Section 5.2.1), are depicted in Figure 6.81. These simulations are complemented by the results of experimentation, which include the pressure and temperature measurements, and the photograph of the screw removed from the extruder, presenting the melting progress of the material.

The effect of the screw speed on the pressure profile is depicted in Figure 6.82. It is seen that the pressure increases with an increase of the screw speed which results in an increase in the flow rate. A maximum pressure is observed at the end of the compression zone. Then it drops in the metering zone. The pressure in the extruder always increases from zero at the beginning of the screw to the pressure enabling the material to be extruded through the die. However, the pressure profile is not always monotonic and sometimes pressure drops occur in the screw, even down to zero, e. g. during the starve fed extrusion. The pressure in the die always drops. In general, it can be said that the screw produces the pressure and the die consumes it.

The effect of the screw speed on the melting profile is depicted in Figure 6.83. It is seen that the melting is slower when the screw speed increases, which results from an increase in the flow rate and a decrease in the residence time of the polymer in the plasticating unit (the heating time), although the energy dissipation increases. It is seen that in the compression zone, the SBP parameter, i. e. the ratio of the polymer solid bed width X to the screw channel width W, decreases more slowly. If the screw is compressed too much, this can even increase. The flow of the polymer may be blocked in the screw when the width of the solid bed reaches the width of the screw channel, i. e. when $X/W = 1$.

The effect of the screw speed on the temperature profile is depicted in Figure 6.84. It is seen that the temperature increases with an increase of the screw speed, which results from an increase in the energy dissipation.

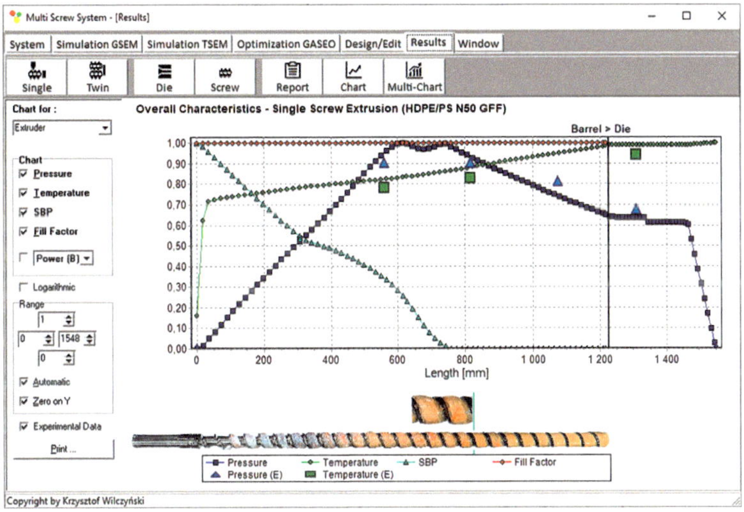

Figure 6.81 Overall characteristics of flood fed single screw extrusion, simulation and experimental data using the Global Screw Extrusion Model (GSEM): *SBP* – solid bed profile, *E* – experiment

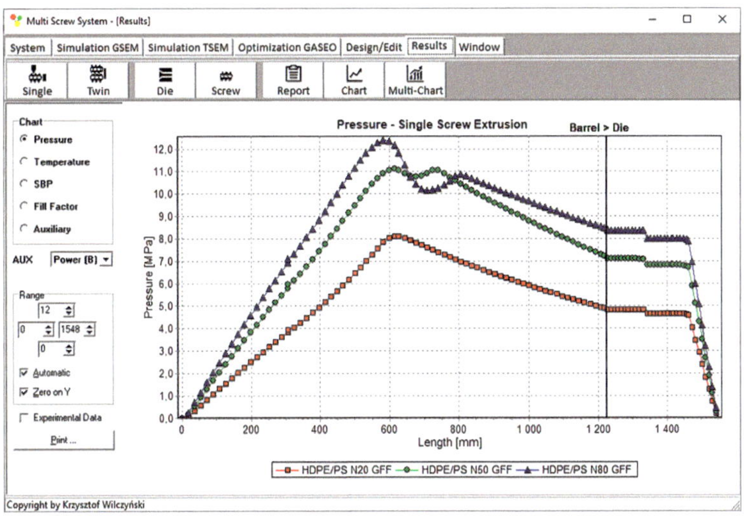

Figure 6.82 Effect of screw speed on the pressure profile, flood fed single screw extrusion: $N = 20$ rpm, $N = 50$ rpm, $N = 80$ rpm

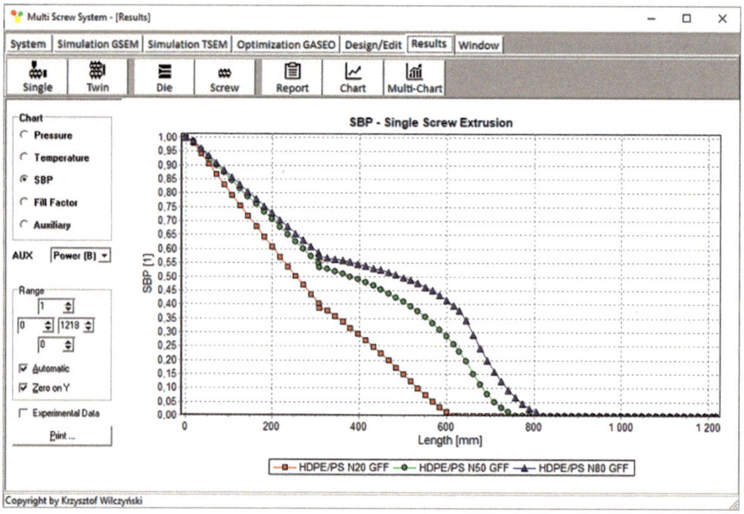

Figure 6.83 Effect of screw speed on the melting profile, flood fed single screw extrusion: $N = 20$ rpm, $N = 50$ rpm, $N = 80$ rpm

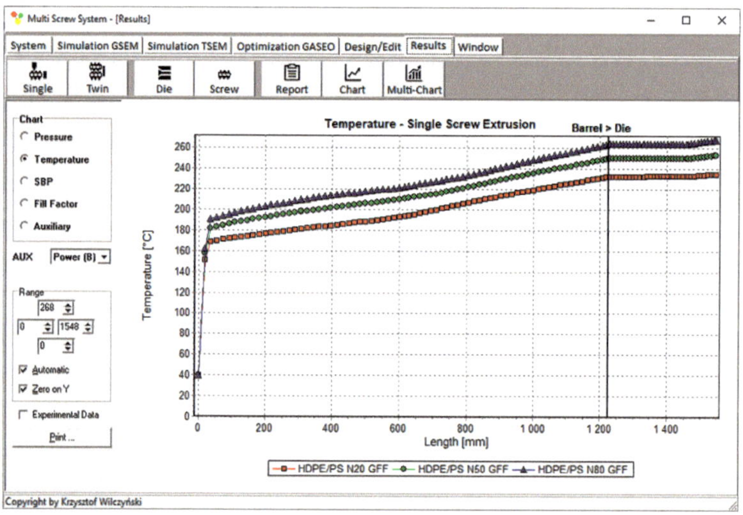

Figure 6.84 Effect of screw speed on the temperature profile, flood fed single screw extrusion: $N = 20$ rpm, $N = 50$ rpm, $N = 80$ rpm

Starve Fed Single Screw Extrusion

Simulations for starve fed single screw extrusion are presented in Figure 6.85 to Figure 6.89. The overall process characteristics (dimensionless), which include the pressure and temperature profiles, the screw filling profile, as well as the polymer melting profile represented by the parameter *SBP* (Section 5.2.1), are depicted in Figure 6.86. It is clearly seen that the pressure falls to zero when starvation begins. It is worth noting that the two stages of melting are observed, In the partially filled region of the screw, the polymer is melted by conduction, mainly. If the polymer is not completely molten until the screw channel is fully filled with the material, the unmolten solid particles of the polymer are suspended in the previously molten material. Such a suspension flows to the die and melting progresses through heat dissipation. The size of unmolten particles continuously decreases until the material is completely molten.

The simulations are complemented by the results of experimentation, which include the pressure and temperature measurements, and the photograph of the screw removed from an extruder, presenting the melting progress of the material and the screw filling with the material. The starve fed region and the fully filled region are clearly seen on the validation pictures.

In Figure 6.86 to Figure 6.89 the simulations are presented for various level of starvation, i.e. for various flow rate and these are compared to the flood fed extrusion.

An effect of the flow rate on the screw filling is depicted in Figure 6.86. It is seen that the screw filling increases with an increase of the flow rate. In the case of flood feeding, the screw is filled with the material along its entire length.

An effect of the flow rate on the pressure profile is depicted in Figure 6.87. It is interesting to note that the die pressure decreases with an increase of the flow rate, which results from an increase in the material temperature and an increase in the shear rate which means a lower viscosity. Comparing the pressure profile with the screw filling profile (Figure 6.86), it is seen that the pressure is developed in the fully filled region of the screw only, and falls to zero in the starved region.

The effect of the flow rate on the melting profile is depicted in Figure 6.88. It is seen that the melting is slower when the flow rate increases, which results from an increase of the material to be melted and a faster mechanism of melting in the case of metered feeding. In dependence on the flow rate (the level of starvation) one or two mechanisms of melting are observed in the case of metered feeding.

The effect of the screw speed on the temperature profile is depicted in Figure 6.89. The temperature increases with an increase of the flow rate, which results from an increase in the energy dissipation.

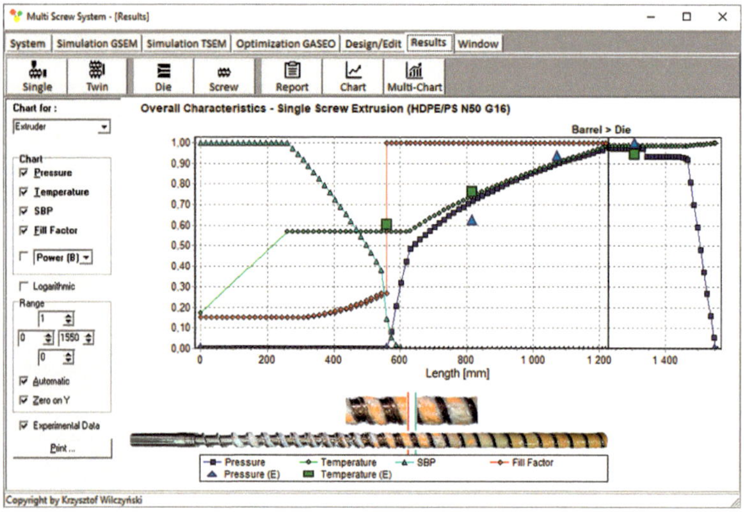

Figure 6.85 Overall characteristics of starve fed single screw extrusion, simulation and experimental data using the Global Screw Extrusion Model (GSEM): *SBP* – solid bed profile, *E* – experiment

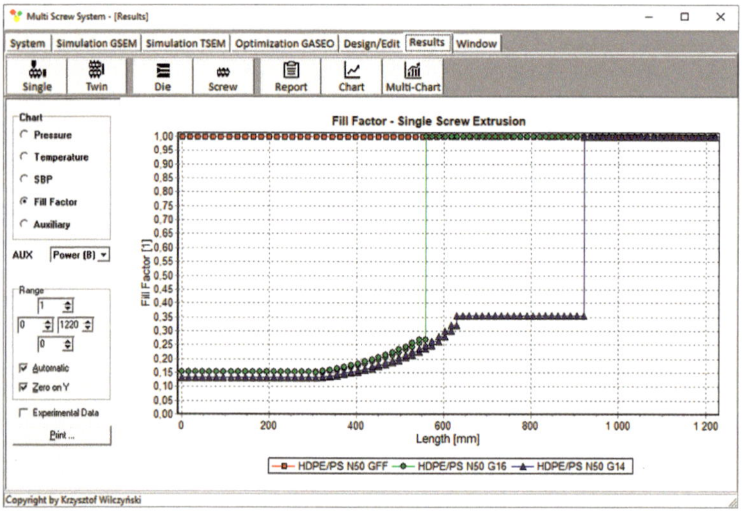

Figure 6.86 Effect of flow rate on the screw filling profile, starve fed single screw extrusion: $G = 14$ kg/h, $G = 16$ kg/h, $G = GFF$ (flood feeding)

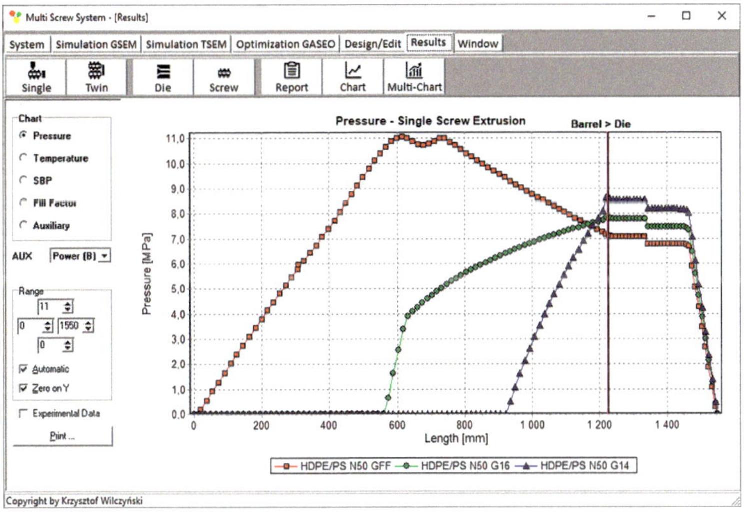

Figure 6.87 Effect of flow rate on the pressure profile, starve fed single screw extrusion: $G = 14$ kg/h, $G = 16$ kg/h, $G = GFF$ (flood feeding)

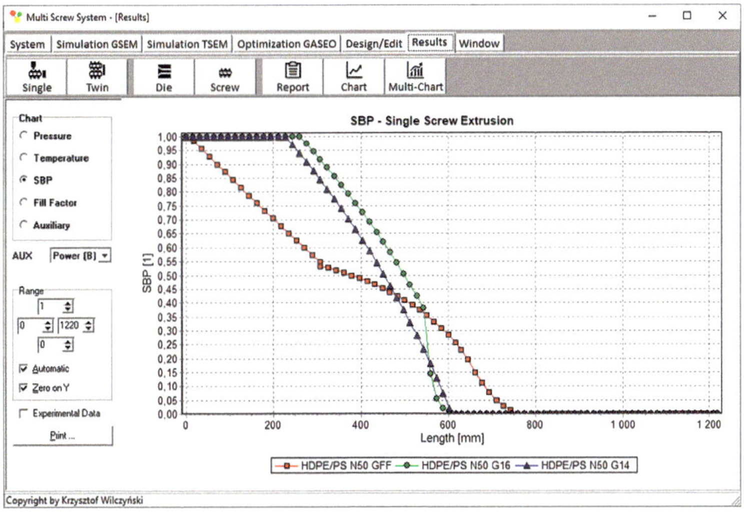

Figure 6.88 Effect of flow rate on the melting profile, starve fed single screw extrusion: $G = 14$ kg/h, $G = 16$ kg/h, $G = GFF$ (flood feeding)

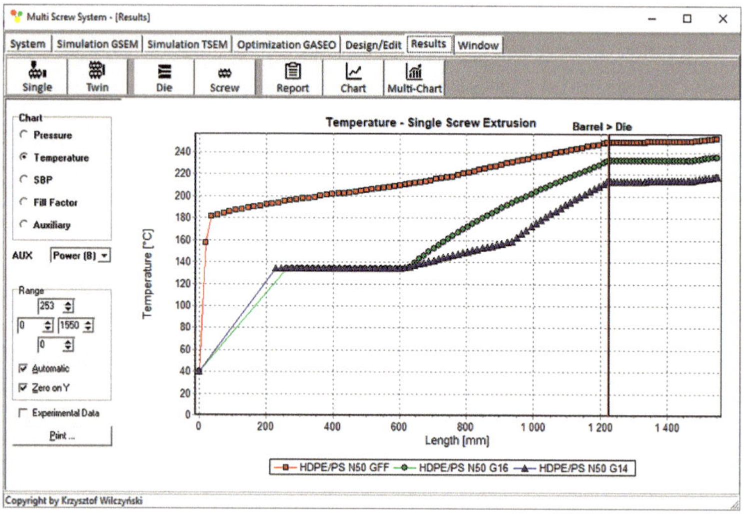

Figure 6.89 Effect of flow rate on the temperature profile, starve fed single screw extrusion: $G = 14$ kg/h, $G = 16$ kg/h, $G = GFF$ (flood feeding)

6.4.3.2 Twin Screw Extrusion

In this section, the simulations for counter-rotating twin screw extrusion are presented. The simulations were performed using a modular, closely intermeshing configuration of screws of diameter $D = 27$ mm with a distance between centerlines $d = 22.5$ mm. Various screw elements were arranged on the shafts to produce a particular screw configuration which is depicted in Figure 6.90. These are thin-flighted elements (FF), thick-flighted elements (FD), elements with variable pitch (KFD) and (KDD), and shearing elements (ZSS).

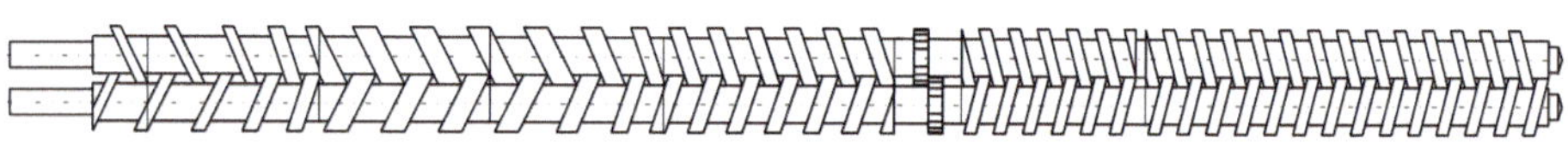

Figure 6.90 Modular, closely intermeshing twin screw configuration: *FF 1-30* - thin single-flighted element of pitch $t = 30$ mm, *KFD 1-30/20* - thin single-flighted element of variable pitch $t = 30/20$ mm, *FD 1-30* - thick single-flighted element of pitch $t = 30$ mm, *KDD 1-30/20* - thick single-flighted element of variable pitch $t = 30/20$ mm, *FD 1-20* - thick single-flighted element of pitch $t = 20$ mm, *FD 3-45* - thick triple-flighted element of pitch $t = 45$ mm, *FD 2-30* - thick double-flighted element of pitch $t = 30$ mm, ZSS - shearing element

Simulations for counter-rotating twin screw extrusion are presented in Figure 6.91 to Figure 6.93. The overall process characteristics (dimensionless), which include the pressure and temperature profiles, the screw filling profile, as well as the melting profile represented by the parameter *SBP* (Section 5.2.1), are depicted in Figure

6.91. These simulations are complemented by the results of experimentation, which include the photograph of screws removed from an extruder, presenting the melting progress of the material and the screw filling with the material. It is clearly seen that the pressure falls to zero when starvation begins. The starve fed region and the fully filled region are clearly seen on the validation pictures.

In Figure 6.92 and Figure 6.93, the simulations are presented for various level of starvation. i. e. for various flow rates.

The effect of the flow rate on the screw filling is depicted in Figure 6.92. It is seen that the screw filling increases with an increase of the flow rate.

The effect of the flow rate on the pressure profile is depicted in Figure 6.93. The pressure increases with an increase of the flow rate, and is developed in the fully filled region of the screws only, and falls to zero in the starved region. When the fully filled region increases the pressure region also increases.

Figure 6.91 Overall characteristics of starve fed single screw extrusion, simulation and experimental data using the Twin Screw Extrusion Model (TSEM): *SBP* - solid bed profile, *E* - experiment

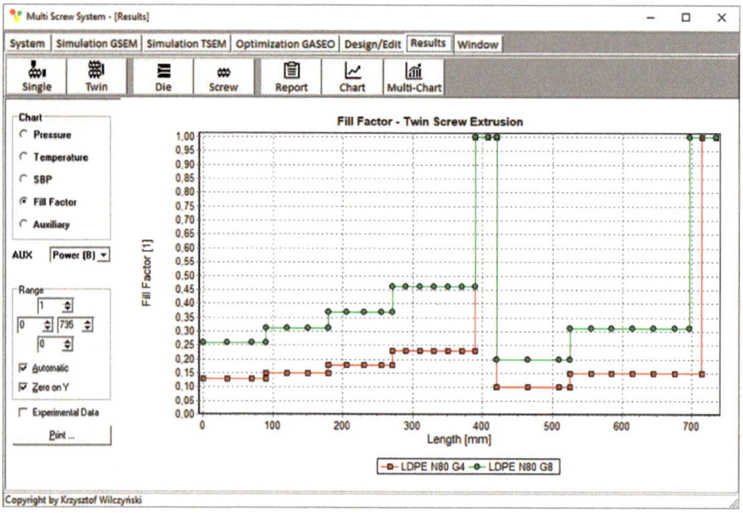

Figure 6.92 Effect of flow rate on the screw filling profile, twin screw extrusion: $G = 4$ kg/h, $G = 8$ kg/h

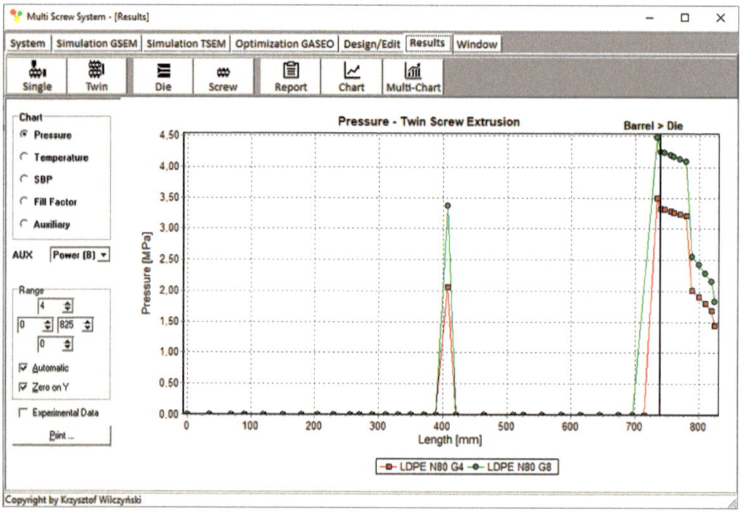

Figure 6.93 Effect of flow rate on the pressure profile, twin screw extrusion: $G = 4$ kg/h, $G = 8$ kg/h

References

1. Tadmor, Z., Klein, I., *Engineering Principles of Plasticating Extrusion* (1970) Van Nostrand Reinhold Company, New York
2. Rao, N.S., *Computer Aided Design of Plasticating Screws* (1986) Carl Hanser Verlag, Munich
3. Tucker, C.L. (Ed.), *Fundamentals of Computer Modeling for Polymer Processing* (1987) Carl Hanser Verlag, Munich

4. Manzione, L. T. (Ed.), *Applications of Computer Aided Engineering in Injection Moldding* (1992) Carl Hanser Verlag, Munich
5. O'Brien, K. T. (Ed.), *Computer Modeling for Extrusion and Other Continuous Polymer Processes* (1992) Carl Hanser Verlag, Munich
6. Baird, D. G., Collias, D. I., *Polymer Processing. Principles and Design* (1995) Butterworth-Heinemann, Oxford
7. Kennedy, P., *Flow Analysis of Injection Molds* (1995) Carl Hanser Verlag, Munich
8. Menges, G., Michaeli, W., Mohren, P., *How to Make Injection Molds* (2001) Carl Hanser Verlag, Munich
9. White, J. L., Potente, H., *Screw Extrusion* (2002) Carl Hanser Verlag, Munich
10. Tadmor, Z., Gogos, C. G., *Principles of Polymer Processing* (2006) John Wiley & Sons, New Jersey
11. Osswald, T., Hernandez-Ortiz, J. P., *Polymer Processing. Modeling and Simulation* (2006) Hanser Publishers, Munich
12. Kazmer, D. O., *Injection Mold Design Engineering* (2007) Carl Hanser Verlag, Munich
13. Manas-Zloczower, I. (Ed.), *Mixing and Compounding of Polymers. Theory and Practice* (2009) Carl Hanser Verlag, Munich
14. Osswald, T., Turng, L.-S., Gramann, P., *Injection Molding Handbook* (2009) Carl Hanser Verlag, Munich
15. Kamal, M. R., Isayev, A. I., Liu, S.-J., *Injection Molding. Technology and Fundamentals* (2009) Carl Hanser Verlag, Munich
16. White, J. L., Kim E. K., *Twin Screw Extrusion* (2010) Carl Hanser Verlag, Munich
17. Vlachopoulos, J., *Polymer Rheology and Extrusion* (2011) McMaster University, Hamilton
18. Rao, N. S., Schott, N. R., *Plastics Engineering Calculations* (2012) Carl Hanser Verlag, Munich
19. Rauwendaal, Ch., *Polymer Extrusion* (2014) Carl Hanser Verlag, Munich
20. Campbell, G. A., Spalding, M. A., *Analyzing and Troubleshooting Single-Screw Extruders* (2014) Carl Hanser Verlag, Munich
21. Vlachopoulos, J., *Fundamentals of Fluid Mechanics* (2016) Polydynamics, Hamilton
22. Agassant, J.-F., Avenas, P., Carreau, P. J., Vergnes, B., Vincent, M., *Polymer Processing.Principles and Modeling* (2017) Carl Hanser Verlag, Munich
23. Hopmann, Ch., Michaeli, W., *Extrusion Dies for Plastics and Rubber* (2017) Carl Hanser Verlag, Munich
24. Wang, M.-L., Chang, R.-Y., Hsu, Ch.-H., *Molding Simulation: Theory and Practice* (2018) Carl Hanser Verlag, Munich
25. Beaumont, J. P., *Runner and Gating Design Handbook* (2019) Carl Hanser Verlag, Munich
26. Vlachopoulos, J., Polychronopoulos, N. D., *Understanding Rheology and Technology of Polymer Extrusion* (2019) Polydynamics Inc., Dundas
27. Chung, Ch. I., *Extrusion of Polymers. Theory and Practice* (2020) Carl Hanser Verlag, Munich
28. ANSYS Polyflow, *CDF for Extrusion, Forming and Molding*; ANSYS Inc.: Canonsburg, PA, USA
29. Autodesk Moldflow, *Plastic Injection and Compresion Mold Simulation*; Autodesk Inc.: San Rafael, CA, USA
30. Wilczyński, K., Nastaj, A., Lewandowski, A., Wilczyński, K. J., *Multipurpose Computer Model for Screw Processing, Polymer-Plastics Technology and Engineering* (2012), 51, 626–633, https://dx.doi.org/10.1080/03602559.2012.659313
31. Wilczyński, K., Nastaj, A., Lewandowski, A., Wilczyński, K. J., Buziak, K., *Fundamentals of Global Modeling for Polymer Extrusion, Polymers* (2019), 11(12), 2106; *https://doi.org/10.3390/polym11122106*

32. Wilczyński, K., Nastaj, A., Lewandowski, A., Wilczyński, K. J., *Modeling of the Polymer Melt Flow in a Twin Screw Counter-Rotating Extrusion Process. Simulation and Experimental Studies, Polimery* (2011), 56, 45–50 (in Polish). https://dx.doi.org/10.14314/polimery.2011.045

33. Wilczyński, K., Lewandowski, A., *Study on the Polymer Melt Flow in a Closely Intermeshing Counter-Rotating Twin Screw Extruder, International Polymer Processing* (2014), 29, 649–659. *https://doi.org/10.3139/217.2962*

34. Lewandowski, A., Wilczyński, K., *Global Modeling of Single Screw Extrusion with Slip Effects. International Polymer Processing* (2019), 34, 81–90. https://doi.org/10.3139/217.3653

35. Lewandowski, A., Wilczyński, K., *Global Modeling for Single Screw Extrusion of Viscoplastics. International Polymer Processing* (2020), 35, 26–36. *https://doi.org/10.3139/217.3866*

36. Wilczyński, K., Nastaj, A., Wilczyński, K. J., *Computer Simulation of Flow Balancing in Injection Molds with Melt Flipper. Mechanik* (2008), 81, 327–330 (in Polish)

37. Wilczyński, K., Wilczyński, K. J., Narowski P., *Experimental and Simulation Studies on Filling Imbalance in Geometrically Balanced Runner Systems of Multi-Cavity Injection Molds, Polimery* (2015), 60, 411–421 (in Polish), *https://dx.doi.org/10.14314/polimery.2015.411*

38. Wilczyński, K., Narowski, P., *Simulation Studies on the Effect of Material Characteristics and Runner Layout Geometry on the Filling Imbalance in Geometrically Balanced Injection Molds. Polymers* (2019) 11(4), 639; *https://doi.org/10.3390/polym11040639*

39. Wilczyński, K., Narowski, P., *A Strategy for Problem Solving of Filling Imbalance in Geometrically Balanced Injection Molds. Polymers* (2020) 12(4), 805; https://doi:10.3390/polym12040805

40. Wilczyński, K., Jiang, Q., White, J. L., *A Composite Model for Melting, Pressure and Fill Factor Profiles in a Metered Fed Closely Intermeshing Counter-Rotating Twin Screw Extruder, International Polymer Processing* (2007), 22, 198–203, *https://doi.org/10.3139/217.2001*

41. Wilczyński, K. J., Nastaj, A., Lewandowski, A., Wilczyński, K., *A Composite Model for Starve Fed Single Screw Extrusion of Thermoplastics, Polymer Engineering and Science* (2014), 64, 2362–2374, https://doi.org/10.1002/pen.23797

42. Lewandowski, A., Wilczyński, K. J., Nastaj, A., Wilczyński, K., *A Composite Model for an Intermeshing Counter-Rotating Twin-Screw Extruder and its Experimental Verification, Polymer Engineering and Science* (2015), 55, 2838–2848, *https://doi.org/10.1002/pen.24175*

43. Wilczyński, K. J., Lewandowski, A., Nastaj A., Wilczyński K., *Modeling for Starve Fed/Flood Fed Mixing Single-Screw Extruders, International Polymer Processing* (2016), 31, 82–91. https://doi.org/10.3139/217.3154

44. Wilczyński, K. J., Lewandowski, A., Nastaj, A., Wilczyński, K., *A Global Model for Starve-Fed Nonconventional Single-Screw Extrusion of Thermoplastics, Advances in Polymer Technology* (2017), 36, 23–35, *https://doi.org/10.1002/adv.21570*

45. Wilczyński, K. J., Nastaj, A., Wilczyński, K., *A Computer Model for Starve-Fed Single-Screw Extrusion of Polymer Blends, Advances in Polymer Technology* (2018), 37, 2142–2151. https://doi.org/10.1002/adv.21873

46. Wilczyński, K. J., *Experimental and Theoretical Study on Screw Filling in Starve Fed Single Screw Extruders, International Polymer Processing* (2018), 33, 171–179. *https://doi.org/10.3139/217.3386*

47. Wilczyński, K., Buziak K., Wilczyński, K. J., Lewandowski, A., Nastaj, A., *Computer Modeling for Single-Screw Extrusion of Wood-Plastic Composites. Polymers* (2018), 10(3), 295; https://doi.org/10.3390/polym10030295

48. Nastaj, A., Wilczyński, K., *Optimization for Starve Fed/Flood Fed Single Screw Extrusion of Polymeric Materials . Polymers* (2020) 12(1), 149; https://doi:10.3390/polym12010149

Index

A

apparent viscosity 112
apparent viscosity curve 113
arrangement of mold cavities 156
Arrhenius law 35

B

Bagley correction 110
Bagley method 109
barrier screws 137
Barus effect 57, 148
blow molding 157
- extrusion blow molding 159
- film blowing 158
- injection blow molding 160

C

calendering 164
capillary rheometers 99
C-chamber 141
CFD modeling 297
- co-rotating twin screw extrusion 332
- counter-rotating twin screw extrusion 339
- extrudate swell 317
- extrudate swell inverse problem 321
- pressure flow 300
- single screw extrusion 325
compression molding 166
- plunger transfer compression molding 167
- pot transfer compression molding 167
compression ratio 135
computer modeling software 295
- generally oriented systems 295
- systems oriented on a specific processing technique 295
cone-plate rheometers 117
conservation laws 8, 21
- equation of continuity 21
- equation of energy 21
- equation of momentum 21
conservation of energy 16
conservation of mass 8
conservation of momentum 11
constitutive equations 8, 21
continuum mechanics 1
corrected shear rate 112
Couette flow 79
Couette method 108

D

Deborah number 56
degree of swelling 58
dispersive mixing elements 257
distributive mixing elements 260
drag flows 91, 207
- isothermal flow between parallel plates 208
- non-isothermal flow between parallel plates 210

E

effect of flight flanks 250
elongational viscosity 44
elongation rate 45
end effects 106
equations of continuity 11
equations of energy 20
equations of momentum 16
Euler's motion analysis 5
extensional rheometers 123
extensional viscosity 115, 124
extrudate swelling 58
extruder feeding system 133
- flood fed 133
- gravitational feeding 133
- metered feeding 133
- starve fed 133
extruder operating characteristics 263
- die characteristics 263
- extruder operating points 263
- screw characteristics 263
extrusion 129
extrusion dies 142, 265
- annular die for extruding pipes 145
- annular dies 278
- circular die for extruding rods 143
- circular dies 273
- flat slit die for film extrusion 144
- flat slit die for sheet extrusion 144
- Newtonian model 268
- non-Newtonian model 270
- profile dies 286
- slit dies 273
- wire-coating die 147
extrusion line 131
extrusion process 224
- melt conveying zone 224
- melting zone 224
- pre-melting zone 224
- solid conveying zone 224

F

flow 22
- one-dimensional flow 22
- steady flow 22
- three-dimensional flow 22
- two-dimensional flow 22
- unsteady flow 22
flow curve 25
- of a Newtonian fluid 26
- of non-Newtonian fluids 26
flow curves 113
flow in the co-rotating twin screw extruder 139
- axial flow 139
- drag flow 139
- leakage flow 139
- pressure flow 139
flow in the counter-rotating twin screw extruder 140
- calendering leakage flow 140
- flight leakage flow 140
- pressure leakage flow 140
- side leakage flow 140
flow in the single screw extruder 136
- drag flow 136
- leakage flow 136
- pressure flow 136
flow parameters 22
- boundary conditions 22
- initial conditions 22
flow rate in the extruder 247, 248
- drag flow 248
- pressure flow 248
- throttle coefficient 249
flow rate of solid bed 234
forces 1
- body forces 1
- surface forces 1

G

generalized Newtonian fluids 47
generalized shear rate 48

glass transition temperature 37
global modeling 292

H

Hagen-Poiseuille law 103
hydrostatic pressure 4

I

infinite shear rate viscosity 30
injection molding 149
injection molding cycle 153
- holding 153
- mold filling 153
- molding ejection 153
- mold opening 153
- plasticating 153
injection molds 153

L

leakage flow 252
length of melting zone 244
linear viscoelasticity 59

M

material deformation 61
- elastic 61
- flow 61
- plastic 61
mechanical rheological models 61
- Kelvin-Voigt 65
- Maxwell 63
mechansim of melting initiation 236
melt conveying in conventional screws 245
melt conveying in non-conventional screws 257
melt flow Index 95
melt flow indexer 96
melting 236
melting course 237
melting process 243
melting rate 243
melt volume rate 97
mixing screws 137
modeling of extrusion 223
modeling with Autodesk Moldflow 349
- filling analysis 351
- flow balancing 352
- gate location 350
- Melt Flipper 354
modeling with the MULTI-SCREW system 361
- flood fed single screw extrusion 362
- single screw extrusion 361
- starve fed single screw extrusion 364
- twin screw extrusion 368
modulus 71
- damping factor 72
- loss modulus 72
- storage modulus 72
mold 153
- clamping and mold platens 153
- cooling system 153
- ejection system 153
- mold cavity 153
- sprue and runner systems 153
Mooney method 111
movement of the solid bed 230
multiphase systems 53
- composites 53
- emulsions 53
- foams 53
- mixtures 53
- suspensions 53

N

Newtonian fluid 25, 40
Newton's law of viscosity 24
non-corrected shear rate 112
nonlinear viscoelasticity 76
non-Newtonian fluids 26
- shear thickening fluids 27
- shear thinning fluids 27
normal stress differences 77

normal stress effects 79
- Barus effect 81
normal stress effects
- Weissenberg effect 79
normal stresses 5, 116
- first normal stress difference 5
- second normal stress difference 5
numerical methods 214
- finite difference method 215
- finite element method 215

P

perfect bodies 61
- elastic Hookean 61
- plastic St. Venant 61
- viscous Newtonian 61
phenomenon of antithixotropy 39
phenomenon of thixotropy 39
physical model of extrusion 224
plasticating 234
plasticating unit 130, 149
polymers 28
- elastomers 28
- thermoplastics 28
- thermosets 28
power law model 27
pre-melting 234
pressure-drag flow between parallel plates 212
pressure flows 91, 172
- disk flow 199
- flow between parallel plates 173
- flow through a circular tube 184
- flow through a cone 193
- flow through an annulus 194
- flow through a tapered channel 192
- Hele-Shaw flow 204
pressure rheometers 90
- capillary rheometers 90
- slit rheometers 90
process characteristics 362
- melting profile 362
- screw filling profile 362
- solid bed profile 362

R

Rabinowitsch correction 105
Rabinowitsch-Mooney equation 105
rate of strain 5
relaxation stress modulus 60
representative shear rate 272
representative viscosity 271
rheologically unstable fluids 29
- antithixotropic 29
- thixotropic 29
rheological models 40, 48
- Bingham 52
- Bird-Carreau-Yasuda 50
- Herschel-Bulkley 53
- Klein 51
- Muenstedt 51
- power law 48
- Vinogradov and Malkin 51
rheological models of viscoelastic liquids 83
- Criminale-Ericksen-Filbey 86
- Maxwell convective 86
- Rivlin-Ericksen 84
- White-Metzner 86
rheometers 90
rheometry 89
rotating rheometers 90
- cone-and-plate rheometers 90
- plate-and-plate rheometers 90
- with coaxial cylinders 90
rule of Cox-Merz 74

S

screw geometrical zones 136
- compression zone 136
- feeding zone 136
- metering zone 136
shear type rheometers 90
simple flows 172
single screw extruder 130
single screw extrusion 133, 224
slip effects 111
solid bed break-up 236

solid conveying 228
solid conveying mechanism 228
special injection molding processes 156
stress 1, 3
- normal stress 3
- shear stress 3
stress tensor 2, 4
- deviatoric part 4
- isotropic part 4
stress vector 2
stretching type rheometers 90
swelling degree 116

T

tensile flows 95
tensile type rheometers 95
thermoforming 161
- negative thermoforming 162
- positive thermoforming 163
time effects 59
- dynamic time effects 59
- static time effects 59
time effects of viscoelasticity 66
- creep deformation 68
- dynamic time effects 68
- relaxation time 67
- retardation time 68
- stress relaxation 66
Trouton's viscosity 47
true viscosity 112
true viscosity curve 113
twin screw extrusion 138
- co-rotating twin screw extrusion 139
- counter-rotating twin screw extrusion 140

V

velocity 228
- longitudinal (down-channel) velocity 228
- transverse (cross-channel) velocity 228

velocity gradient tensor 7
- antisymmetric part 7
- rate of deformation 7
- rate of rotation 7
- symmetric part 7
velocity profile 247
- cross-channel velocity profile 247
velocity profile in the screw channel 250
- axial profile 250
- down-channel profile 250
viscoelasticity 55
viscometric functions 77
- first normal stress difference function 77
- second normal stress difference function 77
- shear stress function 77
viscosity 23, 25
- dynamic viscosity 25
- kinematic viscosity 25
viscosity curves 30, 113
viscosity of dispersion 54
- intrinsic viscosity 54
- reduced viscosity 54
- specific viscosity 54
- viscosity ratio 54
viscosity of polymers 28

W

Weissenberg effect 57
Weissenberg number 79
WLF equation 35

Z

zero shear rate viscosity 30